Springer Studies on Populism, Identity Politics and Social Justice

This book series offers an outlet for cutting-edge research on all areas at the nexus of populism, identity politics, as well as natural and social justice. The series welcomes theoretically sound and empirically robust monographs, edited volumes and handbooks from various disciplines and approaches on topics such as nationalism, racism, populism, human rights, diversity, discrimination, identity of minority groups, LGBTQ rights, gender politics, minority politics, social and environmental justice, political and social effects of climate change, and political behavior. Taking as its benchmark global relevance and research excellence the series is open to different approaches from the case study to cross-cultural and transnational comparisons. All books in this series are peer-reviewed.

Barbara Wejnert
Editor

Democracy in Retreat and the Climate Crisis

Challenges and Strategies for Adaptation in an Evolving Political Landscape

Springer

Editor
Barbara Wejnert
Department of Environment and Sustainability
University at Buffalo, State University of New York (SUNY)
Buffalo, NY, USA

ISSN 2731-894X ISSN 2731-8958 (electronic)
Springer Studies on Populism, Identity Politics and Social Justice
ISBN 978-3-032-02880-8 ISBN 978-3-032-02881-5 (eBook)
https://doi.org/10.1007/978-3-032-02881-5

This Springer imprint is published by the registered company Springer Nature Switzerland AG
The registered company address is: Gewerbestrasse 11, 6330 Cham, Switzerland

Contents

Contributors

T. L. Thurston The University at Buffalo, State University of New York, Buffalo, NY, USA

Lane F. Fargher-Navarro The PAST Foundation, The Ohio State University, Columbus, OH, USA

Barbara Geddes Political Science Department, University of California, Los Angeles, CA, USA

Senay Imre Global Gender Studies, University at Buffalo, SUNY, Buffalo, NY, USA

Neelam Kedia Department of English and Foreign Languages, Central University of South Bihar, Gaya, Bihar, India

Kapila Khemundu Department of Sociology, Central University of Odisha, Sunabeda, Odisha, India

Niranjan Majhi Department of Sociology, Central University of Odisha, Sunabeda, Odisha, India

Nibedita Priyadarsini Department of Philosophy, Ravenshaw University, Cuttack, Odisha, India

Alexander Ruser Department of Sociology and Social Work, Center for Digital Transformation (CeDiT), University of Agder, Agder, Norway

Divya Sharma Faculty of Economics and Business Studies & Convener, Department of Commerce and Social Science, Manava Bharati International School, Patna, India

Muthukuda Arachchige Dona Shiroma Jeeva Shirajanie Department of Public and International Law, Faculty of Law, University of Colombo, Colombo, Sri Lanka

Izabela Skórzyńska Department of History, Adam Mickiewicz University, Poznan, Poland

Maria Anita Stefanelli Department of Anglo-American Language and Literatures, Roma Tre University, Rome, Italy

Jagriti Upadhyaya Department of English, Sardar Patel University of Police, Security and Criminal Justice, Jodhpur, Rajasthan, India

Jyotasana School of Liberal Arts and Humanities, Woxsen University, Hyderabad, Telangana, India

Barbara Wejnert Department of Environment and Sustainability, University at Buffalo, State University of New York (SUNY), Buffalo, NY, USA

Camille Wejnert-Depue Center for Global Sustainability, University of Maryland, College Park, MD, USA

Soe Win Department of Women and Gender Studies, SUNY Brockport, State University of New York, Brockport, NY, USA

Introduction: Navigating Environmental Politics and Policies in an Era of Autocracy Rise

Barbara Wejnert

Abstract The era of profound historical and political change, from the rapid rise of democracy to the ascent of autocracy, calls for attention to the effects of these changes on environmental politics and policies aimed at preserving the natural world and protecting people from extreme weather events for future generations. This volume navigates the changes in environmental politics and policies. The volume's introductory part sets the tone for the included multidisciplinary chapters, which are thematically organized into three clusters addressing (a) *Democracy-Autocracy Conundrum*, (b) *Climate Change Politics in a Polarized World, and (c) Climate Change Politics Through an Interdisciplinary Socio-Humanities Lens*. These clusters examine the broader debates surrounding the conundrum of democracy and autocracy, the effects of pro-autocracy global shifts on environmental politics, and a unique humanities perspective on the impact of environmental politics on societal and environmental sustainability.

Keywords Democracy · Autocracy · Environmental politics · Environmental policies · Sustainability

Introduction

The edited volume *Autocracy's Ascent and Climate Change Politics: Challenges and Strategies for Adaptation in an Evolving Political Landscape* provides a comprehensive analysis of the global resurgence of autocracy and the erosion of democracy, offering a renewed perspective on the implications of these transformations for climate change governance. Divided into three sections, the volume examines the

---

B. Wejnert (✉)
Department of Environment and Sustainability, University at Buffalo,
State University of New York (SUNY), Buffalo, NY, USA
e-mail: bwejnert@buffalo.edu

B. Wejnert (ed.), *Democracy in Retreat and the Climate Crisis*, Springer Studies on Populism, Identity Politics and Social Justice,
https://doi.org/10.1007/978-3-032-02881-5_1

diverse pathways that lead to backsliding democracy and the rise and spread of autocratic regimes, examining the profound impact of these changes on global climate actions, politics, and policies.

Democracy Erosion

Research on democracy reveals that political evolution in some democratic systems often culminates in increasing autocratization, as democratically elected leaders become reluctant to relinquish power. Democratic backsliding is frequently accompanied by a decline in judicial independence, increasing restrictions on media freedom, and the politicization of the education system (Economy 2022; Xia 2021). Concurrently, populist movements gain traction, enabling populist leaders to consolidate power by sidelining, silencing, or bullying and intimidating political opponents (Applebaum 2020; Levitsky and Ziblatt 2019). Under these conditions, elected leaders often adopt autocratic behaviors to sustain their authority.

As illustrated in Table 1, the decline of democracy is evident in the diminishing judicial independence, restricted media operations, and increased state control over educational institutions (Johnson 2019; Xia 2021). The transformation of democratically elected officials into autocratic rulers often occurs through the marginalization or menacing of political rivals, particularly in contexts marked by fragile party systems and the influence of international autocratic networks (Applebaum 2024; Levitsky and Ziblatt 2019).

To entrench their power, these leaders frequently resort to non-democratic decision-making processes, justify or permit violence against political adversaries, and introduce legal frameworks that undermine civil liberties (Mollan and Geesin 2020; Sata and Karolewski 2020). They also restrict media freedoms, penalize civil dissent, and suppress opposition movements (Cianetti et al. 2019). Following electoral victories, such leaders often consolidate power and erode democratic principles, contributing to the emergence of what has been termed an "authoritarian nation" (Geddes et al. 2018). Political enablers further facilitate the regression

Table 1 Autocratic behaviors manifested by elected leaders in eroding democracies

1	Undermine the election result and refuse to accept legitimate ballots (Przeworski 2019)
2	Support laws and policies that restrict civil liberties (Karolewski 2020)
3	Baselessly portray partisan rivals as criminals to disqualify them from political participation (Mollan and Geesin 2020; Johnson 2019)
4	Endorse or fail to condemn the violent behaviors of their supporters (Mollan and Geesin 2020)
5	Praise repressions used by other authoritarian governments (Mollan and Geesin 2020)
6	Restrict criticism of the government (Cianetti et al. 2019)
7	Take legal or other punitive action against criticism of the government by media, civil society, or rival parties (Cianetti et al. 2019; Karolewski 2020; Przeworski 2019)

Source: Table developed by the author following analyses of literary sources

toward autocracy and are frequently accompanied by legal or coercive actions targeting media outlets, civil society organizations, and opposition parties (Cianetti et al. 2019; Karolewski 2020; Przeworski 2019). In adopting these practices, democratically elected governments often align themselves with and even legitimize the repressive strategies of established authoritarian regimes, reinforcing their disregard for civil rights and democratic norms (Mollan and Geesin 2020). In doing so, they erode democratic norms and increasingly mirror authoritarian regimes in both rhetoric and practice (see Table 1).

Democratic erosion frequently aligns with the persistent neglect of marginalized populations. When the socioeconomic and political needs of these groups remain unaddressed, particularly during periods of instability, populist movements often emerge, at times veering toward authoritarianism. In such contexts, populist leaders may serve as catalysts for democratic backsliding, as is evident across several contemporary Western European democracies, including France and Germany (Berman 2019, 2021). The global rise of pro-autocratic and populist forces has further intensified the vulnerability of democratic systems. This decline typically unfolds incrementally, through the gradual weakening of democratic institutions, norms, and practices, eventually giving rise to autocratic governance. Despite regional and national differences in the trajectory of democratic decline, this trend stands in stark contrast to the sustained expansion of democratic governance observed globally until 2005 (see Fig. 1).

Effects of Backsliding Democracy on Environmental Politics

Democratic backsliding presents a particularly urgent concern in the era of global climate change, highlighting the critical need for accelerated progress toward the Sustainable Development Goals (SDGs). A core tenet of democratic governance is its responsiveness to the evolving needs of its citizenry—needs that encompass not only economic equity, civil rights, and access to independent knowledge, but also the protection of public health and environmental sustainability (Campbell and Pölzlbauer 2010; Wilson 2019). Democracies, by virtue of their commitment to the public good, are uniquely positioned to lead efforts aimed at mitigating the impacts of environmental degradation and climate-induced disasters.

However, as Garrett Hardin (1968) illustrated in his seminal work *The Tragedy of the Commons*, individual actors, when operating in the absence of regulation, tend to deplete shared resources for personal gain, resulting in collective harm. Hardin contends that only mutually agreed-upon constraints—effectively enforced—can safeguard common resources from overuse. While such coercive measures may be more readily implemented in autocratic systems, democracies must find innovative, participatory means of instituting environmental protections that align with democratic values. The erosion of democratic institutions thus weakens states' capacity to enact and enforce the collaborative, long-term solutions necessary to address global environmental challenges.

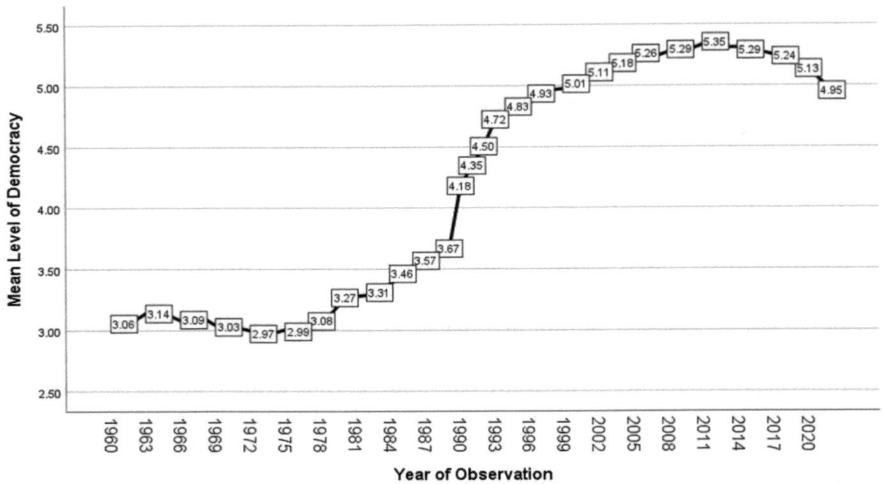

Fig. 1 The mean levels of democracy scores worldwide from 1960 to 2020. *Note: Figure designed by the author using data from V-Dem* (2021*). Varieties of Democracy (V-Dem) project. University of Gothenburg, V-Dem Institute* complemented by data derived from the database Nations, democracy, development: 1800-2005 (Wejnert, 2007)

Crucially, democratic backsliding has significant implications for environmental governance, particularly in the context of climate change. The erosion of democratic institutions often coincides with a state's diminishing capacity to protect populations from the adverse effects of environmental degradation. As democratic governments shift toward authoritarian practices, their focus often turns to consolidating political power—manifested through corruption, patronage, and vote-buying—at the expense of environmental priorities. In such contexts, environmental decision-making and climate policy responses may be deprioritized or weakened.

Research by Povitkina (2018) provides compelling evidence that democratic regimes are generally more inclined to adopt and adapt effective environmental policies, partly by learning from the experiences of other democracies. However, this relationship is not uniform. Analyzing data from 144 countries between 1970 and 2011, Povitkina demonstrates that the efficacy of democratic governance in reducing CO_2 emissions is significantly moderated by levels of corruption. In highly corrupt democracies, the capacity to meet climate targets and implement effective mitigation strategies is severely constrained. As a result, such democracies may perform no better than autocratic regimes when addressing climate change, underscoring the critical importance of transparent and accountable governance in achieving environmental sustainability.

These challenges remain only partially understood and warrant further empirical and theoretical investigation. Addressing the consequences of rising autocratization amid accelerating climate change has become an urgent global matter. This volume aims to examine the underlying drivers and mechanisms that contribute to the rise of autocracy in both historically stable and newly established democracies. By critically examining these dynamics, the volume aims to advance scholarly understanding and inform efforts to safeguard democratic institutions while simultaneously addressing the escalating threats to environmental sustainability and preventing further degradation of democratic values, as well as preserving the natural world as we know it.

Brief Book Description

This volume offers a critical and timely perspective on the global rise of autocratic regimes, reframing the debate in light of contemporary democratic decline and escalating environmental challenges. The emergence of autocratic tendencies within established and emerging democracies has become particularly salient in the context of global climate change, an era that demands responsive, transparent, and accountable governance. As the urgency for political and policy solutions to mitigate and adapt to climate impacts intensifies, the simultaneous erosion of democratic institutions raises profound questions about the capacity of states to address these interconnected crises effectively.

Taking a comprehensive and multidisciplinary approach, the book's first section examines the causes of the rise of autocracy and its global spread. It does so through a set of four studies that conceptualize this rise. Barbara Geddes' study examines the enablers of democracy backsliding in the United States—including politicians, unelected officials, and undereducated in civic education, voters—who contribute to the weakening of democracy, thereby facilitating the rise of authoritarianism. She argues that for politicians and officials whose future well-being depends on the executive's goodwill, acquiescence is the only sensible option even if they sincerely oppose executive's actions. In contrast, officials whose future success depends on their reputations for competence, fairness, and integrity are more likely to stand in the backslider's way. She uses this argument to make sense of the contrast between the supine behavior of elected members of the US Congress, threatened by primary challenges, and the way unelected officials, judges, and military officers blocked various Trump initiatives during his first term. In the last pages, she interprets Trump's barrage of executive orders and attacks on the personnel and funding of the government during the first weeks of his second term as efforts to persuade the kinds of officials who blocked his initiatives during his first term that their future well-being is more dependent on him than U.S. laws and norms might have led them to believe.

Similar dynamics are unfolding in Germany, where democratic backsliding is subtly encouraged by anti-establishment actors and those dismissive of scientific consensus. As Alexander Raser outlines in his chapter, Germany's consensus-driven political system—rooted in coalition-building and corporatist ties between parties, academia, and selected media—has historically limited the influence of radical dissent on climate change. Unlike the United States, Germany lacks strong networks linking political parties with wealthy donors, partisan media, and ideological think tanks. However, the rise of populist parties is beginning to exploit vulnerabilities in this system, creating new opportunities for anti-democratic currents to gain traction.

A more precise understanding of the processes through which democracy weakens and its impact on environmental politics is presented in a significant historical study by Hurston and Fargher-Navarro. This study examines various democratic systems, including direct democracy, which can more effectively promote support for environmental policies. In contrast, representative democracy, where interest group lobbying disrupts voters' intentions regarding environmental protection, can delay or stall crucial environmental regulations. The authors, using an archaeological perspective, suggest that the nature of democracy might hinder effective environmental stewardship. By drawing on numerous historical examples, they offer a new perspective on the increasing involvement of citizens and civic society in environmental decision-making, which adds to the centralized policies of different democratic government types.

While reflection on the dynamics of changes in the political landscape is critical in the context of global climate change, the environmental implications of rising autocracy remain insufficiently understood. The following section comprises several chapters that examine the implications of increasing authoritarianism for environmental politics and human rights. Topics explored include the impact of dictatorial regimes on environmental degradation and disaster response, the challenges of climate governance in eroding democracies, the deterioration of human rights under climate stress, and the evolving roles of women in environmental leadership within democracies undergoing autocratic transition.

The book's second section explores the effects of the evolving political landscape on climate change adaptation. This part starts with an interesting and timely study on the autocracy threat to democratic, independent Ukraine, marked by the February 2022 Russian invasion and persistent war. Wejnert-Depue's investigation of the outcomes of Russian aggression on Ukrainian and global energy security includes various scenarios for rebuilding Ukrainian energy infrastructure that include not only the opportunity to modernize and decarbonize Ukraine's buildings and energy sectors but also to do so in a way that advances energy security (due to a reduction in reliance on foreign-sourced fossil fuels replaced by a domestically sourced green energy transition) and improves citizens' well-being via a decrease in energy service costs and a reduction in air pollution. In another chapter, focusing on war and conflict and their relation to environmental devastation, Soe Win discussed war in Burma, examining the impact of an autocratic takeover on the country's preparedness and susceptibility to environmental disasters. As the author reveals, heightened environmental vulnerability due to widespread deforestation has accelerated the effects of climate change on the environment. However, ongoing internal conflicts—the civil war and political unrest—have led the ruling military regime to prioritize other concerns over environmental concerns, thereby exacerbating the consequences of ecological degradation and the devastating role of military dictatorship in shaping disaster preparedness or lack thereof. The weak preparation is compounded by the broader humanitarian crises emerging in the context of environmental devastation during the authoritarian governance in Burma. The following chapter contributes to this investigation by examining how local governance actors and the central government collaborate or conflict in addressing the climate crisis, navigating tensions between development, political conflicts, and the climate change mitigation. This chapter by Senay attempts to answer these questions by examining climate approaches in Turkey at both the local and national levels. Through a detailed analysis of Turkey's most developed urban megacity, Istanbul, this paper examines the climate policies and rhetoric of various political leaders in both local and central government.

The complex interplay between democratic backsliding, the global ascent of autocracy, and the resulting implications for climate change and population well-being is critically examined in Wejnert's work, "Predicting Climate Vulnerability and Well-being in a Declining Democratic and Increasingly Autocratic World." This study addresses the growing polarization of global governance and the accelerating spread of authoritarian regimes by asking: What are the consequences of declining democratic governance on population well-being? Do the outcomes associated with low levels of democracy differ significantly from those under high-functioning democratic systems? Moreover, how do these political shifts influence countries' progress toward sustainable development?

Complementing this global perspective is an in-depth analysis of the relationship between climate change and human rights, with a particular focus on Sri Lanka. The research explores the adverse effects of climate change on Sri Lankans, highlighting the violations of socio-economic and civil rights that occur in its wake. It further evaluates the extent to which existing legal frameworks address—or fail to address—the protection of climate-affected populations. The findings suggest that the impacts of climate change are gender-differentiated and that current international legal instruments are insufficient in recognizing the variability of population vulnerabilities. Moreover, Sri Lanka's domestic legal system does not adequately safeguard the rights of communities facing climate-related threats, nor does it legally incorporate the implications of climate change mitigation and adaptation. The chapter underscores the urgent need to align national and international legal standards with the evolving realities of climate vulnerability and human rights protection.

The final set of chapters in this section critically examines the political landscape of India, with a particular focus on the governance practices of Prime Minister Narendra Modi. Joysana explores the intersection of populism and climate governance, assessing how the rise of populist politics presents both challenges and opportunities for democratic environmental policymaking. Her chapter evaluates key climate-related initiatives, political rhetoric, and media discourse under the current Indian administration, highlighting the tensions between populist leadership and democratic principles in the context of climate action.

Building on this examination of climate governance in India, the subsequent chapters turn to the role of tribal societies in environmental sustainability and development. Sharma analyzes the Indian government's implementation of the Sustainable Development Goals, with a specific focus on the participation of tribal women in local governance structures. Building on this theme, Kapila examines the significant contributions of tribal communities to environmental conservation. Drawing on examples of indigenous agricultural, cultural, and religious practices, this chapter underscores the importance of traditional ecological knowledge in preserving natural resources—particularly forest ecosystems—and sustaining environmental health. Together, these chapters illuminate the complex interplay between political leadership, indigenous knowledge systems, and sustainable development in contemporary India.

The final section of the volume offers a rare humanities-oriented perspective on climate change amid the rise of authoritarianism, examining how the shifting political landscape and its implications for climate solutions can be understood through historical, cinematographic, and literary frameworks. The contributing chapters critically analyze literary texts that engage with environmental themes, examining how fiction reflects and shapes public discourse on ecological crises. Additionally, this section investigates the historical significance of environmental symbols and their role in shaping collective understandings of environmental protection and resistance.

The section starts with a chapter by the Italian scholar's analysis of cinematography, reflecting on the consequences of the war initiated by the authoritarian Nazi regime in 1939. As the author explains, far from marking an element of the astonishing diversity in sexual characteristics of plants, *red* in the film *Czerwone Maki na Monte Cassino* (1951), directed by Polish filmmaker Krzysztof Łukaszewicz, is the color of the poppy, a poignant icon of loss and memory, sacrifice and remembrance; the sign is associated with the bloodshed of those who fought the final battle on the ruins of the sixteenth-century monastery of Monte Cassino to open the way to Rome for her liberation. Moving back to history, Skorzynska dwells on symbolic forms of environmental memory in the context of the growing climate crisis. She analyzes plant motifs in national and state symbolism from a global perspective, along with the question of their significance in shaping national identity and promoting environmental and pro-climate awareness. My thesis is that plant motifs in national symbolism, which applies to hundreds of countries and national communities around the world, contain the potential to care for the climate, and at the same time, it is being abused and trivialized, the author write. Populists in particular, deprive national symbolism of its historically shaped identity focusing on such values as protection and multiplication of natural resources.

Turning to literary analysis, the chapter titled *"Earth Democracy: Justice, Sustainability and Peace: Healing Ravaged Ecologies through the Concept of Ecosocialism and Eco-Justice: An Ecocritical Reading of Vandana Shiva"* by Jagriti Upadhyaya explores the philosophical underpinnings of ecological justice and sustainable living through an ecocritical lens. Upadhyaya interrogates the consequences of modern materialism, marked by unrestrained consumption, hedonistic values, and a market-driven economy that prioritizes production and profit over ecological balance. This paradigm, she argues, has led to the unchecked exploitation of natural resources, exacerbating economic inequality and producing a distorted model of development. The chapter challenges the dominant cultural narrative that equates material accumulation with wealth, proposing instead a vision of well-being rooted in ecological harmony and minimal desire. Upadhyaya highlights the environmental consequences of this exploitative ethos, including water and air pollution, deforestation, species extinction, climate change, and the encroachment of urban infrastructure on natural habitats—phenomena that deepen social divides and threaten both ecological and human resilience. Continuing this theme, the last two chapters focus on ecofeminism while analyzing literary sources. Authors Priyadarsini and Neelam Kedia draw attention to feminists and philosophers who have emphasized the importance of ethics and practices that connect humans to the Earth, underscoring the abstractness and generality of deontological and consequentialist theories that propose a need for the conservation and sustainability of the environment through feminist ethics. As a social and political movement, ecofeminism seeks to draw attention to how women and nature suffer and are dominated by patriarchy, development, capitalism, and scientific advancement. It examines the resilience and resistance of both disenfranchised populations and nature in the face of exploitation. The central focus of these papers is to present how the beautiful surroundings have become a green prison for its inhabitants, highlighting the necessity of sustainability and empowerment to secure the survival of disenfranchised populations.

This book enriches existing knowledge on the politics of climate change, analyzing how authoritarianism impacts the protection of the environment and the pursuit of a sustainable environment for all. It embraces a multidisciplinary approach, featuring contributions from scholars and practitioners across various countries and fields, including democracy, autocracy, and sustainability. The authors engage in a vibrant debate on the impact of the evolving political landscape and its essential effect on environmental and sustainable living. This multidisciplinary volume features authors from various disciplines, including sustainability studies, political science, sociology, law, communication, anthropology, history, and literature. They represent countries from the United States of America, Europe (Italy, Germany, Poland, and Norway), and Asia (India, Sri Lanka, and Kashmir), further enhancing the breadth of investigations. This book is a valuable source of information on the intersection of politics and climate change, an essential read for scholars and students of climate politics, sustainability studies, democracy, political sociology, comparative politics, international relations, and globalization. It also provides valuable insights for policymakers and activists navigating climate adaptation in an increasingly authoritarian world.

References

Applebaum A (2020) Twilight of democracy. The seductive lure of authoritarianism. Doubleday, New York

Applebaum, A. (2024). Autocracy Inc. The dictators who want to run the world. New York: Doubleday.

Berman S (2019) Populism is a symptom rather than a cause: Democratic disconnect, the decline of the center-left, and the rise of populism in Western Europe. Polity 51(4):654–667

Berman S (2021) The causes of populism in the West. Annu Rev Polit Sci 24(1):71–88

Campbell, D.F.J. and Pölzlbauer, G. (2010) The Democracy Ranking 2009 of the Quality of Democracy: Method and ranking outcome. Comprehensive scores and scores for the dimensions. Vienna, VA: Democracy Ranking. Accessed 17 September 2021. http://www.democracyranking.org

Cianetti L, Dawson J, Hanley S (2019) Rethinking "Democratic Backsliding" in Central and Eastern Europe. Routledge, New York, NY

Economy E (2022) Xi Jinping's new world order. Foreign Aff 101(1):52–67

Geddes B, Wright JG, Frantz E (2018) How dictatorships work: Power, personalization, and collapse. Cambridge University Press

Hardin, G. (1968). The tragedy of the commons: the population problem has no technical solution; it requires a fundamental extension in morality. Science, 162(3859), 1243–1248.

Johnson R (2019) Children of the dream: Why school integration Works. Basic Books

Karolewski IP (2020) Towards a political theory of democratic backsliding? Generalizing the East-Central European experience. Palgrave Studies in European Union Politics (book series). https://link.springer.com/content/pdf/10.1007%2F978-3-030-54674-8.pdf

Levitsky S, Ziblatt D (2019) How democracies die. Crown

Mollan S, Geesin B (2020) Donald Trump and Trumpism: Leadership, ideology, and narrative of business executive turned politician. Organization 27(3):405–418. https://doi.org/10.1177/1350508419870901

Povitkina, M. (2018). The limits of democracy in tackling climate change. Environmental Politics, 27(3), 411–432. Delete: Sata and Koralewski 2020

Przeworski A (2019) Crisis of democracy. Cambridge University Press, New York

V-dem. (2021). Autocratization Turns Viral. DEMOCRACY REPORT 2021. V-Dem Institute: Department of Political Science, University of Gothenburg, March 2021.

Wejnert B (2007) Nations, democracy and development: 1800-2005. Database ICS

Wilson, R. (2019) 'Authoritarian environmental governance: Insights from the past century'. Annals of the American Association of Geographers, 109 (2), 314–23. https://doi.org/10.1080/24694452.2018.1538767

Xia C (2021) The party that failed: an inside break with Beijing. Foreign Aff 100(10):78–96

Barbara Wejnert is a Professor in the Department of Environment and Sustainability at the University at Buffalo, former Associate Professor of Transnational Studies, and the founding Chair of the Department of Global Gender Studies. She has also served as a faculty member and Director of Academic Programs at Cornell University. While her research broadly addresses global issues, particularly economic development, diffusion of democracy, and the effects of democratization on the well-being of women and girls, her current projects examine the impact of democratic and autocratic regimes on global climate change politics, as well as the acculturative experiences of immigrant and refugee women in Buffalo, NY. She is an author and editor of several books, including *Diffusion of Democracy* (2014), published by Cambridge University Press, and a database, *Nations, Development and Democracy* (2007). She is affiliated with the University at Buffalo's School of Social Work through the Immigrant and Refugee Research Institute (IRRI) and the Jaeckel Center for Law, Democracy, and Governing. She holds a Ph.D. in Sociology from Mickiewicz University in Poznań, Poland.

Democracy-Autocracy Conundrum When Facing the Politics of Climate Change

Analyzing the Effects of the Global Rise of Autocracies for a Sustainable Future and Environmental Protection

Barbara Wejnert

Abstract The global rise of autocracy underscores the urgent need to critically assess the implications of these political shifts for a sustainable future. Revisiting the discourse on authoritarian resurgence is essential to understanding how different political regimes shape environmental governance and influence the prospects for long-term global sustainability. In the context of accelerating climate change, evidenced by the growing frequency and intensity of extreme weather conditions, timely and coordinated responses to scientific warnings about the tragedy of the commons are urgent.

Keywords Backsliding democracies · The global rise of autocracy · Environmental sustainability · The climate change crisis

Introduction

Since the turn of the millennium, the stability and prevalence of democratic states have markedly declined. The proportion of democratic countries fell to 67% by 2015, and the diffusion of autocratic regimes gained momentum (e.g., Applebaum 2020, 2024; Cianetti et al. 2019; Geddes et al. 2018). These new trends underscore the broader geopolitical tensions inherent in the struggle between democracy and

Part of this study represents a modified version of a paper published in the *International Affairs Forum* in May 2025 (Wejnert 2025).

B. Wejnert (✉)
Department of Environment and Sustainability, University at Buffalo, State University of New York (SUNY), Buffalo, NY, USA
e-mail: bwejnert@buffalo.edu

B. Wejnert (ed.), *Democracy in Retreat and the Climate Crisis*, Springer Studies on Populism, Identity Politics and Social Justice,
https://doi.org/10.1007/978-3-032-02881-5_2

autocracy, amplifying the trend of democratic backsliding evidenced by the noticeable drop in the average level of democracy among sovereign states from 5.35 in the early 2000s to below 4.95 by 2021. This assessment uses a scale of 0–10, where zero indicates no democracy and 10 refers to a fully developed, stable democratic system (Coppedge et al. 2021; Wejnert 2021).

Consequently, unlike the latter half of the twentieth century, the early twenty-first century has witnessed a growing number of democratically elected leaders embracing increasingly authoritarian practices. Among the most prominent are Viktor Orbán, Prime Minister of Hungary; Andrzej Duda, former Prime Minister of Poland; Jair Bolsonaro, former President of Brazil; Benjamin Netanyahu, Prime Minister of Israel; Recep Tayyip Erdoğan, President of Turkey; and Donald Trump, former and current President of the United States. This global shift has significantly eroded democratic norms and institutions, leading to a marked decline in the proportion of the world's population living under stable democracies, from nearly 50% to approximately one-third. In contrast, the share of individuals living under autocratic regimes surged from 48% in 2010 to 68% by 2020, according to the Varieties of Democracy Project (V-Dem) (Coppedge et al. 2021, 2025).

Of particular concern is the growth of populism and public prioritization of autocratic rule, which is perceived to deliver faster and more pronounced decision-making and changes that benefit societies. Although the common root of the birth of populism in democracies is popular discontent caused by economic and sociopolitical inequalities, people's dissatisfaction with the existing political and economic situation limits the attractiveness of democracy as a means for progress. Considering the vast economic disparities it turns most of the public away from the democratic system. Democracy is viewed as a poor protector of the economic well-being of average citizens, especially among lower economic brackets. Such trends are visible not only in developing economies but also in well-developed economies of Western democracies, including the United States, where the income disparity leads to a massive increase in the earning gap, with the ratio of CEO-to-worker compensation rising from 21 times in 1965 to 351 times in 2020 in the US, and 89% of stock owned by the top 10% of the wealthy individuals, whereas only 11% is owned by the 90% of the rest of the US citizens by 2021 (Galloway 2022). The inequality in earnings is topped by the limited opportunity for advancement, with over 30% of American jobs requiring a completed university-level or MA degree in 2021 (Galloway 2022). Such suitable conditions for the rise of populism are frequently exploited by populist leaders who reconnoiter popular discontent, trying to destabilize the democratic rule of law and democratic governing.

One of the principles of populism and populist leaders is to speak against the principles of political representation, thus constituting a "shadow of representative politics" represented by a strong leader (Müller 2016, p. 101). This creates a façade of democracy, with populist leaders using the language of democratic values to win the support of the electorate (Albertazzi and McDonnell 2008). To solidify public support, populist leaders employ communication forms that resonate with ordinary people, including direct interaction with followers, spectacles and performances, derogatory language and inflammatory comments, along with extensive use of media and social media (the so-called mediatization politics) and "celebrity strategy" to promote their agenda and expand their follower base (Moffitt 2016, p. 70). If populist leaders succeed in garnering support, their claims become moralistic, empty in content, sensational, and symbolic, mainly offering rhetoric, unfounded solutions, and false promises (Freedom House 2018). Interacting with democracy, populism proves destructive, appealing to social divisions between the average citizen ("the people") and the ruling elite, while claiming the need to combat the elite that unjustly gained power by exploiting the people. This peculiar contrast, a dichotomy between friend and enemy, symbolizes a societal conflict reminiscent of a struggle between adversaries (Giraudi 2018). By generating conflict and polarizing society, populism—through its destructive role—obstructs cooperation and political dialogue among diverse social or political groups, hindering the chances to collaborate, reach consensus, make joint decisions, forge political agreements, or create political coalitions (Mudde and Kaltwasser 2012, p. 83). When populist leaders assume the role of policymakers, they seek to control the state, particularly the judiciary and executive branches of government, using corruption and "clientelism," while suppressing civil society and freedom of the press (Przeworski 2018, p. 1). They also aim to limit the representation of minority institutions and employ the concept and practice of majority rule to bypass minority rights (Przeworski 2018, p. 1). Therefore, populism is regarded as a decline in democracy, a potential threat to civil society (Taggart 2004), and a force opposing the global spread of democracy (Arch Puddington and Tyler Roylance 2017).

The growing popularity and allure of populist demagogues among dissatisfied segments of society is linked to the rise of autocracy through the autocratization of democratically elected leaders (Geddes et al. 2018), following their rise to power largely with the help of populism. Accordingly, Levitsky and Ziblatt (2018, pp. 33–53) contend that a weak representative polity coupled with weak and highly divisive political parties cannot safeguard democratic institutions from populism and may lead to the demise of democracies. Unsurprisingly, the observed global decline of democracy and rise of autocracy support the forecast that societal division, powered by populist leaders' rhetoric, agenda, and actions, coincides with the rise of autocracy.

The marked reduction in mass mobilizations supporting democratic governance since 2017 (see Fig. 1 above) aligns with a parallel trend identified in V-Dem data of an increasing number of democratically elected executive branches exhibiting disregard for constitutional norms and constraints (see Fig. 2).

The global decline of democracy alongside the rise of autocracy underscores the urgent need to critically assess the implications of these political shifts for a sustainable future. Revisiting and reframing the discourse on democratic backsliding and authoritarian resurgence is essential to understanding how different political regimes shape environmental governance and influence the prospects for long-term global sustainability. In the context of accelerating climate change, evidenced by the growing frequency and intensity of extreme weather events, timely and coordinated responses to scientific warnings about the tragedy of the commons are more imperative than ever.

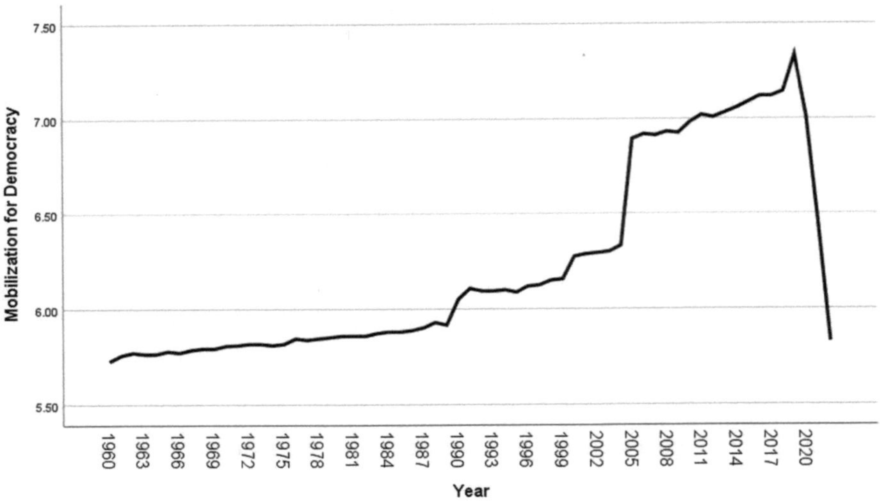

Fig. 1 The average number of mass mobilizations for democracy across all sovereign states, 1960–2020. *Notes*: Events are pro-democratic if they are organized with the explicit aim to advance and/or protect democratic institutions, such as free and fair elections with multiple parties, courts, and parliaments, or if they support civil liberties, such as freedom of association and speech. This question concerns the mobilization of citizens for mass events such as demonstrations, strikes, and sit-ins. *Source*: *Project Manager(s)*: Sebastian Hellmeier. Hellmeier and Bernhard (2022, *V-Dem Working Paper Series* 2022:128); Pemstein et al. (2023, *V-Dem Working Paper Series* 2023:21); Coppedge, M. et al. 2023. *V-Dem Codebook v.13. Varieties of Democracy (V-Dem) Project.* University of Gothenburg, V-Dem Institute, Sweden

Democracy vs. Autocracy and Protection of the Commons from the Tragedy of Unsustainability

In the late 1960s, two pivotal and parallel debates emerged within academic and policy circles focused on environmental and social sustainability challenges. The first, articulated by Paul R. Ehrlich (1968), warned of the dire consequences of unchecked global population growth, framing it as a demographic "bomb" threatening planetary survival. The second, advanced by William Ophuls in 1977 (Ophuls 1977), emphasized the equally unsustainable trajectory of overconsumption, pollution, and the rapid depletion of Earth's finite resources. Both lines of inquiry significantly shaped the understanding of human survival on a fragile planet and contributed to the growing prominence of the concept of sustainability.

A landmark contribution to this discourse was Garrett Hardin's influential 1968 essay, The Tragedy of the Commons, published in Science. Hardin warned that individual self-interest in exploiting common resources, such as air, water, and land, inevitably leads to collective ruin when those resources are overused and irreversibly degraded. Hardin advocated for societal intervention to avoid such outcomes through coercive mechanisms, such as "laws or taxes that make it cheaper for the

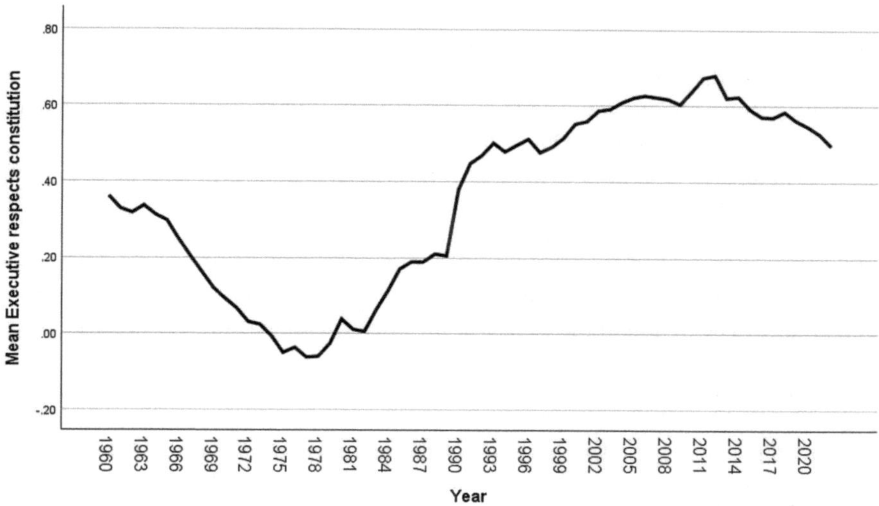

Fig. 2 Change in mean values of elected government leaders (the executive branch) respect for countries' constitutions worldwide, 1960–2020. *Sources*: Figure created by the author using a database V-dem (Coppedge et al. 2021). *Question*: To what extent does the executive respect the constitution and comply with court rulings, and to what extent is the judiciary able to act in an independent fashion? Project Manager: Jan Teorell. Data is coded on a scale: Interval, from low to high (0–1). *Aggregation*: The index is formed by taking the point estimates from a Bayesian factor analysis model of the indicators for executive respect for the constitution (v2exrescon), compliance with the judiciary (v2jucomp), compliance with the high court (v2juhccomp), high court independence (v2juhcind), and lower court independence (v2juncind). Pemstein et al. (2021, *V-Dem Working Paper Series* 2021:21); Coppedge, M. et al. 2023. *V-Dem Codebook v.13. Varieties of Democracy (V-Dem) Project.* University of Gothenburg, V-Dem Institute, Sweden

polluter to treat his pollutants than to discharge them untreated" (Hardin 1968, p. 1243).

The urgency of these early warnings eventually galvanized international policy efforts, culminating in the establishment of the World Commission on Environment and Development. Its seminal 1987 report, Our Common Future, commonly known as the Brundtland Report, popularized the concept of "sustainable development" and emphasized the need to integrate environmental protection with economic and social progress. In the aftermath of the report, both democratic and authoritarian governments were called upon to craft a "foreign policy for the environment," promoting international coordination of national environmental strategies as a vital step toward global sustainability.

Despite decades of global environmental advocacy, international agreements and sustainability goals frequently collide with the national interests of individual states. These interests, often driven by economic and political self-preservation, routinely undermine the implementation and enforcement of international environmental protections. This persistent clash presents a tangible barrier to effective global environmental governance. More than half a century after the Brundtland Report introduced a global agenda for sustainable development and protecting a "common future," meaningful progress on environmental preservation and climate change mitigation remains limited. The irresponsible exploitation of shared environmental resources continues, mirroring the same patterns of self-interested and irrational resource use that Garrett Hardin warned against in his "tragedy of the commons."

For example, across various regions, cattle ranchers leasing national lands often lobby for higher grazing quotas, ignoring ecological consequences until overgrazing results in soil erosion and the spread of invasive species. Similar dynamics unfold in marine ecosystems, where efforts to curb overfishing and protect marine biodiversity, such as those outlined in the High Seas Treaty, are met with limited compliance. The notion of the oceans as an "inexhaustible resource" persists, pushing marine ecosystems toward collapse.

Political leaders have further exacerbated environmental degradation in weakening democracies, prioritizing short-term economic gains over long-term ecological stewardship. Under President Jair Bolsonaro's administration in Brazil (2019–2022), deforestation in the Amazon, commonly referred to as the "lungs of the planet," reached its highest level in over a decade. Weakened environmental regulations and aggressive promotion of extractive industries contributed to a 75% increase in deforestation compared to the previous decade, with rates soaring by 150% in Bolsonaro's final month in office (Bakermans and San Martín 2021). This surge further accelerated the alarming deforestation trend from 2001 to 2023, which saw a 13% loss of forest cover, pushing the Amazon closer to a critical tipping point—its transformation into a drier, savanna-like biome with profound implications for global climate change (Nobre 2019; Roy 2022; Tollefson 2013, 2014).

A similar pattern emerged in Poland under President Andrzej Duda's administration in 2020, which intensified logging in the UNESCO-protected Białowieża Forest, one of Europe's last remaining primeval woodlands. This move endangered the forest's unique biodiversity and ecological value (Blavascunas and Cope 2022; Czajkowski et al. 2009). Moreover, in the United States, the second-term presidency of Donald Trump echoed this prioritization of profit over conservation. In April 2025, an executive order authorized extensive logging in federally protected forests, including ecologically sensitive areas in California, Oregon, Colorado, and the Blue Ridge Mountains in the Eastern United States. This policy shift illustrates the recurring pattern of undermining environmental safeguards in pursuit of economic exploitation, despite the well-documented long-term consequences for global ecological health.

Although humanity's survival hinges on the availability of clean air and water, as well as the continued existence of diverse animal and plant life, many nations continue to exploit global commons—resources that belong to all humankind—for their unilateral benefit. The consequences of such unsustainable practices are dire and far-reaching, including the intensification of the greenhouse gas effect, ozone layer depletion, acid rain, global warming, biodiversity loss, and the depletion of vital nutrients in the soil. Around the world, the cumulative impacts of these behaviors are becoming increasingly visible. Nations are grappling with escalating environmental disasters, deteriorating public health, lost productivity and economic output, infrastructure damage, rising global poverty and inequality, food insecurity, and forced migration linked to climate change. Climate change adaptation and mitigation policies have become the center of public discourse as policymakers and the broader public witness daily that the unregulated freedom to pollute, overconsume, and overpopulate threatens the foundations of collective human survival.

In response to mounting environmental degradation and the continued overexploitation of global commons, the United Nations renewed its call for a coordinated international effort to safeguard the planet, mitigate irresponsible practices, and remediate ecological damage. This global appeal culminated in the establishment of comprehensive sustainability frameworks, first through the Millennium Development Goals (MDGs) for the period 2000–2015 and then the Sustainable Development Goals (SDGs) spanning 2015–2030.

Despite forward-looking goals, the United Nations lacks the enforcement capacity to ensure the realization of these MDGs and SDGs, relying instead on the political will of national governments to implement appropriate policies, legislation, and regulatory frameworks. Consequently, the responsibility for protecting the commons, and by extension, the future of life on Earth, ultimately lies with the political systems of individual states. National political institutions matter regarding the official policies, laws, and regulations that aim to protect environmental sustainability and citizens during global climate change and resulting environmental disasters (Wilson 2020). Responsibility is particularly placed on democratic governments for mitigating environmental degradation and disaster outcomes due to their principal goal of protecting citizens' health and well-being and guaranteeing environmental sustainability. Scholars voice support for democracy as a prerequisite for environmental protection and sustainable development practices, believing that democracies are most capable of implementing the necessary ecological measures (McCloskey 1983, p. 157) and that spreading democracy is essential for achieving better environmental policies (Al Gore 1992, p. 179). Democratic and transparent regulations are easier for citizens to follow, allowing them to monitor government responses to environmental and sustainability problems, whereas authoritarian regimes rarely prioritize environmental conservation (Payne 1995). Countries such as the Soviet Union, Putin's Russia, and Nazi Germany are labeled as environmental criminals (Laakonen et al. 2019). Subsequently, democratic governments are more accountable for mitigating environmental degradation and disaster outcomes because they prioritize protecting citizens' health and well-being while ensuring environmental sustainability. Democracies provide more substantial support to citizens than autocracies during environmental disasters, with protective and rebuilding policies in place (Wejnert 2021).

Margrethe Winslow (2005) presents empirical evidence to support a relationship between democracy and one aspect of environmental quality: urban air pollution. She explores the connection between environmental quality and democracy using a regression analysis of urban air concentrations of three pollutants—sulphur dioxide (SO2), suspended particulate matter (SPM), and smoke—along with two measures of democracy: the Freedom House Index and Polity III. The results suggest a significant and robust negative linear relationship between these pollutant concentrations and the level of democracy: the higher the level of democracy, the lower the ambient pollution level. Furthermore, democracies provide incentives for developing an open market of ideas for environmental problem-solving, learning from policy successes and failures in other democracies, sharing preventive and cleanup technologies, and communicating environmental information and knowledge. In this sense, democracies tend to serve as "policy laboratories" for one another; successful innovations in one spark imitation by others, and the innovations diffuse through international cooperation and support from international organizations, as well as through open markets for environmental protection measures and actions. Consequently, democracies engage in the global exchange of environmental protection ideas, sharing preventive and cleanup technologies and disseminating environmental information and knowledge. In the global economic market, democracies respond more vividly to green consumerism, encouraging corporate conduct towards environmentally friendly production (Povitkina 2018).

In contrast, many scholars argue that autocracies (that is, non-democracies) are better equipped to respond to environmental degradation and disasters. The autocratic regimes provide a more robust response when societies exceed "limits of growth," meaning that when societies overconsume, pollute the environment, and procreate unlimitedly. Autocracies more efficiently and quickly mitigate excessive consumption and growth. Centralized power, authoritarian decision-making, and the imposition of control policies allow a fast and efficient governmental response to the "tragedy of the commons," such as food scarcity or overpopulation (Hardin 1968). In contrast, liberal democracies that stress individual liberty and do not impose control measures can potentially cause an ecological catastrophe. Autocratic governments can impose regulations and taxes that curb consumption and pollution to offset businesses' environmentally destructive practices and private consumption. In contrast, democratically elected decision-makers are concerned about their electorates and are less likely to impose limitations (Schweickart 2010). Accordingly, as Wejnert's (2021) analysis claims, some authors argue that history helps us to understand better the possible social and environmental repercussions when democracy erodes and environmental policies of authoritarian leaders proceed unchallenged (Wilson 2020). Wilson argues that the most egregious examples of disastrous twentieth-century environmental governance were collectivization campaigns in the communist Soviet Union, in China during the Mao period and in Nazi Germany. In the Soviet Union, five million farmers suffered severe famine (the so-called Holodomor), which caused the death of about 3.9 million of them. Agricultural production collapsed, and agricultural soil eroded. In China, rural Chinese people were forced to build thousands of backyard furnaces to melt farming equipment, cooking and eating implements, and other household items to forge steel. Predictably, nearly all the steel was useless, but peasants lost critical tools for farming and cooking food. By the end of Mao's rule, and with the collapse of communism, Russian and Chinese state control, and the lack of political opposition that could control government activity, led to widespread environmental degradation and pollution resulting from the countries' industrialization (Wejnert 2021, pp. 8–9).

Autocracies also fail to disclose information about environmental disasters, endangering other countries and often worsening the results of such disasters. A prime example is the official secrecy and the withheld information by the communist government regarding the 1986 Chernobyl disaster. During the incident, 50–185 million curies of radionuclides escaped into the atmosphere—seven times the radioactivity produced by the atomic bombs dropped on Hiroshima and Nagasaki, a figure that the Soviet regime kept from public knowledge. Moreover, the Soviet regime reported an official count of 31 fatalities. In contrast, the Red Cross claimed that hundreds of thousands of people died from radiation due to the 1986 Chernobyl accident (including from thyroid cancer; Alexievich 2006). Similarly, the Soviet regime kept information secret about another nuclear accident in the Urals (Payne 1995). Another instance is China's nearly 3-month silence and misinformation regarding the surge and rapid spread of the new coronavirus, COVID-19, which contributed to its global transmission. Within the first year, it was responsible for nearly a quarter of a billion deaths, and a year and a half later, the virus remained uncontrolled.

Most academic investigations present a more balanced approach, viewing each political system as both benefiting and harming the commons. Among these investigations is Elizabeth Freedman et al.'s (2005: 216) study of pre-democracy Latin America and the environmental regulations and policies of these countries. As the author concludes, autocratic regimes often use environmental laws as a means to appease opponents of authoritarianism. National environmental agencies were established symbolically during autocratic rule, but specific policies and laws were not developed until the democratic transition. Hence, as described by Wejnert (2021), the 1970s pre-democracy environmental agency in Brazil had only three employees and a very modest budget; however, after the democratic transition, the agency expanded to 6000 staff members. Similarly, in Chile, the transformation occurred after the end of the military junta, with a civilian democratic government increasing environmental agency personnel from 6 to 4758 and its budget from $76,000 to $21 million by 2000 (Freedman et al. 2005).

Although democracy enhances political measures regarding environmental outcomes and strengthens environmental protection through the establishment of protective laws and regulations, these measures are not always implemented. As Wejnert explains, "They include limits on CO_2 emissions, clean water acts, and residents' protection during disasters (Freedman et al. 2005). Freedman et al. (2005) and Hochstetler (2003) studied the environmental protection laws of Eastern European and Latin American countries, comparing the practices of new democracies to those during the autocratic period. The authors concluded that countries that have adopted democracy demonstrate better environmental protection practices and laws than authoritarian regimes. In communist Eastern Europe, governments designated one-third of the land as pristine nature reserves and protected forests while also initiating control over water and air pollutant emissions. However, areas of high pollution were often mixed with well-preserved nature. Moreover, the development of laws and regulations controlling pollution occurred after the democratic transition. During the communist period, anti-pollution regulations were ineffective; for instance, Poland's Silesian region and parts of the Czech Republic were among the most polluted areas in Europe. Notably, after the democratic transition, wealthier countries like Poland, the Czech Republic, Hungary, Slovenia, and Estonia have established a stronger rule of law and more robust environmental protection compared to poorer Eastern European nations" (2021, p. 8).

Most analyses of the effects of politics on the environment recognize that the relationship between political systems and environmental sustainability is much more complex and nuanced than the conventional dichotomy between democracy and autocracy suggests. Rather than following a straightforward path, the outcomes of political systems on environmental issues reflect an evolving interplay of both positive and negative effects, shaped by various factors, including governmental corruption, the frequency of climate-related emergencies, and the balance of power between central governments and local political entities. As Wejnert (2021) describes, Povitkina (2013), in her study analyzing environmental policies of 144 countries from 1970 to 2011, found that governmental corruption affects democratic countries' regulations related to carbon dioxide emissions; corrupted governments in democracies do not meet climate targets for CO_2 emissions and are similar to authoritarian regimes in terms of mitigating pollution. Controversial outcomes of democracies on environmental protection are confirmed by other studies showing that democracies may protect some aspects of the environment from deterioration but neglect the protection of other aspects. For example, Midlarsky's (1998) studies found that democracies are more likely to implement policies aimed at protecting large areas of forest and pristine land but are failing to reduce CO_2 emissions, halt deforestation, or combat soil erosion. Furthermore, despite the ambitious goals outlined in the Renewable Fuel Standards program proposed by the US Environmental Protection Agency during the Biden administration (Bracmort 2019), residents of large US cities, particularly in "sacrifice zones" where communities of color reside, were exposed to higher levels of air pollution than wealthy communities. This disproportionate exposure has led to increased risks of severe illness or death from COVID-19 (Wu et al. 2020) and to a pattern of socially determined environmental inequality that is evident in the Black Belt region of Alabama, where unsustainable practices exacerbate health disparities (Alawaduka et al. 2022).

A similar effect of socio-economic disparities on the variance in the response of democratic countries to mitigating environmental degradation can be observed, particularly in the case of building wind farms in England. According to a report by *The Ecologist* (2010), privileged social strata that have a more significant impact on policymaking hold back the creation of wind farms in their areas. Consequently, developers changed the planned location for new wind farm projects from the proposed prime sites in affluent communities to sites in deprived communities, not involving the local population in the planning (*The Ecologist* 2010). Therefore, looking at the data on the effects of environmental disasters on local communities, Stacia Ryder (2017) proposes to include the roots of social inequality in environmental solutions, and the intersectional framework of addressing the roots of social inequality while focusing on environmental justice would lead to a better understanding of environmental harms and disaster vulnerabilities, argues Ryder (2017). Therefore, it will generate more just and equitable planning, preparedness, response, and adaptation to increasingly extreme heat and cold conditions and climate-related environmental disasters. Such an intersection framework would account for citizens' race, class (income), gender, ethnicity, disability, and other inequalities. In his study on the interaction of global and national social inequality with the effectiveness of governmental response to environmental degradation, Jeffrey Sachs (2015: 45–70) supports Ryder's (2017) suggestion.

In countries ruled by authoritarian regimes, society finds a way to avoid imposed regulations. Jessica Teets (2017), studying modern Chinese policymaking, argues that the autocratic government's strict regulations can be altered by civil society, contradicting popular arguments that policymaking in authoritarian regimes lacks societal participation. The author demonstrates that, when China adopted strict regulations to control civil society by requiring registration of civil organizations with a supervisory agency, such organizations used the supervisory agency as an access point to policymakers and successfully imposed policy changes, including governmental response to environmental disasters. Also, autocratic China, the largest emitter of CO_2 and a significant global air polluter, has invested more heavily than any other nation in renewable energy initiatives, aiming to mitigate the adverse effects of its rapid industrialization and to protect the country from global climate change (Beeson 2018).

In both democratic and autocratic systems, the imperative to protect the commons has evolved to encompass three critical dimensions of state-level governance: (a) addressing the needs of the rapidly growing population of the global poor, who constitute the majority of the world's inhabitants; (b) embedding environmental awareness into everyday practices and fostering concern for well-being of local communities and households; and (c) acknowledging the intrinsic link between sustainable development and crises and war conflicts tied to access to common resources, such as water, forests, and food, as well as the inequalities exacerbated by overconsumption and social and climatic disparities (Redclift 2006).

Together, these findings highlight the need for a more nuanced and differentiated understanding of how regime characteristics influence specific environmental outcomes. The arbitrarily set binary categories of either beneficial or harmful responsiveness of democracies or autocracies to their citizens regarding environmental protection and climate change mitigation could be empirically misleading. Multiple factors shape the effectiveness of democratic versus autocratic regimes in responding to environmental sustainability challenges, underscoring the necessity of an intersectional approach to enhance governmental efficacy in addressing these challenges (Ryder 2017; Sach 2015, pp. 45–70). Consequently, a more precise, data-driven investigation will provide a deeper understanding of the relationship between political systems and environmental sustainability outcomes.

Data-Driven Evidence of the Political System's Impact on the Protection of the Commons

The final section of this study uses two datasets, V-Dem (Coppedge et al. 2023) and the World Bank (2023), to examine the relationship between political regime type and its impact on environmental protection. The V-Dem dataset adopts a multi-dimensional approach to assess democracy and autocracy, measuring these political systems along a spectrum rather than as binary opposites. This method offers a more nuanced and comprehensive understanding of political systems and their evolution. The World Bank dataset provides annual records of individual countries' efforts to mitigate climate change, detailing responses to environmental challenges and the protection of the commons. These records include reductions in air pollution, CO_2, and greenhouse gas (GHG) emissions, as well as the protection of plant and animal species, preservation of forests, and progress toward transitions to clean energy. By combining these datasets, the study enables an assessment of how the strength and nature of the political system influence the effectiveness of countries' efforts to protect the commons and achieve sustainability goals.

Specifically, initially introduced by the Polity IV datasets (Marshall et al. 2014), the 0–10 democracy scale of the V-Dem dataset (Coppedge et al. 2023), where 0 means no democracy and 10 represents a strong, well-established democracy, evaluates a broad range of democratic attributes beyond the mere existence of elections. It includes liberal, participatory, deliberative, and egalitarian dimensions, offering a comprehensive measure of the depth and quality of democratic governance. This approach aligns with prevailing scholarly perspectives that political systems differ in ensuring inclusiveness, equality of rights, and equitable resource distribution among diverse citizen groups. For instance, while some democracies may respond inclusively to the needs and concerns of all citizens, others may limit responsiveness to specific segments of the population. Similarly, access to political expression and public resources may be universally available in some democratic regimes but restricted in others. Held (1995) refers to these variations in form and substance as different models of democracy, highlighting the spectrum on which democratic governance can fluctuate between more and less inclusive and equitable structures.

At the same time, autocratic regimes also vary widely in the extent to which they restrict civil liberties, suppress political opposition, and centralize power, resulting in differing impacts on environmental sustainability. To examine the characteristics of autocratic systems using an autocracy scale ranging from 0 (non-autocracy) to 10 (strong autocracy), this study draws on the V-Dem dataset (Coppedge et al. 2023), assessing not only regime type but also the processes of autocratization (Geddes et al. 2021), including the erosion of civil liberties, the intensification of political polarization, the suppression of opposition, and the spread of disinformation—factors that collectively contribute to the weakening of democratic institutions and the consolidation of authoritarian rule. Although the literature on democracy remains more extensive, the global rise of authoritarianism and the erosion of democratic norms since the early twenty-first century have intensified scholarly interest in studying autocracies. Emerging research explores whether autocratic leaders, particularly those reliant on regime performance for legitimacy or facing long-term governance horizons, are more attentive to improving citizens' living conditions, environmental protection, and human development. This includes studies on elected autocrats in competitive authoritarian regimes and monarchies that may achieve relatively strong outcomes in human development (Cassani 2021). Other research investigates whether the institutional design of autocracies affects their policy effectiveness, particularly in sustaining environmental and developmental goals. Croissant et al. (2014) and later Wuster (2014), challenging assumptions about democratic regimes' inherent environmental policy superiority, question whether democracies consistently outperform autocracies in ecological sustainability. National case studies further highlight the distinctive approaches of democratic and autocratic regimes in responding to environmental disasters and enacting protective environmental policies (Wejnert 2021). By systematically analyzing how autocratic versus democratic features and governance practices influence environmental sustainability outcomes, this study contributes to this fast-growing body of research.

The analyses conducted in this study yield several noteworthy empirical findings. First, they corroborate the mixed results in the literature review (see Table 1), underscoring the complexity of the relationship between regime type and environmental sustainability. Specifically, the findings challenge the conventional assumption that democratic regimes are categorically superior in safeguarding the environment. While stronger democracies demonstrate more robust performance in reducing pollution and protecting biodiversity, particularly in curbing species extinction and mitigating environmental degradation, this advantage does not extend uniformly across all dimensions of environmental protection.

The data indicates that autocratic regimes appears to have made more rapid strides in transitioning from fossil fuels to renewable energy sources despite their often limited transparency and accountability. This suggests that centralized governance structures in autocracies may facilitate a swift implementation of large-scale energy reforms, particularly when regime legitimacy is tied to performance outcomes. Conversely, democracies, while more inclusive and responsive to public environmental concerns, may face institutional constraints, political fragmentation, or interest group pressures that slow energy transitions.

In particular, the correlation and regression of environmental protection indicators with regime type demonstrate that the stronger a democracy is, the larger the area of protected terrestrial and marine regions becomes, as indicated by a statistically significant positive correlation score of +0.281. Additionally, a greater percentage of terrestrial land is protected, reflected by a positive correlation score of +0.338, and decreased deforestation by −0.107% of change annually (albeit the correlation is not significant) (see Table 1, Model 14). Democratic effects are also observed in marine life, with the enlargement of protected areas as a percentage of territorial waters (a statistically significant positive correlation score of +0.399). Democracies also protect threatened mammal (the correlation score is -.1, albeit is also not significant, indicates that fever mammal species are threatened in higher level democracies)and plant species (see Table 1, Models 4 &5), while no data is

Table 1 Correlations of political system and countries' protection of the commons worldwide: 1990–2023

Models	Democracy	Autocracy
1. Terrestrial and marine protected areas (% of total land)	0.281** (<.001) N=1067	
2. Terrestrial protected areas (% of total land)	0.338** (<.001) N=1068	
3. Marine protected areas (% of territorial waters)	.399*** (<.001) N=829	
4. Mammal species threatened	-.091 (.264) N= 153	
5. Plant species (Higher) threatened	.017 (.839) N=153	
6. PM 2.5 air pollution, population exposed to levels exceeding the WHO guideline value (% of total population)	−.431** (<.001) N= 413	.232** (<.001) N=411
7. Methane gas emission (% change since 1990)	−.10** (<.001) N=3319	.087** (<.001) N=1833
8. Total GHG emission (% change from 1990)	−.078* (.001) N=1772	-0.04 (.855) N=1766
9. Renewable electricity output (% of total electricity output)	-.055* (.014) N=1967	−.032 (.159) N=1960
10. Renewable energy consumption (% of total final energy consumption)	−.327 ** (<.001) N=1963	.151** (<.001) N=1956
11. Fossil fuel energy consumption (% of total)	.179** (<.001) N= 5300	−.127** (<.001) N=3727
12. Electricity production from coal sources (% of total)	.278** (<.001) N=5503	−227** (<.001) N=3797
13. Energy use/c (kg of oil equivalent per capita)	.237** (<.001) N=5440	−.202** (<.001) N=3805

Note: Data derived from V-Dem Dataset by Coppedge, M. et al. (2023), & World Bank (2023) Dataset
**Correlation is significant at the 0.01 level (two-tailed)
*Correlation is significant at the 0.05 level (two-tailed)

indicated for autocracies. Similarly, positive outcomes of increased democracy levels are noted concerning air pollution and greenhouse gas (GHG) emissions, where the increased strength of democracy correlates with declines in air pollution, methane, and GHG emissions, with correlation scores of −0.431, −0.1, and −0.078, respectively (see Models 1–3 in Table 1). In contrast, an increase in the level of autocracy correlates with rising air pollution, methane, emissions, with correlation scores of +0.232, +0.087, and with small decline in GHG emission -.04 , respectively (see Models 6–8 in Table 1). The higher autocracy levels also correlated with larger annual deforestation than observed in democracies of −.13% of change versus −.107 % of change, respectively (see Table 1, Model 14).

In contrast to democratic superiority in protecting the natural environment, autocracies are often viewed as more effective stewards of available energy sources and

Table 2 Regression analyses of democracy compared to autocracy effects on environmental sustainability worldwide, 1960–2023

Models	Constant	Democracy	Autocracy
1. Annual deforestation (% of change)	2.418 (*.416*) (<.001)	−.288** (*0.51*) (<.001)	−.397** (*.07*) (<.001)
2. PM2.5 air pollution-population exposed to levels exceeding WHO guideline value (% of total)	96.546** (.563) (<.001)	-1.062** (.077) (<.001)	.999 (.081) (<.001)
3. Methane emissions (% change from 1990)	8.872) 21.912** (2.67) (<.001)	-1.781** (.431) (<.001)	) -1.411** (.407) (<.001)
4. GHG emissions/removals by LUCF	−62.115 (*39.04*) (.112)	2.061 (*4.255*) (.628)	9.703 (*7.42*) (.191)
5. Total GHG emissions per capita (t CO_2 e/capita)	8.059** (.16) (<.001)	-.121** (.022) (<.001)	
6. Renewable energy consumption (% of total energy consumption)	38.3** (.531) (<.001)	-7.03** (.073) (<.001)	.451** (.077) (<.001)
7. Electricity production from renewable sources (excluding hydroelectric)	2.462 (*.345*) (<.001)	) .255** (.02) (<.001)	-.223** (.021) (<.001)
8. Electricity production from coal sources (% of total)	11.951** (.43) (<.001)	) (<.001) .952** (.059) (<.001)	-.918** (.062) (<.001)
9. Fossil fuel energy consumption (% of total)	29.302** (*2.22*) (<.001)	4.91 (*.258*) (<.001)	4.35 (*.313*) (<.001)
10. Total GHG emission (% change from 1990)	37.78** (1.06) (<.001)	-1.248** (.146) (<.001)	1.444** (.154) (<.001)
11. Renewable energy output (% of total energy output)	32.7** (.592) (<.001)	.497** (.081) (<.001)	-.596** (.085) (<.001)
12. Energy use/c (kg of oil equivalent per capita)	−1983.5** (*149.75*) (<.001)	527.85** (*17.42*) (<.001)	454.5** (20.969) (<.001)
13. Electric power consumption/c (kWh per capita)	−3565.18 (*260.2*) (<.001)	876.92 (*30.265*) (<.001)	632.02** (*36.42*) (<.001)

Note: Data derived from V-Dem Dataset by Coppedge, M. et al. (2023), & World Development Indicators (World Bank 2023)

*$p < 0.05$; **$p < 0.001$; coefficient at least twice its standard error. Standard errors in italics and parentheses

use more renewable energy than democracies. As autocracy increases, renewable energy consumption also rises, demonstrated by a significant correlation score of +0.151. This contrasts with democracies, where the decrease in renewable energy consumption correlates with an increase in democracy level, with a correlation score of −0.327 (see Table 1 , Model 9). Furthermore, fossil fuel consumption and electricity production from coal sources decline as the level of autocracy increases, with correlation scores of −0.127 and −0.227, respectively (See Table 1, Models 10–11). Conversely, higher levels of democracy are associated with increased energy use per capita, reflected in a correlation score of +0.327 (see Table 1, Model 12), as well as a higher reliance on non-renewable energy sources, including increased fossil fuel consumption and electricity production from coal sources (by +0.179 and +0.278, respectively) as a percentage of total energy consumption. This is coupled with an increase in renewable energy consumption by +.151% of total energy consumption despite a detectable, but not statistically significant, decline in renewable energy output (see Table 1, models 8–11). In contrast to democracies, a rise in autocracy level correlates with decreased energy use (−0.202 correlation score in Table 1, Model 12) and the output of renewable energy as a percentage of total energy production is declining in autocracies by −0.32% with the increase in autocracy level, and it also declines with increasing level of democratic system (although for both democracy and autocracy, the trends of renewable energy outputs are not statistically significant) (see Table 1, Model 9).

The multiple regression analysis confirms the correlation results, revealing both positive and negative effects of each type of political regime—democracy and autocracy—on protecting the commons. On one hand, the results affirm that democracies, in contrast to autocracies, are superior guardians of environmental sustainability, executing clean water and air acts and securing species protection. With each increasing level of a country's democracy, there is a corresponding higher level of environmental protection. As the regression analysis attests, for each one-point increase in the democracy level on a scale of 0–10, deforestation decreases by −.288%, air pollution declines by −1.06% of population exposed to levels exceeding WHO guideline value, and methane emissions decline by −1.781% as compared to 1990 (see Table 2, models 1–3). Contrarily, in autocracies, with the increase of autocracy level by one on a scale of 0–10, methane emission decreases less (by -1.41%) compared to the 1990 level, and an increase in air pollution by.99% of the population exposed to levels exceeding the WHO guideline value is observed, nonetheless, similarly to democracies, deforestation also reduces in autocratic regime with an increased level of autocracy. Moreover, the level of all GHG attributable to forest and land use (e.g., the increase of biomass stock due to fuel collection and forest management) increases to a larger extent than in democracies (increase by 9.7 vs 2.06 kt of CO_2 equivalent, respectively) (although the indicators are not statistically significant). The total GHG emissions per capita in democracies are reduced contrasting an increase in autocracies (decline of -.121 vs an increase of +. 196 CO_2 e/capita, respectively) (see Table 2, Models 4–5).

Considering energy production and consumption, democracies lag behind autocracies in rapid transitioning to sustainable, green energy. Democracies consume more energy per capita than autocracies (876.9 vs 632.0 kWh per capita, respectively), and energy consumption increases as countries achieve higher levels of democracy; however, the transition, demands, and consumption of alternative energy are lower than in autocracies except for higher energy production from renewable sources excluding hydroelectric)(see Table 2, Models 7, 11 –13). Moreover, the consumption of renewable energy as percentage of total energy consumption decreases with higher energy consumption by a larger percentage in democracies while increases in autocracies (by −7.03% in democracies versus +.451% in autocracies), however, the renewable energy output as a percentage of total energy output increases in democracies whereas declines in autocracies (+.497% compared to −.596%, respectively) (see Table 2, Models 6, 11). Nonetheless, in democracies, fossil fuel energy consumption and electricity production from coal sources exceed the levels in autocracies (+4.9% of total energy from fossil fuel in democracies versus 4.3% in autocracies, and +.95 vs -0.91% of total electricity production from coal sources) (see Table 2, Models 8 and 9). Subsequently, the contrast in renewable energy consumption attests to the less impactful outcomes of democratic ruling compared to autocratic regimes on countries' rapid transition to better for the environment, non-fossil fuel energy, however, more substantial and increasing in higher democracy levels renewable electricity production, contrasting declining renewable electricity production in autocracies, indicate that in the long term, democracies demonstrate more sustainable protection of natural resources.

The study thus supports multilayer outcomes of political regimes on environmental sustainability and the protection of the commons, aligning with scholars arguing for mixed impacts of political systems. Several examples of current policies of democratic and authoritarian systems support these empirical findings. For instance, authoritarian China is the largest emitter of CO_2 in the world and one of the greatest polluters. Nonetheless, it has also invested more than any other country in renewable energy to substantially mitigate the effects of rapid industrialization (Beeson 2018). Likewise, Midlarsky's (1998) examination of the effect of democracy found that the statistically significant effect of democracy on the environment was both positive and negative. Regarding protected land areas, the impact of democracy was positive. However, no significant effect of democracy on the environment was indicated concerning the impact of chemicals on freshwater availability and soil erosion.

Another study underlines that the greater autonomy of local governments in democratic countries provides an opportunity for local environmental decision-making to challenge the policies of a central government. This practice was represented in the pushbacks from the state governments against the first Trump presidency's proposal to substantially reduce the Renewable Fuel Standards program (Bracmort 2019). The response of elected policymakers at the local governmental level oscillates between being more and less responsive to environmental protection and climate change mitigation policies, laws, and practices. For example, Slavin (2011: 188) notes that former New York City Mayor Michael Bloomberg announced on Earth Day in 2007 the MillionTreesNYC Campaign, which aims to plant a million trees by the year 2030. This initiative exemplifies the impact of local policymakers and regional variability within democratic countries regarding mitigating environmental pollution effects. As mayor, Bloomberg increased energy efficiency and reduced New York City's carbon footprint by 13%, helping make its air the cleanest it had been in 50 years. The role of local policymakers in protecting the environment further extenuates the study by Povitkina (2018), which shows that widespread corruption in local democratic governments reduces their capacity to meet climate targets and lower CO_2 emissions.

Therefore, a political system's effectiveness in protecting the commons depends not solely on its democratic or autocratic nature but also on contextual factors such as governance quality, institutional capacity, the impact of local politics, and regime priorities.

Concluding Notes on the Significance of Political Regimes' Protection of the Commons

According to Hardin and others, finding solutions to avoid the "Tragedy of the Commons" is crucial for human survival and environmental sustainability. However, several scholarly investigations show that protecting the global commons is broader than safeguarding nature and human survival; it also involves securing and maintaining world peace. The seemingly divergent topics of environmental and human survival on one hand and peace on the other are intricately linked. Von Weizsäcker (2019) asserts that "there can be no peace with nature without peace among men." The author claims that one of the most significant barriers to a potentially peaceful modus vivendi between the superpowers is their mutual perception of each other, founded on a sense of justice and just distribution and use of scarce goods. Similarly, Blundell and Harwell (2016) claim that failing to improve the governance of natural resources "increases the risk of conflict recurrence because access to natural resources is an especially valuable prize worth fighting for." This position, also supported by Rustad and Binningsbø (2012), may explain why peace is so difficult to maintain. The link between maintaining peace and protecting nature and the commons is most straightforwardly demonstrated by analyzing how democracies and autocracies view the protection of a sustainable future and how often autocracies' expansionist tendencies lead to wars, as exemplified by Russian military attacks on Ukraine, the war between India and Pakistan over Kashmir, the war in Gaza, and other current wartime events that have undermined prolonged European and global stability and peace.

Unlike leaders in democratic systems, heads of major autocracies typically do not prioritize environmental protection in their political agendas. This tendency arises from the high costs of implementing environmental safeguards and the lack of immediate political returns. Since autocratic regimes operate without electoral accountability, their survival depends on maintaining elite loyalty and suppressing dissent, often through extensive control mechanisms and patronage networks. The financial burden of societal control and elite appeasement incentivizes authoritarian leaders to pursue immediate gains, including the increased extraction of global commons resources. This resource-driven strategy promotes expansionist behavior and raises the likelihood of military aggression, as autocracies seek to assert control over scarce resources and enhance regime legitimacy through nationalist ventures (Lake 1992).

The expansionist orientation of autocracies diverts resources away from long-term environmental investments in favor of military expenditures. One possible explanation for the higher performance of some autocracies in transitioning to renewable energy lies in their pursuit of increased energy capacity to support military production, an interpretation that merits future empirical investigation but is supported by China's current policies of rapid transition of its energy toward renewable energy sources. Expansionism also contradicts the ethos of international environmental cooperation, which requires collective restraint and adherence to mutually agreed-upon norms for managing shared resources that are typical for democracies (von Weizsäcker 2019).

Democracies are structurally inclined towards cooperation and sustainability because political leaders in democratic regimes, who depend on electoral approval, are more responsive to citizen demands for environmental protection and are more likely to pursue multilateral agreements aimed at conserving global resources. They are also less prone to initiating unpopular wars and are more likely to form alliances to counter expansionist autocracies—alliances that enhance their chances of prevailing in conflicts (Tangerås 2009; Lake 1992). These dynamics reinforce the characterization of democracies as having a more pacifistic orientation and, by extension, as more reliable stewards of the global commons. Unsurprisingly, despite the advances made by autocracies in transitioning to renewable energy, democratic regimes provide a framework that is more conducive to protecting humanity's shared sustainable future and are thus, unsurprisingly, more frequently cited as stronger preventers of the global tragedy of the commons.

References

Aladuwaka S, Wejnert B, Alagan R, Mishra M (2022) Social determinants of health disparities and COVID-19 in Black Belt communities in Alabama: geospatial analyses. In: Systemic inequality, sustainability and COVID-19. Emerald Publishing, pp 3–32

Albertazzi, D., & McDonnell, D. (2008). Twenty-first-century populism: The spectra of western European democracy. London, UK: Palgrave Macmillan.

Alexievich S (2006) Voices from chernobyl: The oral history of a nuclear disaster. Dalkey Archive Press, Normal, IL

Applebaum A (2020) Twilight of democracy: the seductive lure of authoritarianism. Vintage

Applebaum A (2024) Autocracy Inc. The dictators who want to run the world. Doubleday, New York

Bakermans M, San Martín W (2021) The ultimate demise of the world's lungs: the deforestation of the Amazon. Extinction Stories. Pressbooks. pub

Beeson M (2018) Coming to terms with the authoritarian alternative: the implications and motivations of China's environmental policies. Asia Pac Policy Stud 5(1):34–46

Blavascunas E, Cope B (2022) The wild border of politics: the Białowieża forest and changing ecologies of protest. Ethnologia Polona 43

Blundell AG, Harwell EE (2016) How do peace agreements treat natural resources? Forest Trends Report Series: Forest Trade and Finance Report, January

Bracmort K (2019) The Renewable Fuel Standard (RFS): an overview (Rep. No. R43325). Congressional Research Service (CRS), Washington, DC

Cassani A (2021) Varieties of autocracy and human development. In: Research handbook on democracy and development. Edward Elgar, pp 135–150
Cianetti L, Dawson J, Hanley S (eds) (2019) Rethinking democratic backsliding in Central and Eastern Europe. Routledge, Abingdon
Coppedge, M., Gerring, J., Knutsen, C. H., Lindberg, S. I., Teorell, J., Alizada, N., Altman, D., Bernhard, M., Cornell, A., & Fish, M. S. (2021). V-Dem dataset v11.1. https://www.v-dem.net/data/the-v-dem-dataset/
Coppedge M, Gerring J, Knutsen CH, Lindberg SI, Teorell J, Alizada N et al (2025) V-Dem dataset v15. V-Dem (Variety of Democracies). Global Standards, Local Knowledge. V-Dem Institute, Department of Political Science, University of Gothenburg, Sweden. https://v-dem.net/data/the-v-dem-dataset/
Czajkowski M, Buszko-Briggs M, Hanley N (2009) Valuing changes in forest biodiversity. Ecol Econ 68(12):2910–2917
Croissant A, Kailitz S, Koellner P, Wurster S (eds) (2014) Comparing autocracies in the early Twenty-first Century: Vol 2: The performance and persistence of autocracies, 1st edn. Routledge. https://doi.org/10.4324/9781315752853
Ehrlich PR (1968) The population explosion and ecology. Ballantine, New York
Freedom House. (2018, December 22). Nations in transit, 2017: The false promise of populism. Freedom House. Retrieved from https://freedomhouse.org/article/nationstransit-2017-false-promise-populism.
Friedman EJ, Hochstetler K, Clark AM (2005) Sovereignty, democracy, and global civil society: state-society relations at UN world conferences. SUNY Press
Galloway, S. (2022). Adrift. America in 100 charts. New York: Blackwells
Geddes B, Wright J, Frantz E (2018) How dictatorships work: power, personalization, and collapse. Cambridge University Press
Giraudi, G. (2018). Populism: What is and what we need a multidimensional approach to understand it. European Scientific Journal, pp. 14, 16–29.
Hardin G (1968) The tragedy of the commons: the population problem has no technical solution; it requires a fundamental extension in morality. Science 162(3859):1243–1248
Held D (1995) Democracy and the global order: from the modern state to cosmopolitan governance. Stanford University Press
Hochstetler, K. (2003). 'Fading green: Environmental politics in the Mercosur Free Trade Agreement.' Latin American Politics and Society 45 (4), 349–371. https://doi.org/10.2307/3177129.
Laakonen, S., Tucker, R. and Vuorisalo, T. (2019). The Long Shadows: A global environmental history of the Second World War. Corvallis: Oregon State University Press.
Lake DA (1992) Powerful pacifists: democratic states and war. Am Polit Sci Rev 86(1):24–37. https://doi.org/10.2307/1964013
Levitsky, S., & Ziblatt, D. (2018). How democracies die. New York, NY: Crown
Liu L (2023) Covid-19 excess mortality in China: a regional comparison. medRxiv:2023–2006
Marshall MG, Gurr TR, Jaggers K (2014) Polity IV project: political regime characteristics and transitions, 1800–2013. Center for Systemic Peace, p 5
Midlarsky MI (1998) Democracy and the environment: an empirical assessment. J Peace Res 35(3):341–361
Moffitt, B. (2016). The global rise of populism. Performance, political style, and representation. Stanford, CA: Stanford University Press
Mudde, C., & Kaltwasser, C. R. (Eds.). (2012). Populism in Europe and the Americas. Cambridge, UK: Cambridge University Press.
Müller, J.-W. (2016). What is populism? Philadelphia, PA: University of Pennsylvania Press. Reprinted in 2017 by Penguin Books, Random House, New York, NY
Nobre CA (2019) To save Brazil's rainforest, boost its science. Nature 574(7779):455–456
Ophuls W (1977) Ecology and the politics of scarcity: prologue to a political theory of the steady state. W.H. Freeman, New York

Payne RA (1995) Freedom and the environment. J Democracy 6(3):41–55
Povitkina M (2018) The limits of democracy in tackling climate change. Environ Polit 27(3):411–432
Przeworski, A. (2018). Mozemy lepiej! (We can do better!). Interview with Adam Przeworski. Polityka, p. 1. Taggart, P. (2004). Populism and Representative Politics in Contemporary Europe. Journal of Political Ideologies 9 (3), 269–288.
Redclift MR (2006) Sustainable development (1987-2005): an oxymoron comes of age. Horizontes Antropológicos 12:65–84
Roy D (2022) Deforestation of Brazil's Amazon has reached a record high. What is being done? Council Foreign Rel 24
Rustad S, Binningsbø H (2012) A price worth fighting for? Natural resources and conflict recurrence. J Peace Res 49(4):531–546
Ryder SS (2017) A bridge to challenging environmental inequality: intersectionality, environmental justice, and disaster vulnerability. Social thought & research: a continuation of the mid-American review of sociology, pp 85–115
Sachs JD (2015) The age of sustainable development. Columbia University Press
Slavin MI (ed) (2011) Sustainability in America's cities: creating the green metropolis. Island Press
Schweickart D (2010) Is sustainable capitalism possible? Proc Soc Behav Sci 2(5):6739–6752
Tangerås TP (2009) Democracy, autocracy and the likelihood of international conflict. Econ Gover 10:99–117
Teets, J. (2017). 'The power of policy networks in authoritarian regimes: Changing environmental policy in China.' Governance 31 (1), 125–141.
The Ecologist. (2010) 'Privileged opposition holding back wind farm development.' 10 September. Accessed 10 September 2021. https://theecologist.org/2010/sep/10/privileged-opposition-holding-back-wind-farm-development.
Tollefson J (2013) A light in the forest: Brazil's fight to save the Amazon and climate-change diplomacy. Foreign Aff 92:141
Tollefson J (2014) The case of the missing heat. Nature 505(7483):276
von Weizsäcker CF (2019) Justice, peace, and the preservation of nature. In: International conflict resolution. Routledge, pp 231–259
Wilson R (2020) Authoritarian environmental governance: insights from the past century. In: Environmental governance in a populist/authoritarian era. Routledge, pp 14–23
Winslow M (2005) Is democracy good for the environment? J Environ Plann Manag 48(5):771–783
Wejnert B (2021) Educating on democracy in a time of environmental disasters. Lond Rev Educ 19(1):1–16
Wejnert, B. (2025). Examining the tragedy of the commons in the era of democracy erosion and autocracy rise. Opinion essay. International Affair Forum, May, 63–77.
World Bank (2023) World Development Indicators. https://databank.wordlbank.org/source/world-development-indicators
Wu, X., Nethery, R. C., Sabath, M. B., Braun, D., & Dominici, F. (2020). 'Air pollution and COVID-19 mortality in the United States: Strengths and limitations of an ecological regression analysis.' Science Advances 6 (45), eabd4049. https://doi.org/10.1126/sciadv.abd4049.
Wuster S (2014) Comparing ecological sustainability in autocracies and democracies, ch.8. In: Croissant A, Kailitz S, Koellner P, Wurster S (eds) Comparing autocracies in the early twenty-first century: Vol 2: The performance and persistence of autocracies, 1st edn. Routledge. https://doi.org/10.4324/978131575285

Barbara Wejnert is a Professor in the Department of Environment and Sustainability at the University at Buffalo, former Associate Professor of Transnational Studies, and the founding Chair of the Department of Global Gender Studies. She has also served as a faculty member and Director of Academic Programs at Cornell University. While her research broadly addresses global issues, particularly economic development, diffusion of democracy, and the effects of democratization on the well-being of women and girls, her current projects examine the impact of democratic and autocratic regimes on global climate change politics, as well as the acculturative experiences of immigrant and refugee women in Buffalo, NY. She is an author and editor of several books, including *Diffusion of Democracy* (2014), published by Cambridge University Press, and a database, *Nations, Development and Democracy* (2007). She is affiliated with the University at Buffalo's School of Social Work through the Immigrant and Refugee Research Institute (IRRI) and the Jaeckel Center for Law, Democracy, and Governing. She holds a Ph.D. in Sociology from Mickiewicz University in Poznań, Poland.

How Politicians, Unelected Officials, Judges, and Voters Obstruct or Encourage Democratic Backsliding

Barbara Geddes

Abstract Democratic backsliding occurs when a democratically elected executive grabs more power for himself at the expense of political actors embedded in other institutions and/or extends his time in office beyond his legal term. Institutions can only constrain backsliders if the human beings who embody them refuse to go along with executive power grabs. This essay argues that the choices of these individuals about how to respond to backsliding efforts depend on how their decisions will affect their future careers and welfare, as well as their values. For politicians and officials whose future well-being depends on the executive's goodwill, acquiescence is the only sensible option even if they sincerely oppose his actions. In contrast, officials whose future success depends on their reputations for competence, fairness, and integrity are more likely to stand in the backslider's way. I use this argument to make sense of the contrast between the supine behavior of elected members of the US Congress, threatened by primary challenges, and the way unelected officials, judges, and military officers blocked various Trump initiatives during his first term. In the last pages, I interpret Trump's barrage of executive orders and attacks on the personnel and funding of the government during the first weeks of his second term as efforts to persuade the kinds of officials who blocked his initiatives during his first term that their future well-being is more dependent on him than US laws and norms might have led them to believe.

Keywords Backsliding · Authoritarianism · Autocracy · Dictatorship · Democracy

---

B. Geddes (✉)
Political Science Department, University of California, Los Angeles, CA, USA
e-mail: geddes@ucla.edu

B. Wejnert (ed.), *Democracy in Retreat and the Climate Crisis*, Springer Studies on Populism, Identity Politics and Social Justice,
https://doi.org/10.1007/978-3-032-02881-5_3

Introduction

Why do some backsliders take their countries all the way into dictatorship, while the autocratizing initiatives of most others are blocked? I define democratic backsliders as political leaders initially chosen in free and fair elections who initiate changes in political institutions, laws, norms, and personnel that increase executive power at the expense of other branches of government, undermine institutions that would ordinarily limit arbitrary executive actions, reduce the authority of state agencies to which the implementation of policies chosen by earlier elected governments have been delegated, interfere with election machinery, and/or increase their control over the media and the distribution of state resources to citizens and businesses in order to tilt the electoral playing field. The backslider's repertoire is extensive, and no individual backslider does all the things mentioned here. The changes initiated by backsliders, though sometimes individually inconsequential, move their political systems toward the autocratic end of the democracy-autocracy continuum, while aiming to perpetuate themselves in power.

Usually a democratically elected executive who tried to initiate such changes would be stopped by other political actors, who have incentives to protect their own institutions from loss of authority and funding, and by officials and institutions given monitoring and enforcement powers by the laws, the constitution, and the courts. For backsliding to occur, these other political actors have to acquiesce or side with the executive.

In this essay, I build on what we know about why some fairly elected executives have more success than others in grabbing power from other political actors who, according to democratic laws and norms, should share it. Recent empirical research has shown that new democracies and recently formed ruling parties are more susceptible to power grabs by elected executives than are older democracies and ruling parties with long histories (Svolik 2008; Frantz et al. 2023). The authors argue that the concentration of power in a party leader's hands happens more often in newly created parties. They then show that the absence of constraints from within their support parties contributes to further power grabs if the party leader is elected to national executive office, especially if he can count on a legislative majority as well.

New parties are more common in new democracies because earlier non-democratic regimes often hinder the formation of opposition parties. I further develop these arguments about *why* older democracies and ruling parties resist autocratization better than younger ones, as well as why some democratic institutional configurations impede executive power grabs better than others. None of this means that any democracy is immune to potential backsliding, but executive power grabs have been more seamless and uncontested in some places than others.

My argument begins by noting that political institutions—which include parties, constitutional provisions that assign powers to different branches of government, all the rules that govern elections, and democracy itself—are just the individual human beings embedded in roles that give them certain responsibilities, powers, and perquisites. Institutions are defined by the people who abide by the laws, rules, and norms embedded in them, as well as those who enforce the laws, rules, and norms when challenges occur. When a democratically elected executive tries to grab power for himself at the expense of political actors embedded in other institutions, other political players can obstruct the effort, shirk or slow implementation, or support the executive. Institutions can only constrain the executive if these individuals refuse to acquiesce in the power grab, possibly at the risk of damage to their lives, livelihoods, and futures.

I suggest that the decisions made by these individuals depend on how their choices will affect their own futures in addition to their values. For politicians, unelected officials, judges, and military officers who see their future careers and even well-being as dependent on the executive's goodwill, acquiescence is the only sensible option, even if the individual believes the executive's choice is wrong-headed or damaging to national welfare. In contrast, officials and officers who see their future success as depending on their own reputations for competence, wisdom, and fairness will try to make decisions consistent with existing legal norms and procedural rules that govern the professional spaces they occupy, along with their understanding of the best policy in the circumstances. These incentives mean that they sometimes oppose backsliding initiatives even if they share the backslider's party affiliation.

If we think of backsliding as a strategic game, executives make the first move when they try to grab more power or not. Other politicians, officials, and judges make the next move; they can oppose or support the executive. The paragraph above discusses the costs and benefits that determine these second movers' payoffs. The game ends after the second move for unelected officials and judges. However, voters then weigh in on the futures of elected officials, including the executive. As last movers, voters should in principle have a substantial influence on what political elites do.

In general, voters have not punished backsliding executives for actions that diminish democracy despite telling survey researchers that they support democratic government. Instead, they support backsliders as long as the economy is doing reasonably well, corruption is either no worse than usual or hidden from citizens, the streets are safe, and so on—the same decision calculus as for candidates fully committed to democracy. Earlier research has suggested a number of reasons for voters' surprisingly strong support for backsliders, which I discuss below.

Past research has given little attention, however, to the way voters respond to elected legislators and judges who oppose backsliding initiatives. Most noteworthy, some voters have apparently punished Republican members of Congress who have voted against President Trump's backsliding efforts or criticized his supporters' riot on January 6, 2021 (Bartels and Carnes 2023).

In the remainder of this essay, I first describe two backsliding histories, that of Victor Orbán, prime minister of Hungary, and that of Donald Trump, the past and current US president. The purpose of these fairly extensive descriptions is, first, to be sure that readers can picture what backsliding means in practice. The second goal is to show how different initial democratic institutional contexts affect the kinds of strategies available to backsliding executives and likely to be successful. Hungary's institutional configuration, which included an electoral system that significantly advantaged the plurality winner of elections and allowed constitutional change by a simple parliamentary supermajority, facilitated Orbán's rapid transition to autocracy relative to Trump's mostly blocked backsliding efforts during his first term. Third, by comparing Trump's efforts during his first term with what we know so far about his second-term strategy, I illustrate how the incentives associated with different political roles permitted some officials, judges, and military officers to obstruct Trump's backsliding impulses during the first term and how he seeks to change those incentives as he begins the second.

In the next section, I further elaborate on my argument about why other political actors sometimes obstruct backsliders but often do not. I also describe the way these processes have played out in different political systems. I discuss in more detail the ways that politicians and unelected officials of various kinds are likely to respond to executive power grabs. I interpret a number of Trump's recent actions, which most media reports see as expressions of personal vengefulness, as efforts to change the perceptions of unelected political actors about the likely costs of standing up to Trump. Finally, I describe the role voters have in perpetuating backsliders.

What Backsliders Do

Fairly elected executives who push their political systems toward dictatorship usually do so as part of an effort to change aspects of the status quo that probably could not be changed through strictly democratic means. They may sincerely favor these changes, or they may aim to reward important supporters or reduce the authority of other political elites in order to protect themselves from legal or political challenges. Democratic political systems tend to have a pro-status quo bias, which can frustrate reformers from the left as well as from the right. To name a few of the features that create the bias: separation of powers systems require more widespread support in order to change the status quo than do systems that lack separation of powers; most elected legislatures in the world are malapportioned to some degree in ways that favor rural areas, which tend to be more traditional, over urban; courts in most systems put a strong emphasis on precedent and the preservation of the constitution and current laws. In most countries, supermajorities are required to change constitutions. These features and others specific to particular democratic political systems make it difficult to alter the status quo in dramatic ways.

To make bigger changes, some backsliders begin with attempts to change the constitution in order to facilitate government responsiveness to citizen demands, they claim. When Viktor Orbán's party (Fidesz, the Hungarian Civic Alliance) won Hungary's 2010 parliamentary election, he became Prime Minister for the second time and immediately rewrote the Constitution to increase executive power and enshrine Christian, nationalist, and traditional values. Orbán amended the Constitution 12 times during his first year (Scheppele 2022). This was easy to do because Fidesz controlled two-thirds of the seats in parliament, and the constitution could be amended by a simple two-thirds majority. Furthermore, as one of the founders of his party and its leader during much of its history, Orbán exercised substantial personal control over the party, reducing the likelihood that its members would oppose anything he proposed (Frantz et al. 2023).

At the time of Orbán's election, the Hungarian Constitution and electoral laws, originally written as part of the transition from communism, provided a large seat-share advantage for the party that won the largest plurality in parliamentary elections. When the new democratic political system was being negotiated, the Communist Party demanded the continuation of single-member district electoral constituencies, in which it expected to have an electoral advantage because of its access to party resources and the earlier development of local patronage machines by Communist Party legislators. A compromise between the Communists and opposition-party leaders resulted in Hungary's mixed electoral system, in which a little more than half of members of parliament are elected in single-member districts under first-past-the-post rules, and the rest are elected by proportional representation in a single national district (Geddes 1995). Under first-past-the-post rules, the party that wins the most votes is the only one to gain representation regardless of how small the plurality is or how close the election. Thus, in countries in which the same party wins a plurality in most regions, single-member districts result in a large advantage for the largest party in the translation of votes to seats.[1] Constitutional changes were possible during Orbán's first year back in office because his party, which won 52.7% of the vote, gained a 68% supermajority of seats under these rules (Bánkuti et al. 2012).

If considered as part of an assessment of backsliding, the most important constitutional changes were the ways it limited judicial powers. (During his earlier term, Orbán's backsliding initiatives were obstructed by the courts and parliament, which his party did not fully control.) The new constitution limited the constitutional court's right to review legislation. As a further attack on the judiciary, the retirement age for judges was reduced from 70 to 62 years, forcing the retirement of hundreds of judges, who were replaced with Orbán allies. The law provides for raising the retirement age back to 65 in a few years, thus extending the appointments of the newly appointed Orbán-allied judges. (Other government officials were not required to retire.) New provisions in the Constitution must be implemented by "Cardinal Acts" of the legislature, which require a two-thirds majority to pass or change, making it almost impossible for the opposition to revise them even if it wins a plurality in a future election. Cardinal Acts are not subject to constitutional review, so they cannot be reversed by the courts. These areas of legislation include family policy, the pension system, and taxes (Feher 2011). In short, large areas of Hungarian policy are no longer subject to judicial oversight.

The new Constitution also gave Hungarians living outside Hungary the right to vote. Over 90% of the diaspora votes for the ruling party. Before the next election in

[1] In the United States, where party domination varies by region, first-past-the-post rules for electing members of Congress result in a fairly even distribution of seats between the two parties advantaged by the electoral system.

2014, the government gerrymandered electoral districts to give the ruling party an even greater seat-to-vote advantage (*Economist* 2022). Since 2014, the ruling party, often in coalition with a much smaller party, has continued to win at least two-thirds of the seats in parliament, despite usually winning less than a majority of votes.

In addition to direct changes in the political system, Orbán has taken control of most traditional media. He replaced the leadership of public television and radio shortly after coming to office in 2010. Their news broadcasts now reflect the talking points provided to them by the ruling party and the think tanks and pollsters funded by the party and government. They have been criticized for showing Fidesz campaign videos during news programs. Since taking over public broadcasting, Orbán and the ruling party have also gained control of most private media companies. "Orbán has used media buyouts by government-connected 'oligarchs' to build a true media empire subject to his party's orders" (Spike 2024). Reporters Without Borders estimates that Orbán now controls 80% of traditional media resources in the country. The government also spends heavily on advertising on social media platforms and trains conservative influencers (Spike 2024).

Orbán's domination of media makes it difficult for opposition politicians to attract supporters. Voters usually do not find out about government corruption, policies that fail to deliver, or alternative points of view, which helps maintain sufficient support for Fidesz to retain its control of parliament and thus the prime ministry. Fidesz (sometimes in coalition with a smaller party) has been able to continuously control two-thirds of the seats in parliament and thus to choose the prime minister, with less than 50% of the vote because of the largest-party bias in the formula for allocating votes into seats built into Hungarian electoral law and significant gerrymandering. Freedom House now rates Hungarian elections as neither free nor fair (Meltzer and Buyon 2022).

To sum up the history of backsliding in Hungary, existing democratic electoral rules made it relatively easy for a winning party to achieve the two-thirds majority needed to change the constitution. Orbán was one of the founders of his party, he had been the formal party leader for many years, and those who had challenged his conservative policy priorities early on had been forced out, leaving the party a personalized and disciplined vehicle for supporting Orbán's leadership (Frantz et al. 2023). The large parliamentary majority held by a subservient ruling party meant that parliament never functioned as a check on Orbán during his second term as prime minister, and it provided him with the means of defanging the judiciary and thus eliminating it as a check as well. Because Orbán's backsliding efforts during his first prime ministership had been frustrated by the courts, the media, and Parliament, he moved immediately after his second election to disempower the courts, make other constitutional changes, and grab control of the media. In short, Hungary's political institutions and party history made backsliding relatively easy, and Orbán's personal history of having his initiatives blocked during an earlier stint as prime minister intensified his impulses toward weakening the political institutions and players that might otherwise have been able to thwart him.

In contrast to Hungary, the US political system makes backsliding more difficult and incentivizes a different strategy for executives bent on self-aggrandizement.

The major US parties are very old, and presidents do not automatically control their activities. In contrast to parliamentary systems, both houses of the legislature can be and often are controlled by the opposition in the United States. Because of lifetime tenure for Supreme Court justices, the most powerful part of the judiciary can also challenge the president, even though judges are appointed by presidents (with what is usually almost pro forma consent by the Senate). So most of the time, presidents can face checks from a court that was mostly appointed by their predecessors, at least some of whom came from the other party. In addition, the US process for amending the Constitution is difficult and time-consuming because it requires not only supermajorities in both houses of Congress but also ratification by three-quarters of state legislatures or conventions; it would be hard to complete during a single presidential term even if the change were widely popular.

These characteristics of the US political system mean that a self-aggrandizing executive would be unlikely to start by trying to change the Constitution. Instead, if his party controlled both houses of the legislature, he would try to achieve his goals by collaborating with Congress to pass legislation. During his first term as president, Donald Trump used this strategy to pass a large tax cut that predominantly benefited the wealthy, probably his top priority. However, several of his other efforts to change the status quo, for example, repealing the Affordable Care Act (the national health care reform initiated by President Obama), were blocked by a few Republicans from states that received substantial benefits from "Obamacare," who voted along with Democrats. Generally speaking, little legislation was initiated during the first Trump presidency, and he made few major changes to the status quo. Arguably the most consequential decisions of Trump's first term were the appointments of three conservative judges to the Supreme Court, which then ended the federal right to abortion and made other important changes to established law after the end of Trump's first term.

Many of Trump's backsliding efforts were attempted via executive orders, baseless lawsuits, and legally questionable efforts to get around procedural rules and standard ways of handling issues. The Muslim travel bans were issued as presidential decrees, and the effort to overturn the 2020 election result included numerous frivolous lawsuits, the encouragement of a mob attack on Congress, and a convoluted illegal plan to substitute Trump-loyalist electors in some states for those actually chosen by voters. Congressional Democrats attempted to use the legislature to obstruct these efforts, but their inability to persuade enough Republicans to support impeachment doomed their challenge.

However, important unelected political actors embedded in longstanding institutions stopped these and other Trump power grabs. Judges and high-level bureaucrats—including some Trump had appointed—blocked Trump's most extreme dictatorial impulses. So did state-level election administrators and the high command of the military, which refused to allow the use of soldiers against demonstrators (Goldberg 2023). Strong institutions shielded some of these individuals from retribution (though see below), but others stood up to Trump rather than tarnishing their own reputations, even if they lacked institutional protection. These individual officials, officers, and judges had reputations for probity, expertise, and high competence, on which their futures and their sense of their own worth depended. Their unwillingness to tarnish their reputations helped them to stand up to the verbal abuse, social media trolling, threats to harm or kill them and their families, and other forms of retribution used by Trump and his supporters. Even some unelected officials directly dependent on Trump for their posts, such as Attorney General William Barr, refused to support Trump's effort to overturn the 2020 election.

In contrast, *elected* political actors played little role in preserving democratic norms, though most did ultimately vote to certify Trump's defeat in the 2020 election. Most of the time, most elected officials from Trump's party, including those in Congress, were cowed by his popularity with their voter base and made no effort to block either his legally questionable executive decisions or his false claims about the 2020 election. Trump did not formally lead the Republican party or control campaign funding for Republican candidates, but by the end of his first term, he had established a reputation for ending the careers of Republican legislators who failed to go along with what he wanted by mobilizing effective challenges to them in primary elections and through other actions taken by Republican party leaders and activists (Bartels and Carnes 2023).

Trump ultimately lost his bid for reelection because electoral officials in the states did their jobs despite pressure and harassment from Trump and his supporters: the vote was fair and the counting honest. In short, Trump's autocratic impulses often failed during his first term mostly because of opposition from officials whose power derived from the governance structure of a longstanding democracy and the unwillingness of *un*elected officials, even in Trump's own party, to acquiesce in stealing an election.

Examining the ways Trump was blocked at various times highlights the importance of independent courts and well-established governance institutions staffed by professionals whose futures have depended on their personal reputations for competence and honesty, not on loyalty. When Trump rails against the "deep state," these are the people he means. In countries with limited previous democratic experience, however, or where access to high-level administrative jobs depends on loyalty, the barriers to autocratization are less robust. In new democracies, few people have had the opportunity to build careers based on apolitical competence and integrity because loyalty is required for career success in dictatorships, and strong institutional norms to guide behavior may not have had time to develop since democratization.

As I write during the first few weeks of Donald Trump's second presidential term, we can observe that he is trying to overcome the obstacles to backsliding and other radical policy changes that he faced during his first term. Despite the shock and awe tactics of the new administration, the chaos, and the distracting noise level, we can observe several strategies that suggest that Trump and his advisors are trying to change the incentives facing other political actors as they decide whether to stand up to anything Trump wants to do during his second term.

Most visibly, Trump has nominated individuals to staff the highest offices in the cabinet and executive branch who lack professional expertise in the policy areas they will be guiding, have no independent political base, and hold unusual opinions that run counter to those of most of the experts they will lead. Their only qualification seems to be loyalty to Trump, and even that is a recent development for some of them, increasing their need to demonstrate loyalty now. The most unprepared of these candidates, such as Tulsi Gabbard, Robert F. Kennedy, Jr., Pete Hegseth, and Kash Patel, can probably be counted on to support anything the president wants to do because Trump offers the only path for them to high office. They have reputations for unconventional ideas, spreading conspiracy theories, and hostility to the entities they will be running, not for competence and good judgment. As a result, they will have difficulty developing good working relationships with the professionals in their agencies and will be very unlikely to defend the personnel and budgets of those agencies from cuts or to shield individual civil servants from retribution if doing their jobs conflicts with Trump's goals.

These appointments are a departure from Trump's appointment strategy in his first term. Then he made a few deeply unqualified appointments, e.g., Steve Bannon to the principals committee of the National Security Council, but most of his appointments were conventional conservatives rather than Trump loyalists. Trump won the 2016 election unexpectedly, did not come from a political background, and lacked an extensive network of political friends and partisan experts to call upon. He took advice from influential Republicans about who to appoint. As a result, he appointed some principled conservatives to his administration, who later talked him out of uninformed or illegal proposals and occasionally even opposed him in public. The judges he had appointed sometimes ruled against him.

Trump seems determined that no one he appoints this time around will block power grabs. He is also investing heavily in establishing his reputation for unlimited vengefulness toward anyone who ever disobliged him or followed the orders of her superiors while the agency she worked for helped with investigating Trump or his allies when they were charged with criminal behavior.

His reputation for retribution against Republican legislators who failed to support him was clearly established during his first term. Nearly half of the Republican members of the House of Representatives who failed to support Trump in the key votes on certifying electors for the 2020 presidential election and the second impeachment effort did not return to Congress after the 2022 midterm election. Their careers ended before the general election. Primary challengers supported by Trump and his allies eliminated nearly half of those whose careers ended, and the others chose to retire, often because they expected to be defeated in primaries if they ran. Note that Republican Party activists, donors, and primary voters punished these politicians, not the general electorate (Bartels and Carnes 2023).

Trump's readiness to impose punishments on legislators who fail to back him has given him significant personal power over the Republican Party, despite its age and his lack of long-term engagement with the party. As of February 2025, after a spate of capricious foreign policy initiatives and illegal executive orders, including some that would diminish congressional power of the purse, only a handful of Republican senators and representatives, whose seats are probably safe for idiosyncratic reasons, have spoken out against Trump's initiatives (Blake 2025).

Now Trump is trying to show that he can also impose retribution on people who thought they had institutional protection for standing up to him during his first term or who were just doing their jobs during investigations of Trump or his allies. He has announced publicly that employees of the FBI and Justice Department, as well as top military generals, will be dismissed or lose pensions. He has even threatened some with prosecution.

He plans to dismiss a number of generals and has singled out retired General Mark Milley—who Trump appointed as Head of the Joint Chiefs of Staff in his first term—for opprobrium and harassment. Milley refused to use soldiers against protestors, took a public stand against the involvement of the military in partisan politics, and told the journalist Bob Woodward that "No one has ever been as dangerous to the country as Donald Trump. Now I realize he is a total fascist." Trump's new Secretary of Defense, Pete Hegseth, has ordered the Pentagon's Inspector General to investigate General Milley for wrongdoing to see whether there are grounds for reducing his rank (and thus his pension). A security detail had been assigned to General Milley because of Iranian death threats, but it has been ended along with his security clearance. As a retired four-star general with a strong reputation for accomplishment and patriotism in the highly professionalized US Army, General Milley would normally be shielded from this kind of harassment. The message to other officers is clear: If General Milley can be victimized, so can anyone else who stands up to Trump.

Trump is not just indulging in personal vindictiveness. He is trying to make sure that anyone who contributes in any capacity to challenging him knows they are likely to pay a price for it. Since most federal employees cannot legally be fired without cause, and the people currently being threatened were doing their jobs as they understood them in a competent way, these dismissals would normally be illegal. Thus, Trump's efforts at retribution against ordinary employees in federal agencies will contribute to backsliding if they are successful because they will help to undermine the rule of law.

The effort to dismiss people who should be protected by civil service law and union contracts might be successful because judges previously appointed by Trump may rule in Trump's favor when the cases get to court, as they will. If they are successful, Trump will have established his ability to fire at will, thus undermining the civil service protections that have helped to keep the US government administration nonpartisan for most of the last century.

During the current term, Trump has also issued executive orders aimed at radical changes to the policy status quo at a rapid clip. Backsliding executives attempt to make changes by using executive orders rather than legislation when they lack the support to get what they want from the legislature. Two recent examples are Trump's executive order ending birthright citizenship and his order to end the USAID (US Agency for International Development) program, which provides US funding to support health care, education, and poverty reduction abroad. Trump's party controls both houses of the legislature, but only by a few votes. These changes to the status quo would probably not be supported by enough Republicans to pass, since birthright citizenship is a longstanding right based on the 14th Amendment of the

Constitution, and USAID was created and funded with bipartisan support by previous Congresses. It is unclear whether these executive orders will succeed in making policy changes, but they do change conditions on the ground, especially when implementation is under the control of the executive branch, as is the hiring and firing of people who work for USAID. Even though ending USAID would mean impounding the money Congress allocated for it, which appears to be illegal, a great deal of damage is being done to USAID by dismissing its employees and closing its offices abroad, even if a court ultimately decides the executive does not have the authority to impound funds allocated by Congress. As with vindictive dismissals, if the courts rule in Trump's favor with regard to impounding legislative allocations, Congress's constitutional power over the purse will be seriously damaged, and Trump will have taken a giant backsliding step.

Who Can Block Backsliders in Democracies?

As we observe Donald Trump's second term, the answer to this question is more uncertain than it looked a few months ago. Nevertheless, we can make some observations.

In general, political systems that grant the executive more power over other political actors are more susceptible to executive aggrandizement because they reduce institutional checks on executive power and protections for other political actors who might challenge executive power grabs. Government characteristics that contribute to susceptibility to backsliding include the newness of democracy, the newness of the ruling party, and the concentration of power relative to co-partisans in the ruling party leader's hands. These three characteristics tend to be correlated for historical reasons. There are more new parties in recently created democracies because earlier autocratic governments often limited the organization and activities of opposition parties. Party leadership tends to be personalized in new parties because founders often have outsized influence in their parties, and the parties may not have had time or reason to build intra-party institutions for developing legislative strategies or choosing candidates to run in elections under the party label. When the ruling party's leader enjoys a great deal of power relative to co-partisans, then members of his own party will not talk him out of or obstruct power grabs for fear of losing the leader's favor (Frantz et al. 2023).

Personalization of the ruling party intensifies its tendency to support a co-partisan executive even during power grabs. Elected politicians whose future depends on voters who support the backslider are likely to avoid opposing him, especially if he controls or influences whether they can run in future elections. By showing that he can punish legislators who oppose him, Trump has made up for not having the kind of quasi-automatic influence on ballot placement that party leaders in most proportional representation electoral systems have or the kind of control party founders often exercise. Through a strategy of publicized retribution against anyone who expresses disagreement or criticism, Trump has succeeded in personalizing the very old and established Republican Party. The president's legislative co-partisans are now even less likely to hinder backsliding than in his first term.

In general, self-aggrandizing executives who control a legislative majority will face little obstruction by elected politicians.[2] They will face criticism from opposition politicians, but as the minority, they will be able to do very little to impede executive power grabs. If anyone blocks aggrandizing efforts, it will be unelected high-level officials (part of the "deep state" in Trump parlance), the courts, and military officers.

The age of democracy also affects the likelihood that unelected officials and judges stand up to backsliding executives. New democracies often lack the large cadre of professional bureaucrats, nonpartisan judges, and apolitical military officers that typify stable democracies. The kinds of individuals who might become judges or occupy high, unelected government posts have usually had little opportunity to develop reputations for competence, fairness, and honesty during earlier dictatorships in which office depended on loyalty. These kinds of reputations and the availability of other remunerative and high-status jobs in the private sector or international agencies for people who have such reputations provide unelected officials with some insurance against the amount of damage a backslider can cause. They thus increase the likelihood that such officials will stand up to him.

In contrast to elected politicians, many unelected officials who hold high-level positions in government have legally protected tenure in office as a result of civil service laws that were originally enacted to prevent patronage hiring from undermining government efficacy. The executive appoints the heads of ministries and important government agencies in most democracies, but staff are made up of career professionals with experience and expertise in their issue areas. Democratic governments often create entities such as central banks, security agencies, social security administrations, and the public health service to protect citizen safety and welfare from various threats. These agencies need competent staff in order to handle the tasks they are assigned, and they are supposed to be nonpartisan both in the recruitment of staff and their treatment of citizens. In long-standing democracies, many people build careers and reputations for competence, good judgment, and impartiality in these agencies. These are some of the people who may take the risk of trying to talk a self-aggrandizing executive out of actions he proposes or even to obstruct his efforts.

Non-democratic political systems rarely provide many opportunities for individuals to build the kind of reputation for professional competence, wisdom, and impartiality that helps those who occupy important unelected government posts withstand executive power grabs. For this reason, as well as those noted by Frantz et al. (2023) and Svolik (2008), new democracies tend to be more susceptible to executive aggrandizement.

---

[2] See Haggard and Kaufman for empirical evidence that backsliding goes further when the same party controls both the executive and the legislature.

During President Trump's first term, unelected officials talked him out of a number of illegal proposals and blocked others. Even some of those appointed by Trump and dependent on him for continuing in office, though not for their future success in other roles, obstructed Trump at various crucial turning points. Individuals whose future prospects depend on maintaining reputations for competence and integrity have reasons to stand up to an executive's most consequential power grabs. In the United States, the Attorney General and a number of lower-level officials, including some who had been elected in state elections that are likely to remain outside Trump's control, blocked his most egregious attempt to violate democratic rules, the effort to steal the 2020 election.

Now, Trump is trying to show that he can punish any official who stood up to him during his first term or has voiced any criticism since then, no matter what their institutional or reputational protections. Trump's targets include famous directors of important agencies as well as ordinary government employees who should have civil service protection. One of the most visible of those under fire is Dr. Anthony Fauci, head of the National Institute of Allergy and Infectious Diseases during Trump's first term and the Covid-19 pandemic. Dr. Fauci is retired, which somewhat limits the damage that can be done to him, but his security protection has been ended, leaving him more vulnerable to physical threats from extremist anti-vaxers and anti-mask activists who have been trolling him for years. Meanwhile, Trump allies in Congress have brought legal complaints against him for alleged inconsistencies in his testimony before Congress and his many interviews with the media (Etzel 2025; Herlihy 2023).

Since Trump's first term in office, the FBI has investigated him for possible criminal behavior, including complicity in the January 6 insurrection. In this way, they helped to block backsliding. Since Trump's second inauguration, his allies have forced out all six top leaders of the FBI and many heads of its field offices. "All spent decades working their way up the bureau ranks" (Dilanian et al. 2025). Nonpartisan career civil servants staff and run the FBI. The only political appointments are the director and his assistant. Even directors are supposed to have some protection from politicization. They have a 10-year term to try to insulate them from presidential interference. However, the sitting director of the FBI (appointed by Trump during his first term), whose term had not ended, resigned a few weeks ago after Trump let it be known that he wanted him out. Recent threats to FBI employees extend deep into the ranks. The Acting Deputy Attorney General in Trump's Justice Department has asked for a list of the hundreds of FBI employees involved in the investigation of those who stormed the Capitol on January 6, 2021, along with information about their activities and history, in order to decide whether further actions will be taken against them (Dilanian et al. 2025).

Career officials at the Justice Department also helped block backsliding during Trump's first term. Recently, more than a dozen prosecutors who worked on the January 6 prosecutions were dismissed. A week earlier, senior career lawyers involved in the case were reassigned to less desirable tasks and cities. Emails from Acting Attorney General James McHenry to those being ousted explained, "Given your significant role in prosecuting the President, I do not believe that the leadership of the Department can trust you to assist in implementing the President's agenda faithfully." (Perez et al. 2025).

These are explicit and well-publicized efforts to demonstrate to officials, including those protected by civil service laws, that the costs of obstructing the president will be high. In earlier work, I attributed the failure of Trump's backsliding efforts during his first term to non-cooperation by unelected career officials, judges, and military officers who could afford to oppose him because their futures were not directly tied to Trump's support (Geddes 2025). It seems clear that this time Trump seeks to persuade unelected career officials that their institutions and the laws cannot protect them from retribution if they challenge Trump. We do not know yet whether these efforts will succeed.

Judges have blocked backsliding efforts in many countries; for example, they blocked Boris Johnson's effort to abrogate Parliament. They also blocked many of Trump's policy initiatives during his first term. Judges with lifetime tenure, like those in the US Supreme Court, are not dependent on the executive. They have quite a bit of leeway to develop interpretations of laws and the constitution that allow them to make decisions consistent with their own values, and they also need not worry about the effect of partisan-appearing or thoughtless decisions on their future careers (Rakoff 2024). Lower-level judges may be more constrained if they aspire to higher judicial appointments, but the availability of other lucrative and rewarding careers outside the judiciary increases their options. Whether judges oppose executive power grabs depends on their own values and value-laden interpretations of the law. Research on judicial decision-making has shown their rulings to be correlated with the partisanship of the politicians who appointed them (Widmann 2021), though judges' later decisions are never fully predictable at the time of appointment. If an aggrandizing executive has had, or created, the opportunity to appoint most of the judges who will rule on the legality of his actions, we can predict that they will rule in his favor more often than judges appointed by someone else would, but not all the time. Since many judges' futures do not depend on the executive's favor, they retain considerable ability to rule according to their own values and interpretations of the law (Rakoff 2024).

President Trump had the opportunity to appoint three conservative Supreme Court judges,[3] which created a strong conservative majority in the Court. These three were plausibly qualified judges; Trump was deterred from proposing an unqualified crony by Senate Leader Mitch McConnell's warning that the Senate would not consent to it. Thus, the current Supreme Court in the United States has a strong conservative and partisan slant, but it is not composed of cronies and sycophants. So far, the court has ruled against Trump in several instances, though it has also ruled in his favor in some very important cases. We can predict that it will

[3] The first of these appointments was made possible by Republican Senate leader Mitch McConnell, who refused to allow a vote on former President Barack Obama's proposed candidate for the Supreme Court during the final year of his presidency. This unprecedented and constitutionally questionable action gifted Trump with a Supreme Court vacancy to fill as soon as he was inaugurated.

continue in the same vein. Trump and his allies have subjected lower-level judges who have ruled against them to vicious personal attacks and even threats of violence. So far, we have not seen such attacks against Supreme Court justices, though Trump's allies have suggested that their rulings could be ignored. So far, they have not been. Judges will ultimately decide whether the draconian funding cuts and personnel dismissals Trump has announced will really occur, and thus whether Congress's power of the purse and the legal provisions of civil service law will survive in the United States.

The military and security agencies in most democracies are even more independent of the elected executive than judges and unelected officials. They self-select into military careers and achieve promotion over time in consequence of seniority and the demonstration of effectiveness in carrying out the tasks they are assigned. Officers in democracies swear to uphold the constitution; they gain promotions and respect through competence, patriotism, and sound judgment. Normally, politicians play no role in military careers except that in some systems they appoint the very top officer from among a few high-ranking officers eligible for the promotion. This career experience suggests that officers might resist what they see as unconstitutional or illegal efforts by the executive to use force or spying to defeat civilian opponents.

If an aggrandizing executive has gained control over promotions and retirements, however, officers may be unwilling to risk their futures by opposing executive power grabs. Venezuela during the presidency of Hugo Chávez and Turkey under Recep Tayyip Erdogan's leadership are two of the best-studied cases of backsliding to full autocratization. In both, the backsliding executive was able to purge the officer corps after a failed coup attempt, which gave him full control over what had been an autonomous military institution. Since the purges, military officers have been unwilling to challenge the ruler or even to speak out against him.

Since being inaugurated for his second term, Trump seems to be trying to assert unprecedented personal control over the US military by dismissing top officers. So far, the Chairman of the Joint Chiefs of Staff, the Navy's top admiral, the second in command at the Air Force, the Coast Guard Commandant, and the Judge Advocate Generals in all three services have been dismissed. Judge Advocate Generals are the career military lawyers who determine whether specific proposed military actions are legal, for example, whether the military can be used against civilian protestors and what kind of interrogation techniques can be used. They are not political appointees. The president has the authority to appoint top officers (not Judge Advocates), but in the past, appointments have always been made at the end of the incumbent officer's term. It is unprecedented for the president to dismiss officers without cause (Gonyea 2025).

The stated reason for the recent dismissals is Trump's desire to end diversity, equity, and inclusion (DEI) policies in the military. The Chairman of the Joint Chiefs of Staff is Black, and the commandant of the Coast Guard and head of the Navy are women. The dismissals "[signal] to officials that they can be fired at any time" (Detsch and McLeary 2025). Trump's candidate to replace the outgoing Chairman of the Joint Chiefs is a retired three-star general who has not had the "key assignments identified in law as prerequisites for the job" (Copp and Baldor 2025). Although calling an officer out of retirement has happened at least once before, appointing a three-star general is unprecedented (Copp and Baldor 2025). This choice of a less qualified person to replace the highly accomplished outgoing Chairman makes clear that loyalty, shared opinions, and possibly race are more important to Trump than qualifications.

The dismissals will predictably lead to a cascade of promotions down the ranks of officers who will then owe their promotions to Trump and his new Secretary of Defense, Pete Hegseth. Note that Hegseth, who has actively supported and implemented these decisions, is one of Trump's arguably unqualified cabinet appointments. By all accounts, he sincerely supports these choices, but he would be very unlikely to oppose anything Trump wanted.

Historically, the US military has been highly professional, apolitical, and subordinate to civilian control (Gonyea 2025; Copp and Baldor 2025). Trump's dismissals, along with the attacks on retired General Milley, leave the impression that he wants to politicize the military so it will do his bidding without question. During his first term, he wanted to use the military to suppress protests, but top officers refused on the grounds that the military should not be used for political purposes. In the event of some future insurrection by Trump's followers or other effort to extend his term in office, Trump may want to reduce the chance that the military could be used against him by replacing apolitical military leadership with loyalists. At the very least, he wants to increase the perceived cost to officers of speaking out against him or obstructing his plans. As with other unelected officials, we do not know whether Trump's effort to change perceived incentives for military officers will be successful.

Voters

As noted above, voters are the last players to act if we think of backsliding as a strategic game. If they routinely punished backsliding executives at the polls, backsliding would rarely occur because potential backsliders would expect it to result in election defeat. One of the puzzles of the backsliding phenomenon is that voters not only fail to punish backsliders, but they often seem to reward them despite their support for democracy in surveys (Tai et al. 2022; Treisman 2023).

One possible explanation for voters' support for backsliders is that they want the policies backsliders initiate, e.g., many Hungarians favor Orbán's tough policies on immigration and his support of traditional Christian values. A number of studies have investigated what kinds of people vote for backsliders in different countries and also what objective conditions, e.g., large-scale immigration, are associated with increased voting for backsliders. Bartels' 2023 study of anti-immigrant, anti-democratic, and nationalist attitudes in Europe, however, challenges this literature. His investigation shows that in Europe, the number of citizens with nationalist, anti-immigrant, or anti-democratic attitudes did not rise prior to the election of politicians who later became backsliders. Nor were these attitudes more common in countries that experienced more backsliding than in those that experienced little. Substantial minorities hold such attitudes in all European countries, but they were never a majority, and numbers did not change before the first election of backsliders. Bartels' research thus shows citizen demand fails to account for backsliding.

Frantz et al. (2023) suggest an alternative argument. They begin by noting that increased political polarization among citizens, what they call "affective polarization," is associated with backsliding episodes (Haggard and Kaufman 2021; McCoy and Somer 2019; Horz 2021). Polarization can result in so much enmity toward political opponents that voters tolerate abuses of power by their own side in order to reduce opponents' political influence (Svolik 2020). They argue that personalized leadership of the ruling party—a characteristic of backsliding—is associated with increased polarization, first, because inflammatory speeches and social media posts demonize the opposition, and second, because the leader's attacks on democratic institutions and challenges to democratic norms are in themselves divisive. In personalized parties, the leader's co-partisans know that their access to office depends on loyalty to the leader, so they echo the leader's claims and defend his attacks on institutions. Opposition party elites and citizens who oppose the backslider react with indignation, while the backslider's supporters among citizens rally around him, increasing distance and hostility between the two sides. They show that in the US polarization had been rising since the early 2000s but rose even more steeply in the years immediately before and after Trump's election. In Venezuela, where backsliding turned into autocratization during the Chávez presidency, polarization rose very steeply from the time of his initial election in 1998 until the measurement ends in about 2005, the date often used for the beginning of autocratic rule (Frantz et al. 2023, 167–68).

This argument is plausible and probably explains a good deal about why voters who claim in surveys to support democracy nevertheless vote for leaders who damage democratic institutions and norms. Once a substantial number of voters have signed on to the backslider's team, they can then be weaponized to help prevent elected politicians from voicing criticism of specific actions proposed by the backslider. Legislators who might want to defend their institution and its constitutional role often fail to do so if the backslider controls their access to a place on the ballot and thus their future success in politics. In proportional representation systems, party leaders usually choose candidates and assign them places on the party list. Even in full democracies with this kind of a system, voting against the executive on something important could well lead to a lower place on the list, and thus less chance of being elected, or to being left off the list altogether.

In political systems that hold primary elections to choose legislative candidates, it might seem that politicians would have greater freedom to vote against the party leader, but the threat of a well-financed primary challenge deters many from speaking out. This threat frightens legislators because primary voters (who in most of the United States are the more committed members of the party choosing candidates only) are likely to favor candidates supported by the backslider as long as he remains popular. In this way, voters indirectly help the backslider to get unqualified cronies confirmed by the Senate to high appointive office and to reverse policies that would be supported by most legislators if they voted sincerely.

Larry Bartels and Nick Carnes (2023) have shown that about half of the Republican members of Congress who opposed Trump on key votes finished their political careers before the next general election (a much higher proportion than would normally be expected in the US Congress). This punishment was not delivered by ordinary voters but rather by Republican party leaders, donors, and Republican primary voters who, along with Trump, supported primary challenges against these representatives. Primary challenges have not been widely used to discipline co-partisans during backsliding episodes in other countries because primary elections are not a common means of choosing candidates. In the United States, however, Trump's weaponization of primary challenges has resulted in a supine Republican party despite other features of the US party and political systems that would have been expected to give Republican legislators some ability to pursue their own values or the interests of their constituents rather than toeing the Trump line.

Conclusion

This essay has sought to spell out a general argument about what historical circumstances and democratic institutional features have facilitated democratic backsliding and to explain why some kinds of political actors have been more likely to obstruct backsliding than others. It compared the experience of backsliding in Hungary with backsliding in the United States during the Trump presidency to highlight the way Hungary's specific democratic institutional characteristics eased Prime Minister Victor Orbán's path through backsliding to full autocratization. In contrast, during his first term, President Trump's authoritarian impulses were blocked at many points by other political actors embedded in the US institutional framework. The examination of Trump's first term and the comparison between his first and the beginning of his second term cast light on why elected legislators have not been willing to stand up to Trump, while some unelected political officials, judges, and military officers have.

The Hungarian political system had several features that made backsliding relatively easy once an executive determined to do it was elected. Possibly most importantly, the mixed PR and single-member-district electoral system tended to result in a substantial advantage in seats for the party that achieved the highest vote because the same party tends to win all over the country in Hungary. For Orbán, this meant that his party won two-thirds of the seats with only 52% of the vote in the first, fully fair election of his second term. The second feature is that two-thirds of parliament can amend the Constitution, which allowed Orbán to amend it in his favor almost immediately and then repeatedly. Finally, although Orbán's party, Fidesz, is not new, Orbán was one of its founders, and he had been its leader through much of its history. This history resulted in Orbán's personal control over internal party decision-making, as Frantz et al. (2023) would expect, with the result that party deputies had every reason to go along with Orbán's initiatives regardless of their own opinions.

In contrast, the US party system is very old, and its political system was explicitly designed to prevent the executive from assuming monarchical powers. New laws require agreement from both the president and a majority of the legislators. The courts are powerful and somewhat independent because of lifetime tenure at the top. Amending the US Constitution is difficult and time-consuming. The US single-member-district electoral system does not result in a major advantage to the vote winner in seats because each of the two important parties dominates its own region. Candidates are chosen in primary elections, which gives them some independence from party leaders. Finally, Trump did not rise within the Republican Party, as most US politicians would, and initially had little control over it. He gained control over its congressional delegations by visiting retribution on anyone who stood up to him through primary challenges and public criticism.

In countries where constitutions cannot be readily changed, backsliding entails ignoring norms, challenging established procedures, and breaking or changing laws and getting away with it. The more dependent the future success of politicians, officials, judges, and military officers is on the current executive, the more likely it is that the political actors legally empowered to block executive actions and enforce existing laws, procedures, and norms will side with executives, enabling power grabs.

During his first term, unelected political officials, judges, and even military officers blocked many of Trump's impulses. I suggest that these actors faced lower costs for speaking up than did legislators, both because their current positions were insulated from partisan interference and because their future success depended on their substantive achievements and reputations for competence, fairness, probity, and good judgment, not on the voters or the leader's favor.

As I write, about 6 weeks after Trump's second inauguration, he is attempting to change the perceived costs of obstructing his initiatives for unelected officials by imposing retribution on all who played any role in blocking his backsliding efforts during the first term. He has very publicly attacked and/or dismissed numerous officials in various supposedly nonpartisan state agencies, as well as active and retired generals. Since many of the targets of his retribution should be protected by civil service laws or Congress's power of the purse, we do not know how many of the targets will actually suffer the punishment he has announced. These outcomes will depend mostly on the courts, which Trump has shaped through appointments but not brought under full control.

Most of this essay has focused on elite actors, but ordinary voters also play a role in backsliding. To achieve office in the first place and to win any subsequent competitive elections, backsliders have to promise changes that a substantial proportion (not necessarily a majority) of voters support. In the post-2000 period, backsliders in Europe have tended to be right-wing populists who attract nationalistic, religious, anti-immigrant voters. Meanwhile, backsliders in Latin America have tended to be left-wing populists who advocate redistribution and attract anti-elite, anti-American voters. Although they undermine election fairness with various strategies for keeping themselves in power, many backsliders have been genuinely popular. While it lasts, their popularity with voters helps to keep other politicians, who can benefit from their coattails, allied with them and unwilling to oppose anything they propose. Trump's use of primary election challenges to defeat, and thus punish, elected members of Congress for opposing him is a prime example of this process.

Voters do, of course, sometimes defeat backsliders, seemingly for the same reasons they reject other incumbents: poor economic performance, major corruption scandals, failure to prevent damage caused by natural disasters, and so on, rather than for backsliding per se. Arguably, Boris Johnson in the UK, Jair Bolsonaro in Brazil, and Donald Trump in 2020 were defeated by voters partly because of their failure to handle the Covid-19 pandemic and its economic fallout better. The possibility of election defeat motivates many of backsliders' most brazen power grabs. It is the reason for attempts to tilt the election playing field in their favor by gerrymandering and changing election laws, by interfering with the media to stymie opposition efforts to get their messages out, and by politicizing the distribution of government resources that are supposed to be nonpartisan. If backsliding proceeds to full autocratization, as it did in Venezuela, Hungary, and Turkey, however, removing the backslider becomes much more difficult and may not be possible simply by voting.

Donald Trump has already demonstrated his willingness to steal an election and to use violent supporters against other political elites. His recent executive orders show his disrespect for laws and constitutional mandates. Most disturbing, in my view, is the high priority he has given to taking control of the military, the national security apparatus, and the FBI. The armed agencies of the government are the ones that, in the event of a showdown, would determine whether an attempted self-coup or effort to remain in office beyond the end of his term would succeed. Trump seems to be setting the stage for staying in power more than 4 years. A Trump supporter in Congress has introduced a resolution to allow him to serve a third term. Trump himself muses about a third term in social media posts and meetings with supporters (Mangan 2025). Whether this or some other ploy to extend Trump's time in power succeeds depends on the willingness of unelected officials, judges, and officers to risk obstructing him and on how voters respond in the 2026 midterm election.

References

Bánkuti M, Scheppele K, Halmai G (2012) Hungary's illiberal turn: disabling the constitution. J Democracy 23(2):138–146

Bartels L (2023) Democracy erodes from the top: leaders, citizens, and the challenge of populism in Europe. Princeton University Press, Princeton, NJ

Bartels L, Carnes N (2023) House Republicans were rewarded for supporting Donald Trump's 'stop the steal' efforts. PNAS 120(34). https://doi.org/10.1073/pnas.2309072120. Accessed 14 Feb 2025

Blake A (2025) 'Embarrassing,' 'cruel,' 'absurd,' 'extortion': GOP moderates find voices on Trump. *Washington Post*, February 24. https://www.washingtonpost.com/politics/2025/02/24/

gop-moderates-speak-against-trump/?utm_campaign=wp_the7&utm_medium=email&utm_source=newsletter&carta-url=https%3A%2F%2Fs2.washingtonpost.com%2Fcar-ln-tr%2F414ef59%2F67bdae6efe9d672c2a370232%2F66e87f0a97e30f5a0b373170%2F29%2F102%2F67bdae6efe9d672c2a370232. Accessed 25 Feb 2025

Copp T, Baldor LC (2025) Trump fires Chairman of the Joint Chiefs of Staff and two others. *Time*, February 25. https://time.com/7260646/trump-fires-chairman-joint-chiefs-of-staff-other-military-officers/. Accessed 19 Feb 2025

Detsch J, McLeary P (2025) Trump fires top military leaders in unprecedented shakeup. Politico, February 21. https://www.politico.com/news/2025/02/21/cq-brown-trump-fired-joint-chiefs-general-00205593. Accessed 19 Feb 2025

Dilanian K, Winter T, Reilly RJ, Kosnar M (2025) Trump administration forces out multiple senior FBI officials and Jan 6 prosecutors. NBC News, January 31. https://www.nbcnews.com/politics/national-security/trump-administration-forces-multiple-senior-fbi-officials-rcna190138. Accessed 21 Feb 2025

Economist (2022) A wild gerrymander makes Hungary's Fidesz party hard to dislodge. April 2

Etzel GM (2025) State attorneys general look to prosecute Fauci despite Biden pardon. Washington Examiner, February 5. https://www.washingtonexaminer.com/policy/healthcare/3311832/state-ags-look-prosecute-fauci-despite-biden-pardon/. Accessed 25 Feb 2025

Feher M (2011) Hungary passes new constitution amid concerns. The Wall Street Journal, April 18. Wsj.com/articles/BL-NEB-4147. Accessed 12 Feb 2025

Frantz E, Kendall-Taylor A, Wright J (2023) The origins of elected strongmen: how personalist parties destroy democracy from within. Oxford University Press, Oxford

Geddes B (1995) A comparative perspective on the Leninist legacy in Eastern Europe. Compar Polit Stud 28:239–274

Geddes B (2025) What distinguishes 'successful' democratic backsliders from thwarted ones? In: Wejnart B (ed) The global rise of autocracy, its threat to a sustainable future, forthcoming 2025. Routledge Democratization and Autocratization Studies

Goldberg J (2023) The patriot: how General Mark Milley protected the constitution from Donald Trump. The Atlantic 332(4):14–34

Gonyea D (2025) President Trump fires 6 top-level military officers. A retired rear admiral reacts. National Public Radio, February 22. https://www.npr.org/2025/02/22/nx-s1-5305781/president-trump-fires-6-top-level-military-officers-a-retired-rear-admiral-reacts. Accessed 19 Feb 2025

Haggard S, Kaufman R (2021) Backsliding: democratic regress in the contemporary world. Cambridge University Press, NY

Herlihy B (2023) Fauci referred to Justice Department for criminal investigation for allegedly lying under oath to Congress. Fox News, August 9. https://www.foxnews.com/politics/fauci-referred-justice-department-criminal-investigation-allegedly-lying-under-oath-congress. Accessed 25 Feb 2025

Horz CM (2021) Electoral Manipulation in Polarized Societies. J Polit 83(2):483–497

Mangan D (2025) Constitutional amendment to allow Trump third Term introduced in the House. CNBC, January 24. https://www.cnbc.com/2025/01/23/trump-third-term-amendment-constitution-ogles.html. Accessed 27 Feb 2025

McCoy J, Somer M (2019) Toward a theory of pernicious polarization and how it harms democracies: comparative evidence and possible remedies. Ann Am Acad Polit Soc Sci 681(1):234–271

Meltzer M, Buyon N (2022) From democratic decline to authoritarian aggression. Freedom House, April 21. Freedomhouse.org/report/nations-transit/2022/from-democratic-decline-to-authoritarian-aggression. Accessed 22 Feb 2025

Perez E, Campbell J, Rabinowitz H (2025) Trump DOJ demands list of thousands of FBI agents, others who worked on Jan. 6 and Trump investigations for possible firing. CNN, January 31. https://www.cnn.com/2025/01/31/politics/fbi-agents-who-investigated-january-6-fired/index.html. Accessed 21 Feb 2025

Rakoff JS (2024) The most conservative branch. New York Review of Books, Sep 19:51–52

Scheppele K (2022) How Viktor Orbán wins. J Democracy 33(3):45–61

Spike J (2024) How Hungary's Orban uses control of the media to escape scrutiny and keep the public in the dark. Associated Press, July 31. APnews.com/article/Hungary-media-democracy-orban-magyar-european-parliament-f6315d7cc252f210c360863de403054e. Accessed 22 Feb 2025

Svolik MW (2008) Authoritarian reversals and democratic consolidation. Am Polit Sci Rev 102(2):153–168

Svolik MW (2020) When polarization Trumps civic virtue: partisan conflict and the subversion of democracy by incumbents. Quart J Polit Sci 15(1):3–31

Tai YC, Yue H, Solt F (2022) Democracy, public support, and measurement uncertainty. Am Polit Sci Rev 118(1):512–518

Treisman D (2023) How great is the current danger to democracy? Assessing the risk with historical data. Compar Polit Stud 56(12):1924–1952

Widmann M (2021) Domesticating the international: the uneven enforcement of investors' preferences and its unintended consequences. UCLA dissertation

Barbara Geddes is Professor Emeritus of the Political Science Department, University of California, Los Angeles. She has worked on autocratic politics, regime transition, democratization, bureaucratic reform, corruption, and research design. Her recent research has focused on politics in dictatorships. Her most recent book is *How Dictatorships Work: Power, Personalization, and* Collapse (with Joseph Wright and Erica Frantz), 2018. The original data set on which the book was based, The Authoritarian Regimes Data Set (http://sites.psu.edu/dictators/), was awarded the Lijphart/Przeworski/Verba Best Original Data Set Award given by the Comparative Politics Organized Section of the American Political Science Association. Her other publications include *Paradigms and Sand Castles: Theory Building and Research Design in Comparative Politic* (2003); *Politician's Dilemma: Building State Capacity in Latin America* (1994); "What Do We Know about Democratization after Twenty Years?" in the *Annual Review of Political Science* (1999); What Causes Democratization?" in *The Oxford Handbook of Political Science* (2009). She has chaired more than 50 dissertation committees and was the first recipient of the UCLA Political Science Department's Best Mentor award, which was named after her. The Comparative Politics Organized Section of the American Political Science Association awarded her the Bingham Powell Graduate Mentoring Award in 2014. Until her retirement, she taught the Workshop on Authoritarian Politics, Latin American Politics, and Research Design at UCLA.

From Scientific Illiterates to Anti-Establishment Avant-garde: The Changing Image of Climate Skeptics in Germany

Alexander Ruser

Abstract If it is true that climate change skepticism didn't have a long history in Germany, its future looked even more bleak. Its political system is based on consensus and the ability to form coalitions and its strong corporatist roots that favored close relations between political parties, academic research facilities, and public and selected private media companies made Germany a rather unwelcoming environment for proponents of radically dissenting views on climate change. Unlike in the USA, networks between specific political parties, wealthy donors, political think tanks, and partisan media outlets could never form. Consequently, climate change skepticism was considered a marginal phenomenon in Germany and often explained (away) by the scientific illiteracy of a few politically irrelevant groups. Since the financial crisis of the early 2010s, however, the picture has changed. The emergence of new, populist political parties, first the right-wing "Alternative for Germany" (AfD) and more recently the left-leaning "Sarah Wagenknecht-Alliance," changed the dynamics of coalition-forming and turned radical opposition to the political mainstream into a viable political strategy. Opposition to Germany's politics and the rejection of climate science as an integral part of an (alleged) liberal, "globalist," and patronizing agenda thus became a cornerstone of disruptive "anti-establishment" rhetoric.

Keywords Climate change skepticism · Far-right parties · Populism · Alternative for Germany

A. Ruser (✉)
Department of Sociology and Social Work, Center for Digital Transformation (CeDiT), University of Agder, Agder, Norway
e-mail: alexander.ruser@uia.no

B. Wejnert (ed.), *Democracy in Retreat and the Climate Crisis*, Springer Studies on Populism, Identity Politics and Social Justice,
https://doi.org/10.1007/978-3-032-02881-5_4

Introduction

"In June 1990, the German Cabinet agreed to take internationally unprecedented action to reduce its carbon dioxide (CO_2) emissions. This policy is based on what emerged as a distinctly German environmental strategy called *Vorsorgeprinzip* (the precautionary principle), which demands that action to prevent environmental damage be taken *even before scientific certainty is achieved* (Cavender and Jäger 1993, 4; emphasis added).

In their 1993 paper "The History of Germany's Response to Climate Change," Jeannine Cavender and Jill Jäger not only tell the story of growing cross-party consensus on environmental issues in West Germany since the late 1960s. They also point to specific mechanisms that enabled the transfer of scientific knowledge into political debate and the depoliticization of "hot" environmental issues such as human-made climate change. They point to the role of public scientific bodies, for instance, the "Scientific Committee on Problems of the Environment" (ibid. 6) and professional associations, such as the German Physical Society (ibid. 11), in facilitating the understanding of scientific facts and political options. The history of Germany's response to climate change is presented as an expression of the country's corporatist tradition. As I have described in greater detail elsewhere (Ruser 2018, 114), the German political system can be described as a "closed shop" in which a small number of insiders (political foundations, public research institutes and think tanks, business associations, large corporations, and civil society organizations) maintain close ties to political parties and career administrators in ministries and government agencies. The resulting selective and highly structured ecosystem of political consulting and advice-giving suits political parties in Germany, which, always in need of finding and serving as coalition partners, prefer "neutral," "objective," and dry scientific expertise and bipartisan information over divisive, polarizing advice. While this system is not without disadvantages and tends to foster institutional conservatism and is thus slow to respond to radically new challenges and insights, it helped to marginalize climate change denial (ibid.).

Given the further advancement of international climate research and a growing consensus among climate scientists that human activity is the main driving force for global climate change, Germany surely must be an exception to the broken "science-society contract" (Glavovic et al. 2022, 829) that has fuelled the politicization of climate change in countries like the USA. In their 2022 paper "The tragedy of climate science," Glavovic and colleagues lament that despite "compelling evidence" for anthropogenic climate change, "greenhouse gas emissions (…) rise year upon year" (ibid.), thus demonstrating that scientific expertise is not sufficient to compel decisive political action. Germany, though, should be exceptionally well positioned to "listen to the science"—to borrow a popular demand of the Fridays for Future Movement (Sossdorf and Burgi 2022)—and to overcome the tragedy of climate science.

Three decades after Cavender and Jäger told the early history of its response to climate change, however, the "consensus to work towards prevention" (1993, 14) of climate change is eroding, and climate skepticism is on the rise (Ruser 2022).

In this chapter, I will therefore outline a potential connection between beliefs about climate change and (anti) authoritarian worldviews before I trace the rise of climate change skepticism in Germany and explain how anti-science, "fringe" positions could be transformed into a core element of a self-proclaimed "anti-establishment avant-garde." To develop the argument, I will then introduce Germany's "closed shop" and discuss how the selectivity of its insiders, once considered its biggest strength, could become its political Achilles heel.

In the second section, I will turn to more recent cracks in the closed shop. Especially the electoral successes of the right-wing party "Alternative for Germany," which made the rejection of the closed shop a cornerstone of its identity and built its programmatic core in opposition to the dominant logic of consensus of compromise, made climate skepticism politically relevant. Subsequently, I will turn to the AfD's climate politics to show how it and other actors on the right use climate change to separate themselves from the mainstream, "globalist elites." Ironically, this strategy was, as I will show, enabled by the expansion and strengthening of global networks of climate change skeptics and shows how the AfD borrowed from American playbooks and learned how to use climate change as a political weapon.

The fourth and final part of the chapter will discuss the consequences of this development, most notably the disguising of authoritarian sentiments as anti-establishment and empowering rhetoric.

Climate Change Belief, Skepticism, and the Connection to Authoritarianism

It is tempting to portray climate skeptics as either ill-informed, since they lack crucial scientific information about the impact of human activity on the Earth's climate, or morally flawed if they fail to acknowledge the urgency of the problem or do not care about the consequences of anthropogenic climate change. Climate skeptics seem to carry an aura of pre-enlightenment bigotry and might even harbor anti-democratic sentiments as they use all kinds of tricks to undermine scientific knowledge and democratic opinion formation (Powell 2011; Brulle 2014). However, it is not possible to equate a belief in climate science with an enlightened rational position and climate change skepticism with irrational thinking, nor is the rejection of scientific or political consensus automatically anti-democratic (Ruser 2018, 2021). Given the complexity of climate change, the opaqueness of the underlying science, and the uncertainty of the effectiveness of policy measures, attitudes toward climate change reveal more about a person's perspective on political and scientific institutions.

"The climate" is a statistical construct that describes the "average weather" over a set period of time, while climate change is measured as variation of key predefined variables describing such average weather (Ruser 2018, 15–16). Detecting climate change requires the collaboration of specialists from various scientific disciplines and the inclusion of longitudinal data that spans many years, even decades. The very definition of climate and climate change thus prevents us from experiencing the phenomenon directly. Consequently, it is very difficult to "know" that climate change is happening. The logical opposite of climate change skeptic and climate change denier is therefore the climate change *believer* (Ruser 2018, 37). In consequence, if we want to understand why some people believe in climate change, we need to understand why some people believe scientific experts and elected representatives and others don't. Furthermore, it can be argued that this understanding is also key to making a plausible connection between climate change skepticism and authoritarian sentiments.

Attempts to define modern forms of authoritarianism date back to the rise of fascism in Europe in the 1920s and include work on authoritarian structures and the "authoritarian state" (Horkheimer 1973) and studies of the underlying personality structures that make some people susceptible to authoritarian ideas (Adorno et al. 1950) However, in order to understand better if and how climate change skepticism can support authoritarian agendas, authoritarianism needs to be analyzed as a collective phenomenon rather than an individual personality trait and as a political agenda rather than an element of state institutions. Abraham Maslow describes authoritarianism as a "world view," tries to understand its "basic philosophy" and insists that to its supporters it has "a logic of their own which integrates all life for them in such a way as to make their actions not only understandable, but from their own point of view, quite justifiable and correct" (Maslow 1943, 402). For Maslow, authoritarian worldviews are characterized by the idea that the world is a hostile place, a jungle "peopled with animals who either eat or are eaten, who are either to be feared or despised" (ibid. 403). The authoritarian worldview thus values strength (individual power and solidarity within one's own group) and dismisses "love, kindness and cooperation" as "sentimental and unrealistic" (ibid.). The tendency to divide people into members of one's own group ("us") and rivals and enemies ("them") supports hierarchical thinking and justifies the neglect of "lower" groups, for instance, minorities of foreigners (ibid. 404–405). Finally, authoritarian worldviews assume that everyone and every group is striving for power and therefore tend to mistake kindness, altruistic motivations, and care for others as weakness or a clever ploy to conceal egotistical motives.

This admittedly brief overview of authoritarian worldviews is sufficient to compare it to empirical findings on the connection between authoritarianism and climate change skepticism. Aaron McCright and Riley Dunlap, for instance, have shown for the USA that "white males with a hierarchical cultural worldview would be the most likely to downplay or ignore environmental risks, perceiving them as challenges to the existing social, political, and economic hierarchy" (McCright and Dunlap 2011a, 1166). Moreover, their study revealed that conservative white men are more likely to reject scientific expertise and rely on their own personal knowledge (ibid.

1171). The hierarchical worldview and the confidence in one's individual and collective strength—a core element of authoritarian worldviews—thus explain the rejection of threatening information and the challenging of expertise that is seen as hostile or in cahoots with sinister, liberal-leftist forces (Ruser 2020, 52–3).

The connection between climate change skepticism and authoritarian worldviews is not straightforward, though. McCright and Dunlap do not even explore authoritarianism but identify elements of Maslow's typology within conservative positions. Moreover, climate change skepticism, especially if one distinguishes between science skepticism and policy skepticism (see below), is not necessarily an indicator for authoritarian tendencies. However, climate change skepticism can serve authoritarian worldviews and can speak directly to its understanding of the world as a Darwinist "jungle" and antagonistic struggle between us and them. To understand if and how climate change skepticism in Germany fuels authoritarian sentiments, therefore, requires an analysis of the specific circumstances that led to its growing importance. The case of Germany is interesting not only because climate skepticism became a rather unexpected element of the new Authoritarian National Radicalism of the Alternative for Germany (Heitmeyer 2024, 45), but also because it is presented as an attempt to defend and restore democracy and participation.

No Country for Radical Climate Skeptics: Climate Politics in the Closed Shop

Germany, like many Western countries, was slow to acknowledge human impact on the climate. Despite the availability of research on "human activity as a climate factor" (Cavender and Jäger 1993, 5) since the 1940s, environmental politics and political debate about climate change began only in the late 1960s (ibid. 7). Moreover, the particularities of the German political system affected how research on and social reactions to climate change were included in political debate. The German political system has been described as rooted in corporatist traditions (Fox [1977] 2017, 202), which emphasizes the need to create exclusive spaces for negotiating political, economic, and social interest understood as "a movement toward a corporate representation of interests which, despite a high degree of differentiation of occupations, groups, and strata in the modern economy, would be able to bring about public co-responsibility, national-state feelings, and adaptation and subordination to the interests of the whole as represented by the state." (Dahrendorf 1968, 41–42). This not only includes "the incorporation of societal actors into the policy process, typically by means of nonparliamentary consultation." (Zimmer 1999, 40) but that the political system shows "traces everywhere of the deep aversion of all participants (...) to conflict" (Fox [1977] 2018, 192). The corporatist tradition, with its emphasis on involving social actors (typically business associations, trade unions, NGOs, and large charitable associations) in political debates, had a profound impact on scientific expertise and political consulting, too. The German system favors state universities, research institutes, and public think tanks and values impartial advice that can be used in changing political coalitions (Ruser 2018, 114). In consequence, despite Germany's slow response to climate change and the persistence of positions that called for cautious, more market-friendly, and less ambitious climate politics (Cavender and Jäger 1993), attacks on the credibility of climate science and the integrity of climate scientists, common in more competitive and adversarial political systems such as the USA (Ruser 2018, 66), never became significant talking points or parts of political agendas (Goodbody 2019, 94).

Moreover, the German political system and its tradition of nonparliamentary consultation to achieve consensus effectively establish a space of say- and thinkable positions and legitimate consensus as an institutional goal by emphasizing the prestige of the social actors and scientific experts involved. Political debate on climate change was for long effectively constrained by the expertise of renowned public research institutes and think tanks such as the Potsdam Institute for Climate Impact Research, the Wuppertal Institute, or the (private) Öko-Institute (Ruser 2018, 111).

The institutional environment thus established clear boundaries between insiders and outsiders and set rules for what counts as valid expertise. Unlike, for example, in the USA, climate research was treated as a neutral source of information provided to decision makers by non-partisan academic organizations. Battles over the trustworthiness of scientific data or the integrity of researchers were thus largely absent from political debate about climate change. Conflicts about political responses still occurred, and Germany witnessed setbacks and delays in its response to climate change (Cavender and Jäger 1993). Likewise, scientific expertise itself was not enough to promote more ambitious agendas. However, a growing awareness of environmental problems led to the formation of "the Greens," a party that has been represented in the Bundestag since 1983 and which has shared government responsibilities twice since the late 1990s.

In sum, Germany seems to be a rather hostile environment for dissenters and political activists outside the established "closed shop." The example of the Green Party shows that new ideas and new actors can enter the scene. However, despite their ties to environmental movements and unorthodox political style, the Greens shared a trust in institutionalized science, were determined to work with and through established systems of consensus formation and decision-making (Cavender and Jäger 1993, 10) and didn't attempt to disrupt the corporatist traditions. Proponents of climate change skeptic positions, on the other hand, were increasingly and institutionally marginalized. Moreover, strategies and talking points that had been used to undermine a political consensus on climate change in other countries (Brulle 2014; McCright and Dunlap 2011a, b) seemingly couldn't work in a system that didn't reward polarizing and disruptive rhetoric and politics.

Germany's closed shop effectively limited the two main directions of climate skeptic rhetoric: "climate science skepticism" and "climate policy skepticism" (Ruser 2025). The first strategy encompasses attempts to question the soundness of scientific advice and/or the integrity of scientists. Standard arguments include the concerns about the evidential value of existing data (often resulting in demands before responses to climate change can be discussed) and the trustworthiness of climate scientists (frequently pointing to personal gains, attention seeking, or economic advantages because of "sensationalist" findings). By limiting the number of scientific advisors and emphasizing institutional reputation, climate science skepticism not only played a marginal role but could easily be dismissed as scientific illiteracy. Climate policy skepticism, the second strategy, comes in two variants: the rejection of global responsibility and the need for international cooperation and a hostile attitude toward (state) regulation (ibid.). Both rhetorical ploys were of limited use since the corporatist tradition ensured that rules and regulations were always negotiated and stayed within acceptable limits. Moreover, the overall aversion toward conflict affected Germany's "quest" for a "positive national identity since the Second World War" (Goodbody 2019, 92). Arguments against international cooperation and for national solo-runs were generally ill received.

Political parties did not have much to gain from siding with fringe positions. Nor could corporations, wealthy individuals, or media outlets benefit from supporting counter-expertise, e.g., by setting up and funding climate skeptic think tanks and pressure groups. It shouldn't come as a surprise, therefore, that climate change skepticism and denial were marginal phenomena. However, in recent years, and in particular since the financial crisis of 2007/8 skepticism toward the validity of climate science and the adequacy of climate policies is on the rise. Between 2007 and 2016, the number of self-identified climate skeptics in Germany rose from 7% to 16% (Ruser 2022, 107), and recent studies have shown that, especially in the eastern part of the country, there are high and growing numbers of climate skeptics (Lübke 2022, 160). This raises the question of why the consensus on climate change in Germany was weakened, how climate skeptic arguments could enter the public arena, and what this means for the corporatist tradition of democratic decision-making.

The Outsiders from Within: The Rise of the Alternative for Germany

The recent, unlikely increase in self-identified climate skeptics and the growing prominence of public dispute about climate politics cannot be understood without turning to the rise of the Alternative for Germany (AfD), even though the connection might not be immediately clear. The AfD entered the political stage in 2013 as a reaction to the European Currency Crisis, a direct consequence of the Global Financial Crisis that had brought down Lehman Brothers and pushed the global banking system to the brink of collapse a few years prior. The financial turmoil and the bailouts hastily pushed through parliament had highlighted the downsides and limitations of the closed shop and the programmatic shortcomings of Germany's consensus-oriented party system (Grimm 2015; Ruser 2015) and created an opening for radical new voices and a more populist style. The clue for the spectacular electoral success of the Alternative for Germany, therefore, is in its name. The AfD presented itself as outsiders to the closed shop and thus promised to offer a fresh perspective on things. However, despite these attempts to create the image of political mavericks, initially AfD was the brainchild of political insiders and members of the economic and cultural elite: "In late 2012, Alexander Gauland, Konrad Adam, and Berndt Lucke set up the Electoral Alternative 2013 [*Wahlalternative 2013*], from which the AfD emerged the following year. Gauland was a former Department Head of the Federal Ministry of the Environment, Adam was a former Editor of the influential *Frankfurter Allgemeine Zeitung* newspaper, and Lucke was a tenured Professor of Macroeconomics at the University of Hamburg. They were well informed and well networked, and their joint manifesto was endorsed by an impressive array of economists, journalists, business leaders, and political activists" (Lees 2018, 300). Moreover, the AfD was not only established by political insiders but also lacked a broad anti-establishment agenda. Founded as a single-issue party—to oppose the Euro as the single currency in the EU (Grimm 2015)—the party lacked a comprehensive manifesto and was initially not even interested in engaging with environmental problems and climate change. However, the early successes of the party, the media attention it got, and its ability to shift and polarize otherwise "tame" political debates showed its potential as a magnet for disillusioned voters and led to the trying out of more radical positions and rhetoric (Küppers 2024). In consequence, the party went through a series of leadership changes (Lees 2018), each indicating a further deepening and widening of their opposition to the closed shop and the mainstream, consensual style of decision-making it represents. Moderate criticism of Germany's energy transition agenda (*Energiewende*) would gradually lead to the adoption of climate science and climate policy skeptic positions (Küppers 2024). In the process, the AfD exposed a key weakness of the corporatist model: Traditionally, actors that pushed controversial agendas were excluded and, in consequence, lacked public attention. However, the AFD increasingly managed to turn the party's marginal position into a strategic advantage. The party expanded on its Euroskepticism and became the main voice of opposition to the government's reaction to the civil war in Syria and was instrumental in the making of the 2015 "refugee crisis" that dominated public debate (Beck 2021). Likewise, as the only party willing to challenge the consensus on climate science, the AfD began to portray climate research as an elitist, mainstream venture that served the political agendas of mainstream parties and political elites (Ruser 2022). In the course of the radicalization of its political program, the AfD also went through a round of leadership changes. The political insiders of the founding years were replaced by leaders who moved the party further to the right and embraced increasingly combative rhetoric.

These targeted provocations and changes of political style helped to morph the AfD into an agent provocateur, an untamed alternative to an overly domesticated political landscape. Around 2015, the AfD directed its attacks on then-Chancellor Angela Merkel and her "extraordinary capacity for remaining calm, pragmatic and rational" (Mushaben 2022, 38). In contrast, the AfD became increasingly agitated, imaginative, and emotional, and as it tried to establish itself i as true outsiders to the political and social establishment. The AfD more and more became a textbook example for the "fourth wave of right-wing extremism" (Wondreys and Mudde 2022, 87) as a party that combined a traditional nationalist agenda with "far-right positions, populist rhetoric, and conspiratorial thinking" (Sawyer 2022).

A watershed moment that highlighted the AfDs transformation from an outsider party in name only to a true opposition to the tradition of knowledge sharing and decision-making was the Covid-19 pandemic. The detection of the novel SARS-CoV-2 coronavirus in late 2019 and its global spread forced governments around the globe to implement drastic, unprecedented measures (Jaspal and Nerlich 2023, 750). Germany, like many other countries, went into lockdown, imposed social distancing rules, and mandated the wearing of face masks. Moreover, like most European countries, the German government relied on expert advice from national and international medical researchers to design and justify its policy responses and build a national coalition of political parties and economic and social stakeholders. Initially, the corporatist traditions of Germany seemed to determine the country's response. Policy measures based on expert advice were accepted without strong opposition, and approval for the AfD dropped for the first time since 2013 as people "rallied around the flag" (Lehmann and Zehntner 2024, 1110). However, as the pandemic dragged on, the AfD managed to develop a series of responses that not only popularized their "Covid Sceptic" positions but also cast the corporatist structures and especially the government's reliance on a selected group of medical experts as evidence for "elites" conspiring against "ordinary people." As Pola Lehmann and Lisa Zehnter argue, some of the rhetorical frames were a continuation and adaptation of existing strategies. In Spring 2020, echoing Donald Trump, the AfD was blaming China for having ignored and/or lied about the virus (Lehmann and Zehnter 2024, 1117), thus casting the pandemic as a problem brought to Germany by "foreigners." Moreover, the AfD engaged in demonizing the political and scientific "elites," accusing them of conspiring to fulfill their technocratic and autocratic ambitions (ibid. 1122). As I have argued elsewhere, the pandemic moved the AfD closer to conspiratorial groups such as the anti-lockdown, anti-vaccine "Querdenker" (contrarians) movement and reinforced the rejection of the corporatist decision-making model as an expression of an elite conspiracy (Ruser 2025, 145).

The brief dive into the history and evolution of the AfD was necessary to understand how climate change skeptics in Germany not only could become politically relevant in recent years but also change their image from scientific illiterates to defenders of freedom (Lehman & Zehnter, 2024).

Against the "Globalist Elites": Re-politicizing Climate Change

The AfD is not the only proponent of climate skeptic positions. However, since 2013, the party has slowly gravitated toward the center and is increasingly occupying an important position within networks of climate change skeptics in Germany (Ruser 2022) and proponents of other fringe positions and conspiracy theories, including anti-vaccine activists and subscribers of esoteric worldviews or UFO sightings (Ruser 2025, 146). In the following section, I will argue that these alliances are at the core of the changed image of climate skepticism in Germany. At first glance, it seems that nothing has changed: Advocates of climate skepticism were and still are outsiders and are more likely to find allies on the lunatic fringe than in the respectable inner circles of German corporatism. Upon closer inspection, however, we find that being an outsider, rather than condemning the party to political irrelevance, is the very reason the AfD is appealing to an increasing number of voters. Opposition to climate science and climate politics is becoming emblematic for the AfD's self-image as political mavericks: Challenging the consensus on climate politics is a rejection of the alleged "elite" pride in Germany's "reputation for environmental awareness and progressive green legislation" (Goodbody 2019, 91). Rejecting the neutrality of climate science is a dismissal of the corporatist logic and its equation of selective inclusion in political consultation and decision-making with the bringing about of generally a satisfactory consensus.

Moreover, the electoral successes of the AfD are putting pressure on the established parties and their corporate partners and thus beginning to influence decision-making within the closed shop. Finally, and most importantly, as we will discuss in more detail in the following section, the specific framing of climate change skepticism as opposition against technocratic scientists and autocratic elites allows the AfD to disguise their own anti-democratic and authoritarian views as a defense of personal liberty and democratic participation.

Climate change skepticism played an important role in the process of political radicalization of the AfD (Küppers 2024, 6) and indicated the party's increasing involvement in far-right, market-radical, and climate skeptic networks outside of the traditional Berlin circles of corporatist Germany (Ruser 2022). Especially its ties with the *Europäsiche Institut für Klima und Energie* (EIKE), a German climate skeptic think tank with close ties to conservative think tanks in the USA (ibid.), helped the AfD to develop their climate skeptic profile. The party's 2016 manifesto, for instance, featured a chapter on Energie that opened with some general comments on climate change and stated that "Carbon dioxide (CO2) is not a pollutant but an indispensable part of all life" (AfD 2016, 79). Interestingly, the statement is a direct copy-paste of the catchphrase of the (in)famous "We call it life" campaign initiated in 2007 by the Competitive Enterprise Institute (CEI) to counter the message of Al Gore's popular documentary *An Inconvenient Truth*. EIKE, who not only maintained close ties to the AfD but whose leadership joined the party, had close ties to the CEI and had brokered a series of events in Germany that provided AfD politicians and American Veterans of the climate skeptic countermovement (Brulle 2014) with opportunities to meet (Ruser 2022, 123).

In recent years, the AfD has successfully borrowed from the US American playbook and re-politicized debate not only about the reliability of climate science and the adequacy of climate policies but also climate change in general. In a recent study, Anne Küppers has identified the dominant climate skeptic frames used by the AfD, which closely resemble the strategies used by the climate skeptic countermovement in the USA (Brulle 2014).

In addition to challenging the scientific consensus on climate change (Machin and Ruser 2019a, 222), the AfD attacked regulatory policies as "eco-socialism" and patronizing attempts to take away individual freedom (Küppers 2024, 18), thus following the American example of pushing debate about climate change from "evidence-based" to "principled" discussion (Ruser 2021, 201). Arguments for specific policy measures (e.g., investment in renewables or electric cars) are no longer contested as potentially inappropriate responses to a (potentially exaggerated) problem but taken as evidence to inoculate the German population "to so much climate guilt that they are finally ready to take in so-called climate refugees" (Küppers 2024, 19) and accept the emergence of an "ecological authoritarian planning state" (AfD cited after Machin and Ruser 2019b). These interpretations, although directly related to climate change and climate politics, fit into a wider strategy of the AfD. The accusation that Germans should be inoculated to "climate guilt," for example, speaks to a more comprehensive undertaking that seeks to revise the historic guilt of the Shoah and Second World War. In a public speech in 2017, then co-chair of the AfD parliamentary group, Alexander Gauland, stated:

> If the French are, quite rightly, proud of their Emperor, and the British of Churchill, then we [the Germans] have the right to be proud of what soldiers achieved in the First and Second World Wars (Gauland cited after Kaya 2021, 322).

In a similar vein, Gauland has referred to the 12 years of Nazism as just "bird droppings" of Germany's otherwise proud 1000-year history ("Hitler und die Nazis sind nur ein Vogelschiss in unserer über 1000-jährigen Geschichte," AfD 2018), implying that Germans have every right to be proud of their national heritage. Casting climate change as just another trick to make Germans feel guilty, therefore serves the "heritage populism" (Kaya 2021, 320) and allows the AfD to turn climate change skepticism into an expression of national identity and belonging (ibid. 321).

In other words, the AfD replicated the adversarial, highly polarized, and unforgiving style of American public debate (Smith et al. 2024) not only to repoliticize climate change but also to use the topic to force its antagonistic rhetoric on Germany's political stage. Challenging the soundness of climate science and the competence of climate scientists (AfD Dresden County Chapter 2024) disputes the status of public research institutions and representatives of (inter)national scientific communities as neutral and reliable sources of expertise and information. Much like in the USA, scientific expertise is redefined as partisan advice and can therefore be rejected on principle, not evidence. Analogous to the party's anti-vaccine skeptic positions during the COVID-19 pandemic, scientific knowledge can simply be discredited as "elite-knowledge" that serves the interests of an actual or imagined "political establishment." The AfD therefore, actively exploits the selectivity of the corporatist structures of consulting and decision-making in Germany. A particular weakness is its history of trusting a relatively small number of public research institutions and universities. The restricted access that favored large, public research institutions and kept advocacy think tanks and other partisan research facilities largely at bay can easily be reinterpreted as an inner circle in which hand-picked experts serve the interests of a small elite. Against this background, frequent attacks on climate research institutes such as the Potsdam Institute for Climate Impact Research or, related, the strong challenges to the neutrality of the Robert Koch Institute, the government's leading adviser on the coronavirus, by AfD party leader Jörg Meuthen (AfD 2024), thus serve the same purpose: to cast the traditional political system of political consulting and decision-making as elitist, corrupt, and sinister.

Climate Scepticism as a Tool: From Symbol of Anti-Establishment Opposition to Gateway for Authoritarian Ideas?

Climate change skepticism has become a cornerstone of the AfD's self-image as an anti-establishment party and outsider to the elitist inner circles of German politics. Challenging climate science and criticizing climate politics helped the party move on from evolving from a Eurosceptic single-issue party to a true "alternative" to the political establishment. Most importantly, the AfD, following a wider trend in America and Europe (Lejano 2019; de Nadal 2024), weaponized climate change to fuel a German version of the "culture wars." Unlike, for example, the Spanish populist party VOX, which accepts scientific evidence for human-made climate change and focuses on the political responses (de Nadal 2024, 1193), the AfD has not yet moved on to "post-denial narratives" (ibid. 1187) and maintains a more traditional climate skeptic profile that combines a rejection of climate science *and* political responses to the climate change "hoax" (Machin and Ruser 2019b). The fact that the party is unwilling to accept scientific evidence for anthropogenic climate change, even when urged to change its position by the AfD's own youth organization, should not be conceived of as a strategic weakness nor as a potential source of inner-party conflict (Küppers 2024, 18). Following the example of other right-wing parties in Europe and adopting post-denial narratives that exclusively focus on the "flawed," "globalist" responses to climate change would deprive the AfD of a valuable bogeyman—elitist scientists—and thus weaken their overall attack on the inner circle of policymaking. Instead of lagging behind other right-wing parties in Europe or being out of touch with the concerns of younger voters (including their own youth party organization), the AfD's strategic use of climate science and climate policy skepticism allows the party to target a central aspect of the consensus-oriented policy making in Germany. Challenging the neutrality of scientific expertise not only discredits policies based on such advice but also undermines the legitimacy of the mode of political consulting and decision-making. The consensus-oriented corporatist system of nonparliamentary consulting is not seen as a (imperfect) microcosmos designed to bring together selected, albeit representative, perspectives and to discuss responses based on the best available knowledge. Instead, mainstream parties are portrayed as being surrounded by a camarilla of partisan advisors and political influencers.

Climate change skepticism becomes a symbol for anti-establishment opposition. The rejection of scientific knowledge and policy responses challenges the boundaries of think- and sayable things and thus becomes emblematic for the AfD's status as outsiders and hence a true "alternative" to the political mainstream. Moreover, by demonizing their political opponents and defaming democratically legitimized responses to climate change as "eco-dictatorship" (Küppers 2024), the AfD can cast themselves as defenders of individual freedom and common sense (Ruser 2021, 209). Parallel to the party's opposition to Covid-19 lockdowns or mask and later vaccine mandates, the AfD claims to champion individual responsibility, the ability to think for oneself, and to protest government overreach and politics based on irrational "alarmism" (Küppers 2024).

This strategy speaks directly to a populist mindset "that considers society to be ultimately separated into two homogeneous and antagonistic groups, the pure people versus the corrupt elite" (Mudde 2004, 543). However, as Torben Lütjen argues, it contradicts the assumptions of an authoritarian program (2022, 76). Focusing on the "populist" responses to the pandemic, he argues that one would expect support for top-down measures. "Why would many leaders of a party family described as `authoritarian´ be so resistant to really exercise their authority and to lead the effort to protect public health, even if that meant asking to sacrifice some personal freedom" (Lütjen 2022, 77). Similar things could be said about climate change. Why would an allegedly authoritarian party oppose the top-down steering of the national energy transition or measures to protect one's own country from potentially catastrophic changes in the environment?

For Lütjen, this puzzle is resolved by depicting the AfD as representatives of "modern day populism," which for him "is a strange hybrid between authoritarian and antiauthoritarian elements, and it is because of the latter that it represents a historical break within the tradition of conservative/right-wing ideology" (2022, 77). In other words, populist parties, "[i]nstead of calling for discipline, order, restrain and responsibility (as we would expect from 'right-wing', or 'conservative' or 'authoritarian' parties), they have celebrated civil disobedience and a spirit of resistance. Instead of stressing the need for community and the prerogative of the collective, they have focused on the (alleged) needs and wants of the individual, whose freedom shall not be infringed upon" (ibid. 76). Lütjen further substantiates his claim that parties like the AfD should be analyzed through the lens of modern populism and mark a breach with classical authoritarian movements by pointing to two main differences. First, unlike the worldview of authoritarian parties, populists "fashion themselves as highly independent minds" (ibid.85) and show no desire to submit "under a cult leader" (ibid. 86). Secondly, rather than subscribing to ideas of collective identification and trust in pre-existing hierarchies, modern populist parties emphasize "self-empowerment" and individualism (ibid. 89).

Lütjen's argument offers a convincing framework for understanding climate science and climate policy skepticism as an integral part of this new anti-authoritarian element in modern-day populism. The rejection of scientific authority can be interpreted as the reaction of highly independent minds, and the opposition to political regulation might well be an expression of self-empowerment vis-à-vis the political and social mainstream. At the same time, Lütjen's analysis offers some comfort, as fears of authoritarian tendencies within populist parties such as the AfD might be exaggerated.

However, I believe that climate change skepticism allows us to take a different perspective on the connection between authoritarian tendencies and the populist elements of the AfD's party rhetoric. In addition, it might help us to understand better why Lütjen arrived at his conclusions and how we can sustain the claim that climate change skepticism facilitates authoritarian ideas. While Lütjen's analysis of the elements of "modern day populism" is comprehensive, convincing, and original, his understanding of authoritarianism rests mainly on Adorno's and the Frankfurt School's focus on the authoritarian personality. He emphasizes in particular the willingness to surrender to central authority (Lütjen 2022, 78–9) as a core element and emphasizes that modern (anti-authoritarian) populist parties are rallying against central authorities.

If we choose a different starting point, however, for instance, Maslow's remarks on authoritarian worldviews, we arrive at different conclusions. Rather than seeing the attacks on scientific and political authorities as an expression of self-empowerment and an independent mind, we discover the law of the jungle, the struggle for power, fought over the ability to define problems, suggest solutions, and force social change. Moreover, upon closer inspection we quickly see the limits of the support of empowerment and individualism: climate activists of the "Last Generation" (*Letzte Generation*) are, in the eyes of a deputy party leader, "extremist enemies of the constitution" (AfD 2022a), while the party is hedging plans to stop "climate criminals" (AfD 2022b). Moreover, research shows that climate change skeptic attitudes constitute boundaries between specific "in groups" (conservative white man, Dunlap, McCright 2011; middle American radicals Ruser 2020) and an outgroup of believers in climate change whose trust in science and orientation toward international collaboration are depicted as weaknesses of elements of a "globalist" conspiracy (Ruser 2025). Belief in climate change is not depicted as different but as an illegitimate position which conflicts with the democratic prerequisite to accept diverging points of view (Maslow 1943, 404).

However, I agree with Lütjen that the current rise of right-wing, populist parties like the AfD indicates the arrival of a new "adaptive" variant of authoritarian thought. Rather than dismissing democratic decision-making and presenting an authoritarian alternative, the AfD frames its authoritarian ideas as a democratic alternative. Social challenges like the Covid-19 pandemic and especially the ongoing "climate crisis" provide the party with the opportunity to frame the respective policy responses as an "anti-democratic and unconstitutional power grab by the political establishment" (Hansen and Olsen 2024, 658) and claim to defend democratic participation even though large parts of the "demos" are sidelined. In the view of the self-styled saviors of individual thinking, climate scientists wouldn't be heard, activists would be criminalized, and the needs and demands of those most vulnerable to climate change would most likely be ignored.

Conclusion

The AfD's critique of corporatist structures and the risk aversion of political parties is neither without justification nor without historical precursors. Germany's "consensual politics" (Streit 2008, 31) had been identified as the main obstacle to swift and far-reaching reforms (Zolnhöfer 2008, 458). Moreover, the "closed shop" itself had been criticized for becoming less inclusive with the weakening of trade unions and charitable organizations' position in nonparliamentary consultation shifting, tipping the balance in favor of business interests (Heeg 2014). Nor should we conceive of Germany before the arrival of the AfD as a kind of climate policy utopia in which consensus-oriented politicians listened willingly to proven experts to base their decisions on impartial and indispensable knowledge. Germany, like most countries, has a history of lengthy political debate and unambitious proposals. Vested interests and concerns about the economic impact of climate protection frequently get in the way of environmental agendas, and it is not uncommon to have climate scientists appearing in the media to urge politicians to take the issue more seriously.

However, only recently has climate change skepticism risen to national prominence. Borrowing from the playbook of the American countermovement (Brulle 2014) and building networks with national and international facilitators of climate skeptic messages (Ruser 2022, 2025), the AfD managed to turn opposition to climate science and climate politics into a strategic political resource.

Climate change skepticism allows the party to cast itself as an outsider to the political "establishment," discredit the selective tradition of nonparliamentary consulting, and delegitimize democratic decision-making. There is reason to fear that the portrayal of climate change skepticism as a defense of individual thinking and self-empowerment helps conceal authoritarian ideas that deny dissenters and members of alleged "elites" the right to have their voices heard.

The AfD has undeniably been successful in changing the image of climate change skeptics in Germany. Once considered scientific illiterates and a mere footnote to the national debate, climate skepticism is now an important pillar of anti-establishment and, at its core, anti-democratic positions. Moreover, the successful reinvention of climate skepticism has attracted imitators. The "left populist" Sahra Wagenknecht Alliance (*Bündnis Sahra Wagenknecht,* BSW), founded in 2023 after breaking from the "Left Party," for instance, has embraced climate policy skepticism in its 2025 manifesto (BSW 2025, 9) and launched attacks on the climate consensus.

Finally, climate change skepticism has the potential to be a canary in the coal mine of authoritarianism. While the AfD, claiming to speak for "the people" and defending independent thinking, wants to shut down open debate about anthropogenic climate change and seeks to criminalize climate activists, democratic societies must endure climate skepticism. This is not to say that democratic debate must value climate skeptic arguments. However, if Maslow is right that democratic worldviews rest upon the "perception and appreciation of difference" (1943, 404), the best way to address the authoritarian tendencies that accompany the AfDs climate change skepticism is to engage in debate—not to win the "scientific illiterates" over, but to prove the point debates about climate politics can be democratic, open, and transparent.

References

Adorno T, Frenkel-Brunswik E, Levinson D, Sanford N (1950) The authoritarian personality. Harper & Brothers

Altenrative für Deutschland (2018) Pressemitteilung: Wortlaut der umstrittenen Passage der Rede von Alexander Gauland. https://afdbundestag.de/wortlaut-der-umstrittenen-passage-der-rede-von-alexander-gauland/, accessed: 29 January 2025

Alternative für Deutschland (2016) Programm für Deutschland. Das Grundsatzprogramm der Alternative für Deutschland

Alternative für Deutschland (2022a) Stephan Brandner: Klima-Chaoten der sogenannten "Letzten Generation" sind Verfassungsfeinde. https://www.afd.de/stephan-brandner-klima-chaoten-der-sogenannten-letzten-generation-sind-verfassungsfeinde/, accessed 31 January 2025.

Alternative für Deutschland (2022b) Schluss mit lustig: AfD-Plan gegen Klimakriminelle!, https://www.afd-landtagsfraktion-niedersachsen.de/rede/schluss-mit-lustig-afd-plan-gegen-klimakriminelle/, accessed 31 January 2025

Alternative für Deutschland (2024) Jörg Meuthen: Das RKI hat jedes Vertrauen in die Unabhängigkeit seiner Corona-Prognosen verspielt. https://www.afd.de/joerg-meuthen-das-rki-hat-jedes-vertrauen-in-die-unabhaengigkeit-seiner-corona-prognosen-verspielt/, accessed 29 January 2025

Alternative für Deutschland, Dresden County Chapter (2024) Karsten Hilse: Klimawissenschaftler nennen nicht den Temperaturausgangswert. https://afd-dd.de/karsten-hilse-klimawissenschaftlern-nennen-nicht-den-temperaturausgangswert/, accessed 29 January 2025

Beck M (2021) On the making of the German 'Refugee Crisis': Securitizing Muslim immigrants in 2015 and beyond. J Refug Stud 34(2):1307–1326

Brulle RJ (2014) Institutionalizing delay: Foundation funding and the creation of U.S. climate change counter-movement organizations. Climate Change 122(4):681–694

Bündnis Sahra Wagenknecht (2025) Unser Land verdient mehr! Wahlprogramm für die Bundestagswahl 2025

Cavender J, Jäger J (1993) The history of Germany's response to climate change. Int Environ Aff 5(1):3–18

Dahrendorf R (1968) Society and democracy in Germany. Weidenfeld and Nicholson, London

de Nadal L (2024) From denial to the culture wars: A study of climate misinformation on YouTube. Environ Commun 18(8):1186–1203. https://doi.org/10.1080/17524032.2024.2363861

Fox, Alan [(1977) 2017] Corporatism and industrial democracy: The social origins of present forms and methods in Britain and Germany Hist Stud Ind Relat 38: 171–219

Glavovic B, Smith T, White I (2022) The tragedy of climate change science. Clim Dev 14(9):829–833

Goodbody A (2019) Klimaskepsis in Germany in Greg Garrard. In: Goodbody A, Handley G, Posthumus S (eds) Climate change scepticism. A transnational ecocritical analysis. Bloomsbury, New York, pp 91–132

Grimm R (2015) The rise of the German Eurosceptic party Alternative für Deutschland, between ordoliberal critique and popular anxiety. Int Polit Sci Rev 36(3):264–278

Hansen MA, Olsen J (2024) The alternative for Germany (AfD) as populist issue entrepreneur: Explaining the party and its voters in the 2021 German Federal Election. German Politic 33(4):643–667

Heeg S (2014) The erosion of corporatism? The rescaling of industrial relations in Germany. European Urban and Regional Studies 21(2):146–160
Heitmeyer W (2024) Authoritarian temptations and right-wing threat alliance. The crisis of capitalist societies in an uncertain future. Springer, Cham
Horkheimer M (1973) The authoritarian state. Telos 1973:3–20
Jaspal R, Nerlich B (2023) Social representation of COVID-19 skeptics: denigration, demonization, and disenfranchisement. Polit Groups Identit 11(4):750–770
Kaya A (2021) The use of the past by the *Alternative for Germany* and the *Front National*: heritage populism, *Ostalgia* and Jeanne D'Arc. J Contemp Eur Stud 31(2):318–331
Küppers A (2024) 'Climate-Soviets,' 'Alarmism,' and 'Eco-Dictatorship': The Framing of Climate Change Scepticism by the Populist Radical Right Alternative for Germany. German Politics 33(1):1–21
Lees C (2018) The 'Alternative for Germany': the rise of right-wing populism at the heart of Europe. Politics 38(3):295–310
Lejano RP (2019) Ideology and the narrative of climate skepticism. Bull Am Meteor Soc 100:415–421
Lübke C (2022) Socioeconomic roots of climate change denial and uncertainty among the European population. Eur Sociol Rev 38(1):153–168
Lütjen T (2022) The anti-authoritarian revolt: Right-wing populism as self-empowerment. Eur J Soc Theory 25(1):75–93
Machin A, Ruser A (2019a) What counts in the politics of climate change? Science, skepticism and emblematic numbers. In: Prutsch MJ (ed) Science, numbers, politics. Cham, Palgrave Macmillan, pp 203–226
Machin A, Ruser A (2019b) Nationalizing the climate: Is the European far right turning green? Green European Journal 27 September 2019
Maslow A (1943) The authoritarian structure. J Soc Psychol 18(2):401–411
McCright A, Dunlap R (2011a) Cool dudes: The denial of climate change among conservative white males in the United States. Glob Environ Chang 21(4):1163–1172
McCright A, Dunlap R (2011b) The politicization of climate change and polarization in the American public's view of global warming 2001–2010. Sociol Q 52:155–194
Mudde C (2004) The populist zeitgeist. Gov Oppos 39(4):541–563
Mushaben J (2022) The politics of small steps: Angela Merkel's transformative impact on Germany and the European Union. J Common Mark Stud 60:38–47
Powell JL (2011) The inquisition of climate science. Columbia University Press
Ruser A (2015) By the markets, of the markets, for the markets? Technocratic decision-making and the hollowing out of democracy. Glob Policy 6(1):83–92
Ruser A (2018) Climate politics and the impact of think tanks. Scientifc expertise in Germany and the US. Palgrave, Basingstoke
Ruser A (2020) A mission for MARS: The success of climate change skeptic rhetoric in the US. Res Rhetorica 7(2):47–63
Ruser A (2021) Widening the gap: US think tanks and the manufactured chasm between scientifc expertise and commmon sense on climate change. In: Landry J (ed) Critical perspectives on think tanks. Power, politics and knowledge. Edward Elgar, Cheltenham, pp 195–214
Ruser A (2022) En finir avec le consensus climatique: le rôle des think tanks dans l'émergence tardive du réseau climatosceptique en Allemagne. Politique et Sociétés 41(3):105–127. https://doi.org/10.7202/1089851ar
Ruser A (2025) Networks of climate change skeptics and far-right narratives in Europe. In: Wejnert B (ed) The global rise of autocracy. Its threat to a sustainable future. Routledge, London, New York, pp 132–153
Sawyer P (2022) Conspiracism in populist radical right candidates: Rallying the base of mainstreaming the fringe? Int J Politics Cult Soc 35:305–340
Smith KE, Bognar JM, Mayer AP (2024) Polarisation of climate and environmental attitudes in the United States, 1973-2022. npj Clim Action 3(2). https://doi.org/10.1038/s44168-023-00074-1

Sossdorf A, Burgi V (2022) "Listen to the science" The of scientific knowledge for the Fridays for Future movement. Front Commun:1–16

Streit M (2008) Im Reformstau – oder das Elend des Verbändestaates. *ORDO* 59(1):31–38

Wondreys J, Mudde C (2022) Victims of the pandemic? European far-right parties and COVID-19. Natl Pap 50(1):86–103

Zimmer A (1999) Corporatism revisited – The legacy of history and the German nonprofit sector. Volunt Int J Volunt Nonprofit Org 10(1):37–49

Zolnhöfer R (2008) An end to the reform Logjam? The reform of German federalism and economic policymaking. German Polit 17(4):457–469

Lehmann, Pola/ Zehnter, Lisa (2024): "The Self-Proclaimed Defender of Freedom. The AfD and the Pandemic". In: Government and Opposition, Vol. 59, No. 4, Special Issue "The Populist Radical Right and the Pandemic", S. 1109-1127

Alexander Ruser is Professor of Sociology at the Department of Sociology and Social Work and Director of the Center for Digital Transformation (CeDiT) at the University of Agder (UiA) in Kristiansand, Norway. His research focuses on political and environmental sociology and the sociology of knowledge, with a special interest in climate change politics, political polarization, and digital communication.

Democracy, Autocracy, and Ecology in Archaeological Perspective: Materiality, Not Metaphor

T. L. Thurston and Lane F. Fargher-Navarro

Abstract Recent archaeological inquiry enriches discourse on democracy's origins and processes through cases from around the globe, from prehistory to recent eras, seeking their roots, challenges, failures and resurgence: how leaders with followers transform into rulers and ruled, how types shared governance can be maintained enough to satisfy a populace, but also how this can spawn untenable forms of authority that are met with resistance, protest and uprising. Past peoples had long historical consciousnesses - as in our times, various parties had values deemed worth fighting for, defined acceptable levels of political participation, understood when it had been undermined, and enacted plans to regain control. The strength of the archaeological approach is its multi-scalar research design, merging data from single households to settlements, urban systems and whole landscapes, whose changes through time reveal sociopolitical processes that are otherwise invisible.

Keywords Materiality · Democracy · Autocracy · Ecology in archeological perspective

Introduction: Premodern Societies and Modern Problems

Does democracy help reduce environmental degradation or contribute to it? This question, in one form or another, lies at the heart of many research programs examining the last 25–50 year period and spanning the boundary of the twentieth and twenty-first centuries, in part coinciding with the recent resurgence of authoritarian

T. L. Thurston (✉)
The University at Buffalo, State University of New York, Buffalo, NY, USA
e-mail: tt27@buffalo.edu

L. F. Fargher-Navarro
The PAST Foundation, The Ohio State University, Columbus, OH, USA
e-mail: lfarghernavarro@pastfoundation.org

B. Wejnert (ed.), *Democracy in Retreat and the Climate Crisis*, Springer Studies on Populism, Identity Politics and Social Justice,
https://doi.org/10.1007/978-3-032-02881-5_5

regimes. This political turn has increased stress among environmentalists, given the propensity of such regimes to both cause and deny environmental problems of multiple types. In seeking answers, many environmental advocates have turned to political science, as well as other social sciences, to understand these issues.

Sociologists and anthropologists tend to study how various actors instrumentalize their conditions to force each other into desired stances—the public messaging of politicians to garner control of critical powers, legal wrangling between governments and multinational corporations over policies, discourse at international meetings that reflects state support for one bloc over another, or the strategies of ordinary people in support or defiance of local and global powers that impact their daily lives. Hovering around these issues is the ultimate possibility of major reorganizations, whether by vote in a democracy or by violence in a dictatorship.

Many political scientists, economists, and others also key into these issues, often using regression analyses and other statistical methods to determine which factors are in play and their causal relationships. For example, among many others, Bernauer and Koubi (2009) note that the constant churning of regimes typical in modern democracies creates a lack of long-term commitment to resolving environmental problems, as elected leaders are almost continuously weighing re-election with policies that demand sacrifices from the public or are unfavorable to business concerns. Hillman and Ursprung (1992) differentiate between types of democracy, arguing that in a direct democracy, overwhelming societal concerns with good environmental stewardship could sway policy, but in a representative democracy, interest group lobbying interferes with the translation of goals from voters to representatives and thus the outcomes.

Others have found negative statistical relationships between environmental improvement and certain democratic societies: wealthy democracies using nonrenewable energy appear detrimental to the environment (Charfeddine and Mrabet 2017; Farzin and Bond 2006; Farzanegan and Markwardt 2018; Fredriksson et al. 2005; Kim et al. 2019; Lv 2017; You et al. 2015), but similar nations using renewables display improvement (Adams and Acheampong 2019; Bernauer and Koubi 2009; Buitenzorgy and Mol 2011; Gani and Scrimgeour 2014; Pellegrini and Gerlagh 2006). In a recent example (Akalin and Erdogan 2021), the authors thus endorse the UN's Sustainable Development Goals (SDGs) as a way of pushing democracies toward renewable energy sources, despite the SDGs' internal contradictions with the simultaneous calls for environmental justice and antithetical neoliberal principles (Thurston and Pettersson 2024; Weber 2017).

It is certainly true that high offices change every 4–6 years in most current democracies, but in many political systems, the directors of an administrative state's regulatory agencies, the heads of government departments, and other bureaucrats change far less frequently: Li and Reuveny (2006) critique many of these studies for selecting the wrong variables. Instead of quantifying the espousal of executive goals by their principals, they measured the actual amounts of pollutants which directly affect environmental quality as a result of human activities. They found that in stronger democracies, pollutants enter the ecosystem at a much lower rate than in other polity types: a higher level of democracy leads to less CO_2 emissions per capita, less NOx emissions per capita, and less organic pollution in water. There are also lower deforestation rates and less land degradation, largely due to higher levels of regulation (Li and Reuveny 2006: 953).

Conversely, it's also been suggested that the strong controls and extended terms of leadership in autocracies and dictatorships could give the environmental policies of dictators more traction. Thus, one can alternatively ask if autocracy helps reduce environmental degradation. Prominent dictatorships of the twentieth century—Nazi Germany, Stalin's Soviet system, and the early People's Republic of China—sometimes supported "show" environmentalism—large, centralized projects highlighting a particular area or resource (Wilson 2020), while devastating others in search of economic, military, or social agendas (Midlarsky 1998). Sometimes the environmental initiatives were utilitarian, sometimes ideological; in hindsight, all were at the expense of the laboring classes used to implement them and the resulting catastrophes they often had to live with. While liberal democracies have also undertaken large, centralized, and often destructive projects, and although many might contest this distinction, Wilson emphasizes that dictatorships seem to initiate them much more frequently and ruthlessly (2020: 316). Fischer (2017) has directly addressed this issue, expressing the view that democracy must be preserved for a variety of reasons but may be too slow and inefficient to manage processes of coping with climate change that must be implemented quickly. Instead of pressing for a benign dictatorship or "eco-authoritarianism," he suggests other ways that this could be achieved, which we will return to below.

Why Study Ancient Democracies and Autocracies?

Our thumbnail sketch does not even scratch the surface of a vast and complex literature on contemporary societies; it is presented as a prelude to the topic of this chapter: the archaeology of democracy and autocracy in ancient societies and its relation to the environment. In 1969, Michel Foucault first published *The Archaeology of Knowledge* (*L'archéologie du savoir),* using the term "archaeology" as a metaphor for his then-new method of analyzing the history of systems of thought in pursuit of his research on power, subjectification, and resistance (Foucault 1969). He chose "archaeology" to describe digging into knowledge bubbling under the surface as understood "givens" of the conditions of life, but not yet fully articulated in discourse or action. Throughout the late twentieth century, his metaphor spread across the broader social sciences and humanities, and in the work of many scholars, "archaeology" has become metonymous with both Foucault's method and the idea of investigating the "pre-conscious" of a society's thought-world.

The title of this chapter might evoke an expectation of such an exploration, and it indeed touches on power, subjectification, and resistance, but instead of metaphor, we focus on how current archaeologists use materiality to investigate autocracy, democracy, and other political formations in the prehistoric, protohistoric, and ahistoric past. Archaeologists have probably written as much as any group of social scientists about the relationship between politics and environmental issues, as well as impacts of and responses to climate change using long-term perspectives. Humanists, especially Classicists, have long pondered ancient "democracy" as a phenomenon, and some mainstream historians have recently taken up this topic (e.g., Stasavage 2020), but to an anthropological archaeologist in the social science context, any meaningful assessment of events or practices on a 20-, 50-, or even 100-year timeframe must be couched within a much longer time sequence. In addition to better contextualizing change, these long sequences often reflect changes in core political strategies that can be correlated with or even strongly attributed to shifting environmental strategies. If there are notable patterns in these data, we can often find them within the hundreds of ancient societies that have been subject to intense scrutiny. All polities "end," even though the cultures they grow within usually continue. Be it a quiet fading away or a catastrophic "event," when it is all over, we look back over extensive time periods and assess how they grappled with long-term processes, with much more data than textual records to draw on.

Political scientists and environmental sociologists have long discussed their problems with integrating social and environmental data, as well as problems of matching the scale of published scientific data with the humans they study (Forsyth 2011). Archaeology has less problems with these problems of scale—it is by nature a multidisciplinary science that integrates archaeological methods at multiple scales (e.g., excavations, surveys, artifact analyses) with social science (cultural anthropology, sociology, political sciences, economics), natural sciences (chemistry, geology, paleoecology, zoology, physical and human geography, and computational methods of all kinds), as well as aspects of the humanities (history, art, architecture) to holistically study human societies over centuries and millennia.

Archaeologists make good use of data gathered by other scholars but are further enabled by their ability to collect and analyze new data themselves if extant data presents a scalar or categorical mismatch. All have at least three research foci: theoretical interests, expertise in one or more time periods and areas, and methodological specializations in the analytical, spatial, and computational sciences, often through large teams working over many years to create the fullest possible investigation of our research questions. In using what is basically a hypothetico-deductive method that incorporates test implications, we know what types of new data will be needed before we begin. This is essentially similar to adding specific questions to a survey or selecting new variables for a regression but involves the mastery of many field and laboratory skills beyond statistical analysis.

Thus, in addressing whether the nature of democracy is an impediment to good environmental stewardship, or vice versa, archaeologists can perhaps ask different questions and make different observations—not better, but different—adding some dimension that might be missing from other considerations. Some of the details that studies of current and recent times acquire from the analysis of contemporary minutiae are lost in this stretched-out arc, but the long view also eliminates some of the noise created by annual or decadal churn: we have a large sample where not a decade or two, but centuries or millennia of processes shed light on assumptions about how polity type mediates environmental problems, or doesn't. Some premodern societies were authoritarian autocracies, others were more democratic than ancient Athens or most of our own democratic systems—all with complex economies, multiple social classes—not necessarily economic in nature—and political sophistication. In this chapter, we will only touch on a few of those discussed in a very large literature.

Archaeologies of Democracy

Since the 1990s, several overlapping groups of archaeologists with expertise in premodern and ancient states have pioneered new methods for understanding polity in terms of trends toward or away from collectivity and democracy on one hand, and autocracy and authoritarianism on the other. There were many democracy-like state societies in the past, in addition to the well-known Athenian city-state and the Roman republic, the two examples most cited by non-archaeologists.

The notion that "archaeology is ideally suited for examining the deep roots of [state] urbanism, its materialization and physicality, and the commonalities and variability in urban experiences cross-culturally and temporally" is not new (Carballo and Fortenberry 2015: 542) and there have been many substantive efforts. As noted, paleoecology, geoarchaeology, and bioarchaeology are common specializations among archaeologists, and across these societies, there were some environmental problems that were self-inflicted while others were not. There are also many states that coped, successfully and less successfully, with climate change, although not always human-induced, for which findings have been published as monographs and in broadly comparative frameworks. There are numerous case studies of premodern democracies that offer clear evidence of well-chosen, sustainable subsistence strategies, which ultimately fall victim to problems beyond their control. There are many authoritarian regimes which, at a given time slice, seem socio-environmentally sound, yet attainments in one era led to cumulative ecological chaos further down the time stream.

In the long term, all these relationships with "nature" give way to new or highly differing forms of organization. Is it because of *polity* type, or *policy* type? The reasons behind such choices are key to determining their successes or failures. Are they related to a state's well-developed level of bureaucratization or its lack of administrative and infrastructural power? Is delegation to local experts aimed at better provision of public goods or a search for the cheapest possible way to govern, so that principals may maximize their own enrichment? What drives change in such policies—and what are the outcomes? In fact, the notion of "collapse" has been entirely rethought—often what looks like state "failure" is simply the process of figuring out how to cope with change proactively, by migrating away from a landscape that no longer can support people, adopting new subsistence strategies, or inventing new forms of governance.

It has not been lost on contemporary policy-makers that there are lessons to be learned from the past. Approaching the intersection of state societies with environment, climate and ecology, between 2017 and 2019, the US National Science Foundation-supported National Socio-Environmental Synthesis Center (SESYNC) funded a series of interdisciplinary workshops to explore how the study of past cultures, understood through their historical and archaeological contexts, could shed light on the sustainability and resilience of current and future societies.

SESYNC's mission is to "bring together the science of the natural world with the science of social systems and decision making to solve problems at the human-environment interface." (SESYNC 2024). The workshops focused on sharing knowledge with planners, policy-makers, municipalities, and other actors, to extract sustainability strategies from the resilient cities of ancient states and their hinterlands. The participants came from the disciplines of archaeology, history, ecology, regional planning, and urban and landscape architecture, eventually coauthoring a book with a number of case studies as well as essays on their relevance to current issues (Murphy et al. 2022). As archaeologists with social concerns about our own times, the goal is to unify some of this thinking and further communicate it with hopes of making these connections clear.

We have already outlined the types of methods used to create or retrieve data, but when archaeologists seek to *interpret* the materiality that they reveal with trowels, shovels, scientific methods, and remote sensing technologies, they must be epistemologically valid while also applicable to the types of data we employ, from the analysis of residue on cooking vessels and the chemicals trodden into earthen floors to artifact types, structures, and monuments, to the patterns of regional trade and communication, agricultural systems, and the exploitation of natural resources. What does it all mean in terms of studying a topic like polity type or early democracy? What is it telling us? Sometimes we have highly specific and not always helpful inscriptions, or narrow, patchy bits of textual "history" created internally by the first adopters of writing, or curiosity pieces written by outsiders about non-literate people. These minimally documented cases shed light on the truly prehistoric to "help to identify what archaeological signatures are most appropriate for considering the remains of early civilizations who lacked extensive textual documentation" (Carballo 2022: 3).

Theorizing Polity in the Premodern and Ancient World

Archaeologists have excellent laboratory, statistical, and spatial methods for understanding low- and mid-level connections, but to detect the organizational principles on which a society is built, we must also test out our ideas within theoretical frameworks. Not every theory that can be applied to a contemporary or fully documented society can work with archaeological datasets. Some of the theoretical frameworks useful for making connections between ancient polity types and environmental impacts include historical ecology, political ecology, and heterarchy theory (Ashmore 2018; Crumley 1994, 2016, 2021; Scarborough et al. 2003; Thurston 2010, 2019; and others, *ad infinitum*). Since our space here is limited, for the purposes of this chapter, we will focus mainly on one successful approach, especially because it uses concepts that are familiar to many social scientists—an adaptation of Collective Action Theory (CAT), specifically modified for use in non-Western, premodern, and archaeological contexts. The roots of the theory lie deep in the twentieth century (Hardin 1968; Hirschman 1970, 1974; Levi 1988; Lichbach 1996; Olson 1965; Ostrom 1990, 2000; Popkin 1979; Rothstein 2011), while the archaeological rendering is more recent (Blanton and Fargher 2008, 2016; Carballo and Feinman 2016; Carballo et al. 2014; Fargher and Blanton 2007; Feinman 2023; Thurston 2016, 2019).

Collective Action Theory and Archaeology

CAT, as modified for broad cross-cultural applications, including in archaeology and non-Western historical cases, assumes that all members of society are rational social actors (Blanton and Fargher 2008), unlike much customary thinking in the European intellectual tradition, which limited rationality to a select elite prior to the rise of democracy in the West (Isakhan 2015; Marx 1857; de Montesquieu 1989 [1748]; Sherratt 1989: 164; Trigger 2003). For CAT, successful state building can potentially occur at any time in human history, given only the requisites of overcoming collective action problems and achieving some level of cooperation between those being governed and those involved in governing.

Notably—counterintuitively for many scholars, earlier archaeologists included—ancient politically complex societies do not uniformly "start out" more egalitarian and grow toward centralized authority over time. In some, an initial phase of autocracy seems to have given way to a more participatory system. In fact, almost every early state, and many later ones, had at least some phase in their histories when more distributed forms of power and more horizontally organized institutions were present. Earlier generations of archaeologists, and other social scientists, did not imagine this to be possible and thus never looked for evidence, but in the last three or four decades, this has changed significantly (Thurston 2010; Thurston and Fernández-Götz 2021a, b).

Mainstream CAT (Ostrom 1990, 2000) specifies a number of "markers" for collective versus authoritarian societies: presence or absence of venues for public participation in government or redress for detrimental policies, fair and predictable taxation, and the provision of public goods in return for quasi-voluntary compliance with taxation. These are often clear in contemporary or historical cases, where they are documented, not only for merely existing, but for how functional they truly are. Using archaeological CAT, we can often reveal these markers in the absence of texts, suggesting many and varied forms of ancient democracy, discussed in more detail below.

One kind of archaeological CAT research seeks the roots of political systems: how leaders with followers transform into rulers and the ruled. Other authors focus on how untenable forms of authority are spawned within more collective societies, alternately succeeding or inviting resistance, protest, and uprising. As in contemporary societies, many citizens likely went about their daily lives without ceaselessly thinking about complex political issues, but when deteriorating conditions became noticeable, past peoples, like ourselves, had long historical consciousnesses. As in our times, they had values deemed worth fighting for, defined acceptable levels of political participation, and understood when it had been undermined. Sometimes reclamation of "rights" in participatory government was achieved with outright rejections of state structures through rebellion or uprisings, and sometimes through quieter development of resistant accommodation and other strategies. There are cases where, if one can follow along for many centuries, shared governance and autocratic forms alternate again and again as citizens become less or more vigilant (Fernández-Götz and Arnold 2019; Simonton 2024; Thurston 2007, 2022).

Joint-Production and Collectivity

Let us first look at hypothetical states where Collective Action principles are in effect. An important aspect of understanding premodern and archaeological forms of organization is CAT's notion of *joint production,* which characterizes processes that signify functioning, working relationships between people and their governments. Joint production occurs when heterogeneous members at different levels of an organization consciously collaborate with the shared goal of synergistically creating some kind of value. Theorists have emphasized both the "interests" of all parties and the returns they gain, but also important ethical factors that facilitate the necessary willingness to collaborate (Bernacchio et al. 2024: 155), involving a microfoundation of moral factors such as justice and fairness in determining work effort, but also honesty and integrity between parties, and a climate of general friendship and sociability rather than internal division and hostility. The relationship between this and considerations of public goods and fair and predictable taxation in CAT is clear, and in terms of collaboration, the venues for communication and negotiation between governors and governed..

The term "taxes" in relation to many ancient states may sound incongruous to students of current and recent societies, but other terms are generally euphemisms. State-like societies first appeared between 7000 and 6000 years ago, while "coins" only appeared 2500 to 3000 years ago. Taxes were paid in goods, services, or labor prior to the invention of coinage but also afterward. Early anthropologists often labeled such contributions as "tribute," "levies," "corvée," or "tithes," a false distinction implicit in nineteenth- and early twentieth-century views that "civilizations" had organized systems of taxation while others were administratively "primitive." These biases are rare today, and these labels are only synonyms for taxation (in money or in kind), a term which we prefer.

In societies where political architects develop fiscal constitutions dependent upon jointly produced resources to fund the state, they face specific constraints. These constraints arise because tax-paying citizens that jointly produce state revenues constantly monitor the behavior of governing officials, including principals (e.g., kings and queens, presidents, prime ministers, rajas, *tlatoques*, etc.), and how they are disposing of their taxes.

Where tax-paying citizens are dissatisfied with the behavior of their governments, they may exercise "voice" to complain or protest through institutional channels (including voting out officials in societies with voting institutions), withholding tax payments, rebellions, and/or emigrating: "exit" (Blanton and Fargher 2008; Scott 2009). These strategies, if employed by a broad section of the base, have the potential to impede the flow of revenue to the state and debilitate the capacity of political actors to achieve their goals: it behooves leaders and rulers to promote some level of public satisfaction, as it facilitates their governing power. While the base can be violently coerced or punished for failing to pay, the institutions needed for enforcement—prisons, policing, enhanced surveillance—are extremely expensive in terms of money and labor.

Accordingly, ancient principals, like current heads of state, often made concessions to achieve quasi-voluntary compliance with revenue demands and ensure the flow of revenue into state coffers from jointly produced resources (Blanton and Fargher 2008; see also Levi 1988). These concessions include:

- Accepting limits on their agency, including on their personal power and on their use of jointly produced resources for personal consumption.
- Putting in place institutions and infrastructure capable of controlling the agency of political officials: one of the commonest ancient and modern points of contention is the tax-farmer or official who brutalizes the public with impunity.
- Justly and fairly monitoring taxpayer compliance to acknowledge amenable taxpayers and fairly penalize free-riders.
- Actually providing public goods at levels acceptable to tax-paying citizens—public security, national defense, fair and just judicial services, roads, water for domestic and agricultural consumption, market security, public education, etc. (Blanton and Fargher 2008).
- And finally, leaders and rulers must participate in one or more fora where ordinary people can communicate with the government, seeking recourse for problems, justice, or to debate issues in the public domain—the provision of "voice."

Principals achieve these conditions by *bureaucratizing* and establishing what anthropologists call "corporate power," which is analogous in some ways to Weber's rational-legal authority (Blanton 1998; Fargher 2016; Weber 1947). While principals give up *personal* power and agency to achieve cooperative state-building, they are, in turn, seated at the head of an organization that possesses extensive *infrastructural* power (Mann 2012). Importantly, this type of power endows the state with the mechanisms necessary to penetrate deep into the populace's social fabric to facilitate and ensure cooperation and collective action (Blanton and Fargher 2008; cf. Mann 2012). States, even ancient ones, that achieve such conditions have what comparative political scientist Bo Rothstein (2011) calls "good government."

Under these circumstances, which are acceptable to both people and their government, CAT predicts that tax-paying citizens will quasi-voluntarily comply with revenue demands and provide governing officials with jointly produced resources (Blanton and Fargher 2008; cf. Levi 1988). Moreover, considering that thoughtful subjectivity (self-awareness) and rational behavior (within a society's own social context) is a characteristic of all humans—not just Westernized societies and elites—the outcome of collective state-building in non-Western and premodern contexts results in governments that are highly comparable with "modern nation states" (democratic states) in the West, as described by Tilly (1975: 608–609) and Rokkan (1969: 63–68). Importantly, Blanton and Fargher (2008: passim but see especially pp. 295–298), using systematic cross-cultural research, demonstrate empirically that highly collective premodern states exhibit the following: penetration, integration, identity, legitimacy, and distribution—like many modern states (cf. Rokkan 1969; Tilly 1975).

A final note about collectivized, bureaucratized states: they need only provide enough collectivity, or even the convincing illusion of collectivity, to elicit cooperation—up to a point. Archaeologists do not use CAT to imagine that even the strongest ancient democracies were political utopias governed only by earnest altruists.

"Seeing" Collective Action and Autocracy

In many cases, premodern and protohistoric states are partially illuminated by textual records. Literate states, such as Mesopotamian societies, the Maya, the Romans or Greeks, and many others, left their own records. They are never complete: sometimes we have quantities of information, but only about topics that political actors were interested in recording, reflecting views that may be opposed to reality, and sometimes they are partial because only a handful have survived the ravages of time. In either case, archaeologists have the ability to "see" what was *happening* as opposed to what is *written*—often the most revealing part of characterizing a textually documented state. Of course, with prehistoric states, we operate without the textual component.

In textually documented societies, the conflict between authoritarian, non-collective governments and ordinary people often culminates over time in conquest of previously autonomous neighbors, such as the Roman conquest of Gaul, Germany, Britain, and the Mediterranean littoral, the English conquest of Ireland and many subsequent regions, the Spanish, English, and French conquests of many civilizations in the Americas—and numerous other examples. Often written specifically to glorify the chroniclers, there are typically many hidden transcripts (Scott 1990) that some historians and many anthropologists and archaeologists plumb for understanding of the conquered society. A number of these have revealed descriptions of what are essentially democracies, often described as "barbarian" or "primitive" by conquerors.

Another type of conflict comprises various degrees of *internal* protest and rejection mounted by the populace when an extant government, colonial or non-colonial, does not meet an acceptable level of collectivity. Similarly, written records reveal the views of the authors, and archaeological data can measure the distance between reality and the inscribed account.

Finally, there are "prehistoric" situations where we know almost nothing about either governments or governed, or even the "rules of the game" in a particular society. Here, we must use archaeological strategies to examine the livelihood, well-being, and health status of heterogeneous levels of the society, the "access" revealed by urban design or rural networks, monumentality and public arts and what they depict, the natural environment, the supply and control of commodities, and many other types of information.

In such opaque cases, archaeologists actively seek indicators of centralization and authoritarianism versus those associated with *heterarchy* (Crumley 1995, 2003; Thurston and Fernández-Götz 2021a). Heterarchy is another way of describing the outcome of a collectivized state society: what modern societies call a separation of powers with checks and balances. An alternative to a vertical, pyramid-like arrangement of power, heterarchies display a horizontal spreading of power across different but equal social institutions, each of which is internally hierarchic, but where none has precedence over others. This does not mean they are egalitarian, although they may have an egalitarian ethos as a state ideology, nor does it mean they are peaceful (Vidale 2018: 30). In many early states, for example, heterarchic "silos" comprise a complex religious institution, a hierarchy for legal jurisprudence, a warrior aristocracy with defensive and offensive concerns, and an assembly of "all" people—defined by each society in terms of who was eligible to attend and participate.

Ancient heterarchies for which we have some textual confirmation indicate powerful checks and balances: in Iron Age Europe, a warlord was required to appear before the assembly to plead a case for going to war. If the vote went against it, there was no war—an exercise of public veto power over militaristic plans (Fichtl 2012; Wickham 2017; Young 2015). In other societies, those who broke the rules of the collective state were exiled for a decade or more (Forsdyke 2000; Simonton 2024), and there were carefully considered rules about term limits and nepotism. In others, the legal codes used at the voting assembly were committed to memory by "law speakers" so that no person could alter a written law without detection (Brink 2008, 2017; Korpiola 2014; Olsen 2014). In terms of such voting rights, as in recent times, the composition of who qualified for this privilege varied with different models of sociopolitical organization.

Seeing Venues for "Voice"

Assembly institutions are remarkably common in ancient states. These form the material expression of the CAT principle of "voice." Their function often includes the election of principals and debates on issues of public concern. Archaeologists have become astute at detecting the presence of "assembly" (Baker and Brookes 2013, 2015; Dickenson 2024; Gleeson 2015, 2018; Iversen 2013; Sanmark and Semple 2008; Semple and Sanmark 2013; Semple et al. 2020; Skinner and Semple

2016; Smyth 2022). Among the most ubiquitous evidence are persistent placenames indicating political meeting places, outdoor areas marked by certain monuments, structures indicating large gatherings for public fora, and even apparatuses for casting ballots (yea or nay) have been discovered, some of which were continually in use for centuries (Olsen 2014). Evidence of commensality—public feasting with political functions and overtones—often accompany such infrastructure in the form of massive ritual deposits of the feasts' remains (Bray 2003; Neuman 2024; Mainland and Batey 2018; Whitley 2018).

Seeing Joint Production, Public Goods, and Fair Taxation

These markers, archaeologically speaking, are "seen" going hand in hand. Joint production can be "seen" when public works are archaeologically visible: where government spending and the labor of the population come together. For example, in many societies, harbors and ports, temples and shrines, roads and canals, irrigation and drinking water systems, city walls, and large marketplace facilities were built by public labor but funded and maintained at levels too great for any entity but the state to have provided (Baker and Brookes 2013; Carballo 2013; Crabtree 2021; Chase and Chase 2014; Fenton 2020; Kenoyer 2017: 263; Roach 2013). At such sites, there is often evidence for the taxation of production and trade, such as deposits of trade tokens and weights and measures just outside a city gate or entrance to a merchant's quarter or marketplace, where economic activity could not have been successful without partnerships between facilitating governments and producers (Freidel et al. 2017; Kenoyer 2006; Schmandt-Besserat 2010). Public goods can be categorized as those which support both taxpayer activities and the state's political economy and those that are primarily for public benefit. The type and level of investment tells us a good deal about whether the public was getting its money's worth of public goods and services.

Seeing Authoritarian States

In identifying non-collective societies such as autocracies, oligarchies, dictatorships, and absolute monarchies, numerous scholars have confirmed pre- and protohistoric cases where "rulers tended to be flamboyant, bedecked by elaborate trappings, [and] inequality was materially expressed and exaggerated … power largely was unchecked" (Feinman 2018: 7). The "theatre" of rulership was part of a suite of authoritarian strategies. We can see such signaling in murals, sculptures, and other images and representations, as well as exclusionary architectural design and urban layouts. Frequently, inscriptions in public places or recordings of public proclamations state plainly the godlike nature of the ruler and their supreme and unlimited power. In societies that have no written records, we can extrapolate such rulership from these and other data. Few public goods are detectable, and those that are primarily benefit the state by facilitating the economic activity of commoners to pay taxes or contribute labor to state projects. Extreme inequality in terms of material wealth and the skeletal indicators of ill-health, hunger, and excessively hard labor are all among the most accessible types of archaeological data.

Conversely, in some societies, rulers' trappings of power are minimal and balanced with indicators of collectivization or heterarchy (Brellas 2016; Campbell et al. 2009; Moore and González-Álvarez 2021), and in yet other cases, governing "elites" are almost invisible, presumably the material signals of elite-ness subsumed within the ordinary "well-off" class of the society, living in relatively modest housing with few trappings of wealth. This can be accompanied by overt evidence of public meeting places and other aspects of heterarchic institutions and patterns (Froese et al. 2014; Froese and Manzanilla 2018; Green 2022; Hepp 2022; Pauketat and Emerson 2007). "In these cases, which we referred to as corporate [societies] … we observed indicators in these contexts that wealth and welfare were dispersed more evenly. There seemed to be greater opportunities for citizen voice to be expressed in public spaces, and the power of principals was seemingly more distributed, checked, and to a degree balanced" (Feinman 2018: 7).

A classic example of how archaeologists specifically use materiality in an "atextual" context is exemplified by the ancient Harappan civilization of the Bronze Age, 5300 to 3300 years ago, which lay in the Indus River valley. The civilization was spread over an extensive area, with numerous large cities of 40,000 to 80,000 people, as well as towns and villages. They shared the same system of weights and measures, religious practices, and architectural styles, but with no indications that they existed in any urban hierarchy under one authority or state (Eltsov 2016: 303). Internally, the cities display a number of demarcated but not "gated" urban neighborhoods, with no "rich versus poor" neighborhoods (Eltsov 2016: 302). While there are multiroom houses and smaller apartment-like dwellings, indicating households of varied means and sizes, the separation between them is modest.

The cities display a complete lack of "palaces" and "temples." Instead, they feature large, open, raised platforms with what appear to be speaker podiums and large gathering halls with unrestricted public access and indications of vast seating installations. Despite being the antithesis of elite signaling, these spaces and halls had elements and facades that displayed grandeur and monumentality, symbolizing a shared urban and cultural identity. Similarly, despite fantastic technologies for the production of jewelry, furnishings, and other artisanal items, there are no public monuments, architectural reliefs, statues, or murals that depict clear "leaders" or "rulers" (Eltsov 2016: 303). Despite many excavations, no indications of major socioeconomic gulfs can be found in burials, which are all relatively modest, with no discernible "elite" variations. The "ideology requiring this modesty must have been formed prior to the transition from the Early to the Mature Harappan period, because the cultural sequence … displays the gradual *simplification* of burial rites notwithstanding the growth of complexity" (Eltsov 2016: 305).

Finally, the state, and whoever was in charge of it, contributed a massive provision of public goods: urban centers had many large, regularly spaced public water wells, running water, toilet facilities within houses which were connected to extensive covered sewers that emptied outside the city walls, well-maintained avenues and side streets, and large public facilities for what was probably ritual bathing (Eltsov 2016: 303). Indication of tax collection outside some of the city's gates (trade tokens, seals, weights, and measures) shows evidence of taxation as part of joint production.

Thus, connecting the features of collective action and heterarchic organization to the material record is possible, and archaeologists have convincingly argued that, as today, ancient complex sociopolitical organizations were clearly able to exist for long periods without authoritarian government hierarchies: "While hierarchy undoubtedly characterizes power relations in some societies, it is equally true that coalitions, federations, and other examples of shared or counterpoised power abound … forms of order exist that are not exclusively hierarchical and that interactive elements in complex systems need not be permanently ranked relative to one another. In fact, it may be in attempts to maintain a permanent ranking that flexibility and adaptive fitness is lost" (Crumley 1995: 3). In other words, more hierarchical states are fragile and easily break, while heterarchies can more easily reconfigure themselves and thus persist.

Coercion, External Revenues, and Collectivity

One attribute of the bureaucratized state that we have already alluded to is that while, ultimately, its collaborative relationship with citizens makes it responsive (as compared with autocratic systems that lack rational-legal structures), establishing it is expensive: many investments in infrastructure, civil service, and human resources must be made. Throughout time and space, many principals have not wished to give up their power and lavish lifestyles, preferring to expend resources on themselves and thus employ more "selfish" strategies.

In one scenario, selfish principals may forego all or most collective strategies, which are costly, and opt to employ *coercive strategies* to achieve their revenue goals. In this scenario, CAT predicts that principals do not accept limits on their power and agency and do not bureaucratize to achieve collaboration with the populace. Instead, they use punishments like fines, imprisonment, and violence to forcibly extract revenues from the subaltern. As Marc Lichbach (1996) points out, employing such strategies, while producing intended outcomes in the short term, tends to result in social disorder, chaos, and reduced revenue flows in the long term (see also Blanton and Fargher 2008; Thurston and Fernández-Götz 2021a, b). Given that archaeological and historical research is ideally suited to studying long-term social processes, they can provide empirical verification of this postulate in terms of resistance, exit, and chaos.

In a second scenario, principals focus revenue strategies on resources that they own and/or directly control—*external revenues*—such as land-based assets: farmland, mines, and quarries, monopolies on long-distance trade, taxes on foreigners, import/export tariffs on international commodities, attached laborers (e.g., serfs, forced laborers, enslaved people, and indentured servants) among others. Accordingly, compliance is offered in exchange for access to resources, permission to engage in an activity and/or be in a specific location, and other rewards. Cooperation is required from only a limited set of individuals—loyal, allied, or otherwise self-interested elites—and the costs of compliance are greatly reduced for rulers; meanwhile, the benefits of these resources and monopolies flow to principals. Thus, in this scenario, CAT predicts that rulers will not accept limits on their power and agency and will not invest in bureaucratization.

While principals may retain personal power if they opt for either coercion or external revenue sources, the organizations they direct do not possess *infrastructural power* (Blanton and Fargher 2008; Fargher 2016; cf. Mann 2012). Therefore, they are unable to control official malfeasance and corruption or limit/reduce conflict and strife within their political territory. Polities built around these strategies tend to exhibit characteristics consistent with what comparative political scientists refer to as "failed states" (Rothstein 2011), including violent oppression of the subaltern, internal, often sectorial, strife, and environmental degradation.

These findings are behind some of the greatest "shocks" to traditional archaeological thinking and revolutionized the study of ancient states beginning in the late twentieth century. Before this, archaeologists generally viewed historic or prehistoric rulers of state societies where principals had great coercive power or control of external revenues as *powerful*. Their rulers inhabited fabulous palaces, commanded ruthless armies, controlled temples, and had access to vast treasuries and luxurious lifestyles.

We *now* understand that despite the *perception* that such a leader is the "strong head" of a "strong state," in wielding this kind of power over a population vastly differentiated in wealth and privilege, such polities are actually remarkably *weak* states. Not only do they spend more on pursuing and punishing dissidents and anti-state movements, but the rumblings of the masses are opaque to them, since they are not participants in any level of collaborative, jointly produced efforts with their populace. The pressure among the subaltern grows and organizes in ways that rulers cannot always penetrate, and unexpected and violent overthrow can and does occur.

Environmental Impacts Through A Collective Action Lens

Let us now look at a few polities of different types, for which we also have corresponding knowledge of their environmental impacts and response to climate and ecological change. These impacts are generally seen through industrial processes, such as the pursuit of raw materials like ores or energy, and agricultural strategies, which can be sustainable or lead to degradation. State leaders display various levels of concern with how the extraction of industrial or agricultural resources affects ordinary people or impacts the environment and the resources themselves. While the case studies presented here seem to tell a story about autocracy and democracy only anecdotally, the statistical test of the application of CAT to premodern and protohistoric states has been performed already, using a cross-cultural comparative approach (Peregrine 2001; Ember and Ember 1995) of thirty premodern states, carried out to evaluate the role of collective action in state formation (Blanton and Fargher 2008, 2009; Fargher and Blanton 2007). For brevity's sake, we will focus primarily on agricultural production and land degradation, although there is plenty of evidence for other types of impact.

Non-Collective Societies

We start with an ancient autocracy with negligible collectivity: the *Neo-Assyrian empire.* Its image as a degraded paradise has long been rife in the popular imagination. Already a century and a half ago, in 1879, British historian Philip Smith described ancient Babylonia and Assyria in his *Smaller Ancient History of the East.* He says:

> Long lines of mounds, it is true, mark the courses of those main arteries which formerly diffused life and vegetation along their banks; but their channels are now bereft of moisture and choked with drifted sand ... All that remains of that ancient civilization—that 'glory of kingdoms, the praise of the whole earth'—is recognizable in the numerous mouldering heaps of brick rubbish which overspread the surface of the plain. Instead of the luxuriant fields, the groves and gardens, nothing now meets the eye but an arid waste—the dense population of the former times has vanished, and no man dwells there (Smith 1879:136).

While many types of societies rose and fell in southwest Asia through time, we will take a look at this strongly autocratic and dictatorial state in light of Collective Action Theory. The protohistoric Neo-Assyrian Empire (911–609 BCE) was likely the first *empire* to ever exist: a state that conquers and incorporates other states. A somewhat lopsided cuneiform record is based on royal inscriptions and decrees, legal texts, royal correspondence, palace archives, a land census, and archaeological data, mainly due to the focus of excavations on elite establishments; all texts come from archaeological sites. We do not have as much information about its structure as we do for later pre-modern states, but enough to exemplify what Eric Wolf described as "long-range strategies of expansion, intensification, and regulation" (Wolf 1972: 201), which introduce "competing pressures between the ecological limitations of the local environment and the extractionary demands of the powerful ruling from afar" (Rosenzweig 2018: 39), a truly predatory state (cf. Levi 1988). Data from site excavations, however, is not the main strategy used by archaeologists to assess the overall arc of socioenvironmental change. This is done through archaeological surveys that closely walk over and record material over large areas, often taking several years or more than a decade, combined with various kinds of airborne remote sensing with satellite and lidar imaging, sediment, ice, and deep-sea cores reflecting climate conditions, pollen profiles, and other paleobotanical data, then comparing this with periods of considerable reorganization of the landscape of Upper Mesopotamia during the Neo-Assyrian period (Wilkinson et al. 2005: 340).

Anthropological archaeologists, as social scientists, have characterized this era as "the archaeology of desolation," where a core group prospered while the provinces were devastated (Faust 2021: xi). The era's environmental catastrophes, whose repercussions are still obvious, are linked to numerous "scalar mismatches" between short- and long-term goals, core and periphery demands, and local and centralized control (Rosenzweig 2018: 45). These dilemmas echo the current environmental consequences of land commodification and neoliberal development (e.g., Kelly and Peluso 2015), a topic many archaeologists have reviewed.

Empires can be organized in different ways, some moderately collectivized. What was the Neo-Assyrian form of governance? Some information comes through fortuitously preserved cuneiform documents. Before the Assyrian era, under the Babylonians, kings owed certain obligations to the public, and the institutions of accessible courts and the city assembly were still extant, but by the Neo-Assyrian era, this was no longer the case (Postgate 2007b: 355). Postgate's (2007b: 358) immense knowledge of cuneiform records led him to state that "the Neo-Assyrian administration was not bureaucratic, and depended on a sense of institutional loyalty and personal interaction up and down the system … the administrative ethos was nevertheless well-developed, with well-formulated concepts of responsibility and authority, and of appointment to and dismissal from, offices … the hierarchy of posts within the system is largely invisible to us because of the combination of the non-bureaucratic ethos, and the tantalizing usage of the phrase 'Masters' House'."

The Master's House trope analogizes the state's administrative hierarchy with the common laborer's relation to the owner of a business or workshop. "Doing the work of one's Master's House" refers to receiving and executing imperial orders (Postgate 2007b: 356–357); likewise, the "despoiling of one's Master's House" refers to actions taken by others to disrupt the "business" of the state. It renders some aspects of government invisible because it was an ancient euphemism for all the details we would love to know but don't.

The monarchs of the Neo-Assyrian empire conquered with extreme brutality, using both fighting forces and new technologies (siege engines, battering rams, controlled fire, destabilization of defensive walls, etc.) to destroy rather than occupy cities. Archaeologists have found that in some areas, "almost all the excavated sites in the region"—large, planned cities—were "cataclysmically destroyed" by the Neo-Assyrians (Katz and Faust 2012: 49). "The ruin and demographic decline generated by the campaign apparently prevented full restoration" (Katz and Faust 2012: 49). In one famous inscription from 882 BCE, the ruler Aššurnaṣirpal writes, "I captured many troops alive. From some I cut off their arms and hands; from others I cut off their noses, ears, and extremities … I hung their heads on trees around the city. I burnt their adolescent boys and girls. I razed, destroyed, burnt and consumed the city" (Frahm 2017: 169).

Outside the home provinces where native Assyrians dominated, the people who survived the conquest were often used as portable assets: forced labor in a process called *agricultural colonization.* In order to create a preferred political landscape, conquering further and further from their core region, the government deported conquered people into newly founded villages far from their original homes, which appear archaeologically like a massive simultaneous wave of new small settlements. The deportees were then used as an agricultural labor force. "The program divorced resettled subjects from their ties to land, polity, and family; enforced new modes of agriculture based on surplus production; and imposed taxes in kind that fettered agriculturists to the empire" (Rosenzweig 2018: 31). This strategy has been used by countless dictatorships and empires through time (Barbieri-Low 2021; Cooper 1996; Davidson et al. 1997, Davidson et al. 2021; Kasaba 2011; Wernke 2013); the Neo-Assyrians seem to have invented it. From the areas they conquered or defeated, huge amounts of plunder were sent back to the capital region.

We can categorize this as a coercive approach to revenue goals in which principals do not accept any limits on their power and agency and do not bureaucratize to achieve collaboration with the populace. Instead, they use punishments, like fines, imprisonment, and violence, to forcibly extract labor and revenues from the subaltern. Furthermore, Postgate concedes that conquest of Levantine port cities may not have been driven by state economic goals at all: "The dignity of the Assyrian king demanded that his capitals, his temples and his palaces should display his wealth, and the latest luxury items from the west, if not forthcoming as booty or tribute … the incorporation of these cities … provided a new source of skilled craftsmen for the royal court to draw on" (Postgate 2007d: 77), illustrating CAT's acknowledgement that personal desire for luxury is frequently behind autocratic excess, and that many public works facilitating joint production (ports, harbors) can be seen as aimed not toward public wellbeing but satisfying elite demands.

The theatre of elite luxury, wealth, and power was expressed in massive palaces, state-styled "arts," and unheard-of wealth, not only in view of other rulers, elites, and urban dwellers but also to "rural farmers, [for whom] the canal network would have been a highly visible symbol of the might of the ruler, who could remake nature by moving rivers. Local offtakes on the Faida canal were associated with reliefs depicting the king and the gods; thus, even a farmer would daily have been confronted by images of royal legitimacy when opening the sluices. Other reliefs near the hydraulic features … would have been powerful royal symbols in an ideologically contested landscape" (Wilkinson et al. 2005: 32). These monuments would also have been a daily reminder that farmers' immense efforts were directed toward enriching the ruler and not for the benefit of their households or communities.

The ruler set governors over the conquered provinces, and there were law-codes applied during disputes, grievances, or conflicts, but Postgate, from his readings of the cuneiform record, concludes that a subject was only "entitled to question the authority of a superior officer or a local official by appealing … directly to the king." Indeed, this was the preferred method for redress (Postgate 2007b: 338). The ruler "interferes in a great variety of matters, and in various ways, and we frequently come across cases where, for one reason or another, the king's permission, influence or authority is sought" (Postgate 2007b: 339). Sealed documents with orders from the king are referred to in numerous cuneiform letters (2007b: 340). If a representative of the ruler was needed, this task fell to a member of his family or his entourage (Postgate 2007b: 341).

In other words, there was no structured venue for "voice" or reprieve for private citizens with legal problems. "Some device through which the private citizen might reach the ears of the king and by-pass the officials directly superior to him was an obvious necessity … writing letters was one means of reaching the king directly, but even so, those letters we have ask not so much as an ultimate source of *justice*, but as the ultimate source of *power or authority*" (Postgate 2007a: 49). The only difference between the appeal of an elite and that of a peasant was that the former would make a direct request and the latter would use a "cringing and obsequious" tone (Postgate 2007a: 50). In terms of civil administration of justice, almost all rulings came from the central administration: anyone designated as a judge was directly subordinate to the king and only in a small village could a village head (mayor) and a group of elders act as judges in local matters or as go-betweens for a tax collector and a debtor (Postgate 2007a: 48).

The empire's agricultural policy was the vector of its primary impact on the environment. Among free people, the collection of agricultural taxes or military conscription was the responsibility of each province's administration; the central government simply received the levies of in-kind taxes or laborers from them (Postgate 2007c: 31). Most grain, a vital crop, was grown in northern Mesopotamia's agricultural plains, but this was expanded over time to decrease risk when any one region experienced a failed harvest (Postgate 2007d: 75). Hay and straw for animal fodder and brickmaking were equally important, as were wood and leather from pastoralism. A large proportion of villagers were enslaved or forced migrant cultivators for elite landowners, while those who were free often lost their land to moneylenders from whom they borrowed in bad years, as there was no state-sponsored safety net, and sometimes they became unfree themselves.

Both through the acts of agricultural colonization and through preserved proclamations and inscriptions, rulers like Sargon II claimed publicly to have tamed nature: for example, formal declarations that the king could make crops spring from bare rock with the godlike power of his Imperium. Using the metaphor of environmental domination and control of nature, they claimed that their agrarian reform would bring prosperity. They rewrote the prior received wisdom about farming and broadcast new ideas about land tenure, legitimation, and rulership principles to both their own and neighboring peoples (Rosenzweig 2018: 31).

A key problem that many empires and states have faced is the failure to access local expertise in favor of centralized decision-making by actors divorced from agricultural or ecological knowledge. In the Neo-Assyrian case, while differences in the types of land colonized resulted in different colonization processes, an uncanny uniformity can be distinguished, in which rural settlements were highly dispersed and the limits of cultivable lands were stretched beyond sustainable capacity: "colonization of former steppe lands in the rain-fed north, the settlement of massive areas of subdesert lands by irrigation along the southern Khabur," as well as intensification by irrigation in areas where rain-fed farming was possible (Wilkinson et al. 2005: 44). In other words, the population was spread into small, dispersed settlements over a larger land area in order to extend the number of hectares under production. In many regions of forced colonization, the low rainfall and marginal conditions were only suitable for grazing, yet the state encouraged high-risk cereal farming using large irrigation systems (Rosenzweig 2018: 33) in order to produce and ship food back to the empire's core, where it was hoarded for the use of elites.

Farmers, and almost certainly administrators, knew that the imperial approaches were poor and risky choices. The first signs of environmental degradation emerged as over-irrigation led to salinization of the soil. The potential problem was not unknown to Neo-Assyrian rulers in the seventh century BCE (600–700 BCE), but it was ignored: in what is today southern Iraq, in 3500 BCE, wheat and barley were major crops, about equally planted. After a century of irrigation, wheat production fell to one-sixth of the yield, and by 2100 BCE, wheat formed only 2% of the crop. By 1700 BCE, it was completely phased out (Shahid et al. 2018: 45). Centuries later, Neo-Assyrian farmers, acting on their own ecological knowledge, tried to respond by planting less preferred but more salt-tolerant crops, barley and vetch instead of wheat and lentils. Imperial rulers and the elite class also demanded more and more meat, but ecological signals indicate that surplus-based pastoralism—simply putting too many animals on an amount of land that will not sustain them—led to overgrazing and massive erosion (Rosenzweig 2018: 45).

Cities within the empire became highly stratified; while elites consumed the majority of foodstuffs, creating inequality and deprivation, common people responded by using what CAT predicts: "diverse, unique and covert" food procurement strategies that were not part of imperial agriculture itself (Rosenzweig 2018: 39). Among these, visible in paleobotanical studies, were small-scale gardening and forays into areas outside of state control—foothills and outlands where the hungry could forage like hunter-gatherers (Rosenzweig 2018: 39).

Thus, while imperial rulers proclaimed their conquest not only of people but of nature, with "their land modification systems, their crop choices, and their pastoral practices, Neo-Assyrian land managers ultimately had to respect thresholds established by plants, animals, and soils, regardless of the empire's agricultural prowess and bombast" (Rosenzweig 2018: 45). Given the precarity of the system, politically and ecologically, decades-long megadroughts over the course of the seventh century BCE were a strong contributor to political and economic collapse (Sinha et al. 2019).

The era was plagued by intra-dynastic problems pitting brothers against brothers and sons against fathers in a variety of power struggles, as well as a period of interregnum when "magnates" from the military class ruled; an eventual return to royal leadership followed. The empire collapsed suddenly between 614 and 609, possibly because the Neo-Assyrian form of rule had "killed loyalty" that might have once existed (Postgate 2007d: 95). Textual records indicate this weakness during a final conflict over Assyrian succession, plus the death of the Neo-Assyrian's puppet ruler in Babylonia, the last region to be conquered, where a new leader rose to organize the assault on the empire, joined by a new polity on the Iranian plateau, the Medes.

This coalition began with a violent and demoralizing symbolic destruction of royal tombs and temples and, a couple of years later, with a 2-month siege of Nineveh, with looting and an "orgy of destruction" in which the reigning

Neo-Assyrian emperor was killed, seemingly fomented by hatred of the highly oppressive system (Frahm 2017: 192). The Babylonian leader, in his later inscriptions, although he was from an upper-class family, calls himself a "son of a nobody" (Frahm 2017: 192)—perhaps to stress a contrast with those dynastic authoritarians he overthrew. But he did not seek to conquer: he did not place himself on the Neo-Assyrian throne; the empire collapsed, its cities were not rebuilt but largely abandoned, and the polity came to an end. It was not until 500 years later that the Parthians came to rule the region as part of a state again (Frahm 2017: 194).

The plains food-producing region was administered by elites in urban centers; cultivators who were conquered, forced labor, enslaved, or debtors would suddenly have been free from central control (Postgate 2007d: 95). Both government revenue and private wealth would have dissipated rapidly. To put this in context, as Postgate notes, "As the first rulers of an empire of this kind and on this scale, the Assyrian kings met the problems it created for the first time, and their solutions were often those adopted by later generations of rulers too" (Postgate 2007a: 56). This included the mismanagement of the environment as part of a compulsion to display symbolic superhuman power over humans and nature alike.

High Collectivity Societies

While there are many examples of partly or semi-collectivized societies, for our purposes, we restrict our discussion to a contrasting example of high collectivization. An example of high collectivity is found at the protohistoric Late Postclassic state of *Tlaxcallan* (CE 1300–1521) just east of the Basin of Mexico, contemporary with the neighboring Aztec Triple Alliance empire. In a recent comparative study, Fargher et al. (2022) concluded that numerous features of the Tlaxcaltecan political regime rank it among the most collective of known premodern states ever constructed, comparable with the democratic ideals developed and implemented in Classical Athens, and therefore a form of ancient democracy that developed outside of Europe and long prior to European contact or colonization.

Covering about 2500 km^2 of the modern state of Tlaxcala, at its peak in CE 1519 (when Cortés and the Spanish *conquistadores* arrived), the Tlaxcaltecan state governed approximately 300,000 citizens that shared a unified or "national" identity despite being from diverse ethnolinguistic groups, including Nahua, Otomi, and Pinome (Fargher et al. 2010, 2011a, b). Apical political authority was vested in a governing council of up to 250 titled political officials, elected to membership by members, based on merit achieved through military, commercial, or religious service to the state (Fargher et al. 2010, 2011a, b). In turn, they appointed and supervised 4000-5000 political agents charged with the daily operation of the state, a ratio of 1:9 to 1:15 agents to households (Fargher 2019). Archaeological and historical evidence suggests that these agents periodically assembled with the ruling council to form a supernumerary decision-making body (Fargher 2019). Broad participation prevented any individual from centralizing power or wealth and ensured citizens' voices (Fargher et al. 2020).

The Spanish described how the state promoted an egalitarian ethos, recognizing equal rights of citizens (cf. *isonomia*) and freedom of speech (cf. *isegoria*) (Fargher et al. 2022). Archaeological evidence from both Tizatlan, the hilltop seat of the Tlaxcaltecan government, and Tlaxcallan, the largest city, indicates that jointly produced revenues were invested in the construction of many large and well-built public spaces (plazas) accessed via an elaborate network of paved streets (Fargher et al. 2022).

The downward penetration of the state deep into Tlaxcaltecan social fabric, promoting cooperation and collective action, is seen in the city's division into three administrative wards and more than 20 highly uniform neighborhoods centered on one or more plazas. Archaeologically, both residential architecture and the distribution of consumer goods show comparatively low inequality (Fargher et al. 2011a, b, 2020). Early Spanish visitors and later chroniclers noted state administration of public security, market security, commercial rights, and property rights protections, accounting for the wide access to luxury goods across social classes documented archaeologically (Fargher et al. 2010, 2020; Marino et al. 2020) and the comparatively humble character of the apical governing officials' residences. Also evident were large public assemblies, where ordinary citizens made political speeches.

In the sacred sphere, the primary god Tezcatlipoca ("Smoking Mirror") was associated with an egalitarian ethos and equal rights in prehispanic central Mexico, carrying a magical obsidian mirror that could reveal the true moral character of individuals, identifying persons worthy of governing and others with corrupt souls, regardless of social position at birth (Fargher et al. 2010, 2011a, b). He also manifested as Moyocoyotzin, or "he who creates himself"—a reference to egalitarianism—and Ixquimilli, or "blind justice"—a reference to equal rights (Fargher et al. 2010).

Consistent with Tezcatlipocan ideology, the most revered cultural hero, Tlahuicole, was a low-born individual from an ethnic minority (Otomi) who became Tlaxcallan's greatest defender against attempts by the Aztec Triple Alliance to end Tlaxcaltecan autonomy. He, as well as numerous other Otomi, resisted Aztec attempts to bribe them to turn traitor (Fargher et al. 2010). Conversely, the high-born official, Xicotencatl el mozo (the younger), was reviled as a traitor after he defied the ruling council and attacked Cortés and his Spaniards (Fargher et al. 2010), who were allies of the indigenous state.

There is also evidence that this egalitarian ethos was enforced by the courts across the political spectrum, and no individual was above the law. All political officials were monitored for malfeasance, even the highest-ranking individuals (e.g., Xicotencatl), and punished after being fairly judged by the courts or by the ruling council. Importantly, no single individual could pass judgment and invoke punishments (Fargher et al. 2010). A key strength of the Tlaxcallan economy and political independence was joint production in agriculture, in which landscape modifications to increase production were coordinated through state policies that bolstered cooperation among farmers while preserving their autonomy to manage production locally rather than centralizing all farming decisions. This was especially important in terms of coping with the region's geology, topography, and climate. The hilly to mountainous terrain necessitated some form of farming on sloping land. In Tlaxcala, earlier attempts proved highly destructive as they were implemented on an individual and piecemeal basis (Mountjoy and Gliessman 1988). Archaeological research demonstrates that earlier occupations of the region, especially in the period 400 BCE and CE 1000, long before the Tlaxcallan state era, the farming of sloping land was not implemented effectively and caused sequences of high erosion and topsoil loss. This further complicated the task facing later Tlaxcallan farmers. According to Antorcha-Pedemonte et al. (2023:8), "… alternating layers of duripan (tepetate) and clay horizons covered with a thin A horizon are especially prone to erosion, and once stripped away, the underlying duripan hardens into an impermeable layer that supports only sparse vegetation" (Borejsza 2013: 15; Wilken 1969; Mountjoy and Gliessman 1988).

By the Tlaxcallan state era, after 1300 CE, state policy supporting joint production and cooperation provide the framework for the coordinated transformation of hillsides through terrace-*cajete* systems—transforming *entire hillsides* made this system sustainable—which provided extensive environmental recovery (Antorcha-Pedemonte et al. 2023; see also Alcántara et al. 2024: 2). This sophisticated hillslope management strategy combines sloping terraces (held in place by stone revetments), canals, and settling tanks (*cajetes*) with integrated agricultural strategies. The "… terraces … maintain soils in place … channels [direct] excess rainfall … along with suspended sediment … into *cajetes* [in which] water … slowly [seeps] downward, reducing its erosive potential and … [allowing for] the recovery of soils …" (Antorcha-Pedemonte et al. 2023:8–9). Fertility and productivity is maintained on terraces via intercropping, crop rotation, fallowing, planting edges with agaves, nopal cactus, fruit trees, and a variety of shrubs and other annuals, and maintaining managed forest patches (Antorcha-Pedemonte et al. 2023: 9; Mountjoy and Gliessman 1988).

Excavations in the protohistoric city of Tlaxcallan found that residential terraces were built atop areas highly degraded during earlier periods (Antorcha-Pedemonte et al. 2023; Fargher et al. 2020; see Borejsza 2013: 15 for other areas of Tlaxcala). Excavations reveal the tremendous labor needed for the Tlaxcaltecans to construct terraces directly on erosion-exposed bare bedrock. They slowly broke up bedrock into stone debris and used it to create flat living surfaces behind the low terrace walls (Antorcha-Pedemonte et al. 2023: 9; Fargher et al. 2020, 2011a, b) after which compost—food waste, trash, and possibly introduced sediment—was added to create "an entirely new landscape with patches of comparatively deep soils (1–2 m or more) covering anywhere from approximately 1000 m^2 to 12,500 m^2 that could be colonized by diverse forms of vegetation" (Antorcha-Pedemonte et al. 2023: 9). Emphasizing the extent of the built landscape, beyond these areas, places that had not been modified "continued to be nearly barren bedrock hillsides, covered only by dispersed shrubby vegetation and a few trees" (Antorcha-Pedemonte et al. 2023: 10).

In addition to state-supported terrace-*cajete* systems, other types of locally directed and implemented coordinated farming strategies contributed to long-term stability. These included massive drained-field systems in wetlands on valley floors, intercropping, and planting allelopathic species along field edges to provide weed control and green manure, along with silviculture in agroforestry mosaics (Antorcha-Pedemonte et al. 2023; see also Anaya et al. 1987; Wilken 1970; Mountjoy and Gliessman 1988). Such intercropping, generally referred to as "milpa agriculture" in Mesoamerica, provided the " … plant biodiversity that was key to community resilience during times of political or ecological crisis …" (Alcántara et al. 2024: 1).

State-supported coordination of cooperative milpa farming on terraces, in drained fields, and in silviculture patches at the suprahousehold scale would have provided a primary source of food for households along with the joint production that funded democratic governance in Tlaxcallan, including road systems, market security, and the ability to resist onslaughts by the Aztec Triple Alliance (cf. Alcántara et al. 2024: 9). Paleobotanical research demonstrates that these large horticultural plots involved over 40 species: not only maize, beans, and squash but also agave and nopal cacti, amaranth, greens, root vegetables, fruits, and nuts, many of which are resistant to short-term and moderate-term droughts.

The milpa institution still survives in Indigenous areas across much of modern Mexico (Alcántara 2023; Antorcha-Pedemonte et al. 2023; Benítez et al. 2021; Rivera-Núñez et al. 2020). While it has been credited with actively defeating the worst impacts of neoliberalism on the region's poor, working class, and traditional populations (Alcántara 2023), it has been debilitated not only by recent globalization but also by historic eras of authoritarian and corrupt governance in Mexico, Guatemala, and beyond. This is not merely a recent issue: it began shortly after the Spanish conquest, when smallpox and other European-introduced diseases raced through the population. Much like the Neo-Assyrians, the starving, weakened Indigenous population was violently physically abused by Spanish imperial authorities, and the survivors, about 1% to 10% of the preconquest population of about 20 million people, were forced onto the now depopulated flatlands to take up Spanish-introduced (and very different) farming practices and/or work on massive agricultural estates (*haciendas*) (Borah and Cook 1969; Martín Gabaldón et al. 2021; Gerhard 1977; Herzog 2018; Lockhart 1969). In the ubiquitous hilly zones, unattended and unrepaired terraces eroded and, in some cases, were intentionally destroyed to provide access for Spanish-controlled livestock to graze, leading to another sequence of erosion (Borejsza et al. 2017; Butzer and Butzer 1995; Fisher et al. 2003; González-Arqueros et al. 2017; Melville 2021).

This process of post-depopulation abandonment and livestock introduction was repeated in many regions. Before archaeologists were able to precisely date and characterize these specific erosion sequences, they were typically attributed to "poor farming practices" of both prehistoric and colonial-era Indigenous people. In fact, it has turned out to be quite the opposite circumstance (Borejsza et al. 2017; Fisher et al. 2003; González-Arqueros et al. 2017), and areas where pre-Hispanic terraces were preserved retain superior soil fertility and show lower erosion (e.g., central Tlaxcala, which had limited Spanish incursions during the Colonial Period, Gibson 1952). Ethnoecological and historical-ecological researchers have used the concept of "creating a second nature" to describe this: by introducing certain types of disturbances, the areas engineered by pre-Hispanic Tlaxcaltecans increased biodiversity, biomass, and sustainability in ways that have endured for centuries or millennia, long after the state was gone (Antorcha-Pedemonte et al. 2023), in direct contrast to the legacy of the Neo-Assyrians and many others.

The Roman Empire

It's worth taking a brief look at a parallel case: the failed terracing efforts of the Roman Empire. While Mesoamerican terrace systems degraded due to depopulation from disease, landscape modification to accommodate livestock, and forced abandonment under Spanish colonialism and enslavement, the Romans actually managed and oversaw both the construction and abandonment of terraces, essentially watching agricultural regions erode and wash away without any attempts to prevent it. This and similar issues caused human-made crises that spelled great trouble for the empire (Franconi 2021).

Classical scholars and historians frequently refer to the early imperial Roman "agricultural revolution" as an ingenious program of sustained, long-term increases in agricultural productivity. However, in almost every case, Roman-induced productivity is followed by "periods of decline, which often left the land less productive than it had been at the starting point of the development. Moreover, the same cycle of growth and reversal [is] … observed in regions that were totally dissimilar in terms of geology or climate" (Liebeschuetz 2015: 55). Liebeschuetz, as a classical historian, could not explain this phenomenon except to guess that the Romans applied novel technical innovations and that "the return to a more 'primeval' form of exploitation of the landscape will have been an adaptation to changed social conditions linked to the disappearance of the Empire" (Liebeschuetz 2015: 64–65). His guess about changed social conditions was in some ways correct, but it had nothing to do with any resurgence of the post-Roman Celtic or Germanic conquered peoples, who in fact successfully returned to pre-Imperial agropastoral strategies in the face of political and especially climate change (Crumley 1994). As a historian, he lacked any way to find the mechanism. The true cause, overlooked by many Romanists, was revealed some time ago by archaeologists as related to changing conditions *during* the period of empire, not after it disappeared.

The ARCHAEOMEDES research program, funded by the Directorate General XII of the European Commission, ran from 1992–1999 and was aimed at understanding the deep origins of "poverty" in southern Europe, resulting from desertification, land degradation, and land abandonment (van der Leeuw 1998, 2003; van der Leeuw et al. 2005). The entire team was directed and led by Sander van der Leeuw, an archaeologist, involving 65 researchers from seven countries in mathematics, physics, geology, modeling, geochemistry, soil science, life science, geography, sociology, anthropology, and archaeology. As a case in point, the Roman era was one of several important time periods for answering these questions, and research in the Rhone valley, one of the most repeatedly degraded landscapes in

Europe, was subject to study. This turned out to explain much of the Roman sustainability fiasco, which is in line with other "Imperial" or authoritarian conditions.

The Rhone Valley's Roman "environmental crisis" was not of natural origin. The settlement pattern was reorganized after 100 years of increasingly intensive agro-industry, related mainly to political events and a "rapid economic expansion … which depended heavily on continued investment to maintain its extensive infrastructure" (van der Leeuw 1998: 220). For example, in Tricastin, a geological and historic region in the southern Rhône valley, "the Emperor undertook a vast land reclamation project in the first century BCE. He used it to retire soldiers from the many legions that were active around the time of the birth of Christ" (van der Leeuw et al. 2005: 10) and who had carried out a brutal conquest (Fernández-Götz et al. 2020; Raaflaub 2021). Here, these men, who had survived 20 years of military service, were encouraged to settle and work growing grapes for wine, with the aid of an extremely complex irrigated and terraced landscape created by Roman engineers. "However, the 'peace dividend' (post-conquest) of the early decades of the first century CE abruptly reduced the number of soldiers to be retired. Hence, the orthogonal Roman drainage system could no longer be maintained sufficiently well to ensure that it functioned correctly. As a result, the surface available for cultivation, and the yield per hectare, fell drastically, and many farms were deserted" (van der Leeuw et al. 2005: 10). The erosion of the valley's slopes created a "fragilization" of the landscape which underlies today's valley, making it extremely susceptible to modern degradation—"the slightest climatic or anthropogenic event today can cause erosion" (van der Leeuw et al. 2003: 330). When the process of active conquest and the retirement of vast armies was no longer an issue, administrators looked away. Similar issues plague other regions of the Roman realm, and as in modern Central America, underlying problems created centuries or millennia ago persist in today's world, molding the ability of current inhabitants to cope with ecological challenges.

Conclusions

To conclude this summary tour of ancient civilizations, we briefly review some recent large-scale visions of how planetary problems should be managed, as well as what futurists imagine will be the solutions and challenges of the near and more distant future.

We return to Fischer (2017), who argues that successful environmental and ecological policies and solutions are far less about *polity type* and far more strongly related to *who makes decisions*. It has long been recognized that among the components of contemporary "good governance," decentralization and the devolution of decision-making to include mid- and low-level management are key. This creates collaborations between state, regional, and local governance and strikes a balance between who determines the parameters of joint production and all its constituent values. In many archaeological case studies beyond those discussed here, the evidence points to this strategy as one that is successful not only for recent and current "5-year plans" or "10-year plans" but for much longer periods of time. And, when people are prevented by law or custom from managing their own production to produce sufficient well-being, they will pursue those covert, deviant, and non-legal eco-management and food procurement strategies discussed above in order to do so: the foraging Neo-Assyrian forced migrants, or the Mexican farmers slipping away to tend remaining milpas when their back-breaking labor for their hacienda was done—and a multitude of examples from throughout history and the current world.

Given the loss of local and even regional autonomy to control environmental conditions and agricultural production in recent times, how can the precarity of a completely globalized system (let alone a state-level system) be dealt with? Over the last several decades, many have argued that there has been both a "technocratization" of environmental policymaking (Fischer 2000) as well as a spreading notion of public-private management of "sustainable governance" promoted by the United Nations, the World Bank, the EU, and USAID among other bodies, which includes both human-nature and human-human interaction (Utting 2015). This is simply an iteration of neoliberalism, a system which purports to be economic but is in reality an opaque form of governance by non-elected wealth-seekers, who can be capitalists, communists, or of any other political-ideological stripe. As "a polycentric network of inter-relationships among public and private actors in which the boundaries between the two sectors are fuzzy and blurred" (Fischer 2017: 140), neoliberalism lacks both transparency and accountability and in fact threatens rather than promotes democracy (Brown 2019; Cayla 2023; Harvey 2005; Slobodian 2018) not only in its current practice but also in similar earlier models of rule (Thurston and Pettersson 2024). At their most basic, neoliberal principles seek to commodify everything into profit-making enterprises, even the most basic human rights, such as access to water, food, and shelter. Clearly, in a time of ecological and environmental peril, this is a catastrophic approach. Many authors have pondered what should replace it.

Machin (2022) describes three current types of scenario-building for the future: skeptical, rational, and radical:

> the *skeptical imaginary*, found in the 'eco-authoritarianism' of the 1970s that is echoed by much sociopolitical analysis today, [that] casts doubt on the possibilities of democratic mechanisms to respond adequately and swiftly to the problem of climate change. Those who resist such skepticism often defend democracy by arguing that institutions and processes of democracy can be made more 'ecologically rational'—*the rational imaginary* of climate democracy involves improvements in political representation and participation. Finally … the alternative *radical democratic imaginary*, in which the crisis of climate change provides a moment for the rupture of existing sociopolitical structures and the formation of alternatives (Machin 2022: 1).

Machin's conclusion is that although imperfect, the radical imaginary is the best prospect, as it "does not see wicked ecological problems as exposing the limits of democracy as much as driving its renewal" (Machin 2022: 7). She, and similarly minded authors, suggest that this does not require the "redesign of democratic institutions and the creation of new ones (from the inside/above) but rather suggests that it already exists, and calls for the entry of new subjects and conflicting demands into the political realm (from outside/below) to make themselves visible and proffer real alternatives" (Machin 2022: 7).

Fischer (2017) sees a coming conflict between these imaginaries: contemporary environmental theory is "speaking to a reality that is unlikely to exist" in coming times of crisis (Fischer 2017: viii), and given the trending authoritarian revival, "the door will open wider for the eco-authoritarians. And they will have less and less trouble finding support from political and economic elites and worried publics. Casting their arguments in terms of emergency, perhaps even 'survivalism', they will in all likelihood push for crash technology programs to be governed by technocratic decision-makers. Closely related would be calls for a strong police and military role to quell increasing civil unrest in the face of dire circumstances" (Fischer 2017: viii).

Fischer, too, argues for something like the radical imaginary: to rectify *who makes decisions,* we should look to what is already a growing but little-discussed movement—participatory environmental governance embodied in a growing 're-localization' movement, creating a global network of eco-localism (Fischer 2017: 13). He also raises the theory of associative democracy, which posits that voluntary associations are the basic units of political activity—"neighborhood groups, social clubs, arts and crafts associations, political interest groups, professional societies, or eco-transition movements" formed by people sharing specific views, goals, and concerns. Such groups usually have a diversity of membership that helps prevent the emergence of tyrannical government, and "as basic units of civil society … associations … see their well-being in ways that are independent of money, power, or status" (Fischer 2017: 253). He also argues for the "democratization of policy expertise" in order to counteract the dominance of "industry-oriented experts and environmental counter-experts arguing over the validity of competing assessments (Fischer 2017: 46)" and make room for participatory policy expertise that gives equal voice to engaged citizens. While these goals may seem aspirational and unlikely, they are in fact already growing (Dickel et al. 2019; Jalbert 2016; Kimura and Kinchy 2016; Wylie et al. 2014).

Re-localization in terms of agricultural policy and food webs has been studied for some time, partly as policy-oriented and partly as "identity-marking" and culturally meaningful (Hinrichs 2003; Wilhelmina et al. 2010). While some have stressed complete re-localization, others have found that strict local food supply chains are not always optimal and that a mixed strategy might be best (Gava et al. 2018).

In fact, these findings also align quite well with the observations of archaeologists. Certainly, among the key mistakes of ancient states has been the assignment of agro-environmental decision-making to distant and urban managers, with no or negligible inputs from local producers in highly varied environments. Yet it can also be argued that a combination of centrally facilitated but locally managed as well as fully locally controlled sources of food is an ecologically and sociopolitically sound strategy, illustrated in our cases by the Tlaxcallan state. Sometimes with intent, and sometimes purely by trial and error, ancient societies exhibiting these properties have fared best in providing a tenable lifestyle for their citizens, even if it could not last forever.

We hope that our focus on the materiality of the past, rather than its metaphoric power, has provided some insight through a more detailed look into what we understand about how past societies have both caused and managed their environmental problems. These cases, and many others, have led some scholars to charge that archaeological knowledge should be included at the forefront of socioecological research on present times as well as the past (Fitzhugh et al. 2019; Redman et al. 2004; Sabloff 2016; Stephens et al. 2019; van der Leeuw and Redman 2002). Following Fischer, we do not advocate that archaeological knowledge is a superior source of data or that it provides 'the' model for imagining the future. "Rather the argument here is that they provide some useful starting points for beginning to rethink political theory for a time that lies ahead" (Fischer 2017: 17).

References

Adams S, Acheampong AO (2019) Reducing carbon emissions: the role of renewable energy and democracy. J Clean Prod 240:118–245

Akalin G, Erdogan S (2021) Does democracy help reduce environmental degradation? Environ Sci Pollut Res 28(6):7226–7235

Alcántara K (2023) Milpa ecologies: transgenerational foodways in Tlaxcala, Mexico. Elementa: Sci Anthrop 11(1)

Alcántara K, Corral AL, Tung TA (2024) A milpa isotopic foodweb in Tlaxcallan: more than just maize. J Archaeol Sci Rep 60:104819

Anaya AL, Ramos L, Cruz R, Hernández JG, Nava V (1987) Perspectives on allelopathy in Mexican traditional agroecosystems: a case study in Tlaxcala. J Chem Ecol 13:2083–2101. https://doi.org/10.1007/BF01012873

Antorcha-Pedemonte R, Rivera-Núñez T, Fargher-Navarro LF (2023) The impact of ancient Mesoamerican cities on long-term environmental sustainability: the view from historical ecology. Front Ecol Evol 11. https://doi.org/10.3389/fevo.2023.1237953

Ashmore W (2018) Why the archaeology of political ecology matters. Archeol Pap Am Anthropol Assoc 29(1):175–184

Baker J, Brookes S (2013) Monumentalising the political landscape: a special class of Anglo-Saxon assembly site. Antiqu J 93:147–162

Baker J, Brookes S (2015) Identifying outdoor assembly sites in early medieval England. J Field Archaeol 40(1):3–21

Barbieri-Low AJ (2021) Coerced migration and resettlement in the Qin imperial expansion. J Chin Hist 中國歷史學刊 5(2):181–202

Benítez M, Rivera-Nuñez T, García-Barrios L (2021) Agroecología y sistemas complejos: Planteamientos epistémicos, casos de estudio y enfoques metodológicos. CopIt ArXives

Bernacchio C, Foss NJ, Lindenberg S (2024) The virtues of joint production: ethical foundations for collaborative organizations. Acad Manag Rev 49(1):155–181

Bernauer T, Koubi V (2009) Effects of political institutions on air quality. Ecol Econ 68:1355–1365

Birkelbach AW (1973) The subak association. Indonesia 16:153–169

Blanton RE (1998) Beyond centralization: steps toward a theory of egalitarian behavior in archaic states. In: Feinman GM, Marcus J (eds) Archaic states. School of American Research Press, Santa Fe, pp 135–172

Blanton RE, Fargher LF (2008) Collective action in the formation of pre-modern states. Springer Science & Business Media

Blanton RE, Fargher LF (2009) Collective action in the evolution of pre-modern states. Soc Evol Hist 8(2):133–166

Blanton RE, Fargher LF (2016) How humans cooperate: Confronting the challenges of collective action. University Press of Colorado

Borah W, Cook SF (1969) Conquest and population: a demographic approach to Mexican history. Proc Am Philos Soc 113(2):177–183

Borejsza A (2013) Village and field abandonment in post-Conquest Tlaxcala: a geoarchaeological perspective. Anthropocene 3:9–23

Borejsza A, McClung de Tapia E, Vázquez Selem L, Adriano Morán C, Castro Govea R, Rodríguez López I (2017) Changing rural landscapes of the last three millennia at Santiago Tlalpan, Tlaxcala, Mexico. Geoarchaeology 32(1):36–63

Bray TL (2003) The commensal politics of early states and empires. In: The archaeology and politics of food and feasting in early states and empires. Springer, Boston, pp 1–13

Brellas DJ (2016) By the rivers of Babylon: patterns of heterarchy, sustainable wetland agroecology, and urban dynamics in old Babylonian Mashkan-shapir. (Doctoral dissertation. Boston University

Brink S (2008) Law and society: polities and legal customs in Viking Scandinavia. In: The Viking World. Routledge, pp 47–55

Brink S (2017) Law, society and landscape in early Scandinavia. In: Comparative law and anthropology. Edward Elgar Publishing, pp 319–337

Brown W (2019) In the ruins of neoliberalism: the rise of antidemocratic politics in the West. Columbia University Press

Buitenzorgy M, Mol APJ (2011) Does democracy lead to a better environment? Deforestation and the democratic transition peak. Environ Resour Econ 48:59–70

Butzer KW, Butzer EK (1995) Transfer of the Mediterranean livestock economy to New Spain: adaptation and ecological consequences. In: Turner BL, Sal AG, Bernáldez FG, di Castri F (eds) Global land use change: A perspective from the Columbian encounter. Centro Superior de Estudios Científicas, Madrid, pp 151–193

Campbell R et al (2009) Toward a networks and boundaries approach to early complex polities: the Late Shang case. Curr Anthropol 50(6):821–848

Carballo DM (2013) The social organization of craft production and interregional exchange at Teotihuacan. Merchants, markets, and exchange in the pre-Columbian world, pp 113–140

Carballo DM (2022) Governance strategies in precolonial central Mexico. Front Polit Sci 4:797331

Carballo DM, Feinman GM (2016) Cooperation, collective action, and the archeology of large-scale societies. Evol Anthropol Issues News Rev 25(6):288–296

Carballo DM, Fortenberry B (2015) Bridging prehistory and history in the archaeology of cities. J Field Archaeol 40(5):542–559

Carballo DM, Roscoe P, Feinman GM (2014) Cooperation and collective action in the cultural evolution of complex societies. J Archaeol Method Theory 21:98–133

Cayla D (2023) The decline and fall of neoliberalism: rebuilding the economy in an age of crises. Routledge

Charfeddine L, Mrabet Z (2017) The impact of economic development and social-political factors on ecological footprint: A panel data analysis for 15 MENA countries. Renew Sust Energ Rev 76:138–154

Chase DZ, Chase AF (2014) Ancient Maya markets and the economic integration of Caracol, Belize. Anc Mesoam 25(1):239–250

Cooper, F. 1996, Decolonization and African Society: The Labor Question in French and British Africa, Cambridge: Cambridge University Press

Crabtree PJ (2021) Taxation in Anglo-Saxon England, fifth–ninth centuries CE. Ancient taxation: the mechanics of extraction in comparative perspective, vol 11, p 160

Crumley CL (1994) The ecology of conquest: contrasting agropastoral and agricultural societies' adaptation to climatic change. In Historical ecology: cultural knowledge and changing landscapes, pp 183–201.

Crumley CL (1995) Heterarchy and the analysis of complex societies. Archeol Pap Am Anthropol Assoc 6(1):1–5

Crumley CL (2003) Alternative forms of social order. In: Scarborough VL, Valdez F Jr, Dunning N (eds) Heterarchy, political economy, and the ancient Maya: the three Rivers Region of the East-Central Yucatan Peninsula. Tucson, University of Arizona Press, pp 136–145

Crumley CL (2016) Historical ecology: integrated thinking at multiple temporal and spatial scales. In The world system and the earth System. Routledge, pp 15–28

Crumley CL (2021) Historical ecology: A robust bridge between archaeology and ecology. Sustainability 13(15):8210

Davidson R, Fehren-Schmitz L, Llamas B (2021) A multidisciplinary review of the inka imperial resettlement policy and implications for future investigations. Genes 2021(12):215

de Montesquieu C (1989 [1748]) Montesquieu: the spirit of the laws. Cambridge University Press, Cambridge.

Dickel S, Schneider C, Thiem C, Wenten KA (2019) Engineering publics: the different modes of civic technoscience. Sci Technol Stud 32(2):8–23

Dickenson CP (2024) Agoras and forums. In: Flohr, Zuiderhoek (eds) Companion to cities in the Greco-Roman World. Wiley, pp 111–125

Eltsov PA (2016) The ghost of the state in deep antiquity: a closer look at the Harappan Civilization from the viewpoint of Sanskrit literature. In: Abraham SA, Gullapalli P, Raczek TP, Rizvi UZ (eds) Connections and complexity: new approaches to the archaeology of South Asia. Routledge, pp 299–314

Ember M, Ember CR (1995) Worldwide cross-cultural studies and their relevance for archaeology. J Archaeol Res 3:87–111

Fargher LF (2009) A comparison of the spatial distribution of agriculture and craft specialization in five state-level societies. J Anthropol Res 65(3):353–387

Fargher LF (2016) Corporate power strategies, collective action, and control of principals: A cross-cultural perspective. In: Fargher LF, Espinoza VYH (eds) Alternative pathways to complexity. University Press of Colorado, Boulder, pp 309–326

Fargher LF (2019) El Malinche and Tlaxcallan: a field guide to taking down Democracy. Paper presented that the 84th annual meeting of the Society for American Archaeology, Albuquerque, New Mexico

Fargher LF, Blanton RE (2007) Revenue, voice, and public goods in three pre-modern states. Comp Stud Soc Hist 49(4):848–882

Fargher LF, Blanton RE, Heredia Espinoza VY (2010) Egalitarian ideology and political power in prehispanic central Mexico: the case of Tlaxcallan. Lat Am Antiq 21(3):227–251

Fargher LF, Blanton RE, Heredia Espinoza VY, Millhauser J, Xiuhtecutli N, Overholtzer L (2011a) Tlaxcallan: the archaeology of an ancient republic in the New World. Antiquity 85(327):172–186

Fargher LF, Espinoza VYH, Blanton RE (2011b) Alternative pathways to power in late Postclassic Highland Mesoamerica. J Anthropol Archaeol 30(3):306–326

Fargher LF, Antorcha-Pedemonte RR, Espinoza VYH, Blanton RE, Corral AL, Cook RA, Millhauser JK, Marino MD, Rojo IM, Alcántara IP, Costa A (2020) Wealth inequality, social stratification, and the built environment in Late Prehispanic Highland Mexico: a comparative analysis with special emphasis on Tlaxcallan. J Anthropol Archaeol 58. https://doi.org/10.1016/j.jaa.2020.101176

Fargher LF, Blanton RE, Heredia Espinoza VY (2022) Collective action, good government, and democracy in tlaxcallan, Mexico: an analysis based on demokratia. Front Polit Sci 4(832):440

Farzanegan MR, Markwardt G (2018) Development and pollution in the Middle East and North Africa: democracy matters. J Policy Model 40(2):350–374

Farzin YH, Bond CA (2006) Democracy and environmental quality. J Dev Econ 81(1):213–235

Faust A (2021) The Neo-Assyrian empire in the southwest: imperial domination and its consequences. Oxford University Press
Feinman G (2018) The governance and leadership of prehispanic Mesoamerican polities: new perspectives and comparative implications. Cliodynamics 9(2)
Feinman GM (2023) Reconceptualizing archaeological perspectives on long-term political change. Annu Rev Anthropol 52(1):347–364
Fenton A (2020) Royal authority, regional integrity: the function and use of Anglo-Saxon writ formulae. In: The languages of early medieval charters. Brill, pp 412–439
Fernández-Götz M, Arnold B (2019) Internal conflict in Iron Age Europe: methodological challenges and possible scenarios. World Archaeol 51(5):654–672
Fernández-Götz M, Maschek D, Roymans N (2020) The dark side of the Empire: Roman expansionism between object agency and predatory regime. Antiquity 94(378):1630–1639
Fichtl S (2012) Places publiques et lieux de rassemblement à la fin de l'âge du Fer dans le monde celtique. In: Le forum en Gaule et dans les régions voisines. Ausonius, Bordeaux, pp 41–53
Fischer F (2000) Citizens, experts, and the environment: the politics of local knowledge. Duke University Press, Durham
Fischer F (2017) Climate crisis and the democratic prospect: participatory governance in sustainable communities. Oxford University Press
Fisher CT, Pollard HP, Israde-Alcántara I, Garduno-Monroy VH, Banerjee SK (2003) A reexamination of human-induced environmental change within the Lake Patzcuaro Basin, Michoacan, Mexico. Proc Natl Acad Sci USA 100(8):4957–4962
Fitzhugh B, Butler VL, Bovy KM, Etnier MA (2019) Human ecodynamics: a perspective for the study of long-term change in socioecological systems. J Archaeol Sci Rep 23:1077–1094
Flannery KV (1972) The cultural evolution of civilizations. Annu Rev Ecol Syst 3:399–426
Forsdyke S (2000) Exile, ostracism and the Athenian democracy. Class Antiq 19(2):232–263
Forsyth T (2011) Politicizing environmental explanations: what can political ecology learn from sociology and philosophy of science. In: Goldman MJ, Nadasdy P, Turner MD (eds) Knowing nature: conversations at the intersection of political ecology and science studies. University of Chicago Press, pp 31–46
Foucault M (1969) L'archéologie. du Savoir. Éditions Gallimard, Paris
Frahm E (2017) The Neo-Assyrian Period (ca. 1000–609 bce). A companion to Assyria, pp 161–208
Franconi TV (2021) The environmental imperialism of the Roman Empire in Northwestern Europe. In: Climate change and ancient societies in Europe and the near East: diversity in collapse and resilience. Springer International Publishing, Cham, pp 321–345
Fredriksson PG, Neumayer E, Damania R, Gates S (2005) Environmentalism, democracy, and pollution control. J Environ Econ Manag 49(2):343–365
Freidel DA, Masson MA, Rich M (2017) Imagining a complex Maya political economy: Counting tokens and currencies in image, text and the archaeological record. Camb Archaeol J 27(1):29–54
Froese T, Manzanilla LR (2018) Modeling collective rule at ancient Teotihuacan as a complex adaptive system: communal ritual makes social hierarchy more effective. Cogn Syst Res 52:862–874
Froese T, Gershenson C, Manzanilla LR (2014) Can government be self-organized? A mathematical model of the collective social organization of ancient Teotihuacan, Central Mexico. PLoS One 9(10):e109966
Gani A, Scrimgeour F (2014) Modeling governance and water pollution using the institutional ecological economic framework. Econ Model 42:363–372
Gava O, Galli F, Bartolini F, Brunori G (2018) Linking sustainability with geographical proximity in food supply chains. An indicator selection framework. Agriculture 8(9):130
Gerhard P (1977) Congregaciones de indios en la Nueva España antes de 1570. Hist Mex 26(3):347–395
Gibson C (1952) Tlaxcala in the sixteenth century. Yale University Press, New Haven

Gleeson P (2015) Kingdoms, communities, and óenaig: Irish assembly practices in their northwest European context. J North Atlantic 8(sp8):33–51
Gleeson P (2018) Gathering communities: locality, governance and rulership in early medieval Ireland. World Archaeol 50(1):100–120
González-Arqueros ML, Mendoza ME, Vázquez-Selem L (2017) Human impact on natural systems modeled through soil erosion in GeoWEPP: A comparison between pre-Hispanic periods and modern times in the Teotihuacan Valley (Central Mexico). Catena 149(Part 1):505–513
Green AS (2022) Of revenue without rulers: Public goods in the egalitarian cities of the Indus civilization. Front Polit Sci 4:823071
Hardin G (1968) The tragedy of the commons: The population problem has no technical solution; it requires a fundamental extension in morality. Science 162(3859):1243–1248
Harvey D (2005) A brief history of neoliberalism. Oxford University Press, New York
Hepp GD (2022) Heterarchy and the emergence of social complexity in early formative period Coastal Oaxaca, Mexico. In: Framing complexity in formative mesoamerica. University Press of Colorado, pp 19–54
Herzog T (2018) Indigenous reducciones and Spanish resettlement: placing colonial and European history in dialogue. Ler História 72:9–30. https://doi.org/10.4000/lerhistoria.3146
Hillman AL, Ursprung HW (1992) The influence of environmental concerns on the political determination of trade policy. In The greening of world trade issues, pp 195–220
Hinrichs CC (2003) The practice and politics of food system localization. J Rural Stud 19(1):33–45
Hirschman AO (1970) Exit, voice, and loyalty: responses to decline infirms, organizations, and states. Harvard University Press, Cambridge, MA
Hirschman AO (1974) Exit, voice, and loyalty: further reflections and a survey of recent contributions. Soc Sci Inf 13(1):7–26
Isakhan B (2015) Democracy: critiquing a Eurocentric history (Version 1). Deakin University. https://hdl.handle.net/10536/DRO/DU:30079037
Iversen F (2013) Concilium and pagus—revisiting the early Germanic thing system of northern Europe. J North Atlantic 2013(sp5):5–17
Jalbert K (2016) Building knowledge infrastructures for empowerment: a study of grassroots water monitoring networks in the marcellus shale. Sci Technol Stud 29(2):26–43
Kasaba R (2011) A moveable empire: Ottoman nomads, migrants, and refugees. University of Washington Press
Katz H, Faust A (2012) The Assyrian destruction layer at Tel ʿEton. Isr Explor J:22–53
Kelly AB, Peluso NL (2015) Frontiers of commodification: state lands and their formalization. Soc Nat Resour 28(5):473–495
Kenoyer JM (2006) The origin, context and function of the Indus script: recent insights from Harappa. In Proceedings of the pre-symposium of RIHN and seventh ESCA Harvard-Kyoto Roundtable (pp. 9–27).
Kenoyer JM (2017) Trade and technology of the Indus Valley: new insights from Harappa, Pakistan. In: High definition archaeology: threads through the past. Routledge, pp 262–280.
Kim Y, Kim S, Baek J, Heo E (2019) The linkages between democracy and the environment: evidence from developed and developing countries. Energy Environ 30(5):821–832
Kimura AH, Kinchy A (2016) Citizen science: probing the virtues and contexts of participatory research. Engag Sci Technol Soc 2:331–361
Korpiola M (2014) 'Not without the consent and goodwill of the common people': the community as a legal authority in Medieval Sweden. J Legal Hist 35(2):95–119
Levi M (1988) Of rule and revenue. University of California Press, Berkeley
Li Q, Reuveny R (2006) Democracy and environmental degradation. Int Stud Q 50(4):935–956
Lichbach MI (1996) The cooperator's dilemma. University of Michigan Press, Ann Arbor
Liebeschuetz JHWG (2015) Unsustainable development: the origin of ruined landscapes in the Roman Empire. In: East and West in late antiquity invasion, settlement, ethnogenesis and conflicts of religion. Leiden, Brill, pp 54–65

Lockhart J (1969) Encomienda and hacienda: the evolution of the great estate in the Spanish Indies. Hisp Am Hist Rev 49(3):411–429
Lv Z (2017) The effect of democracy on CO2 emissions in emerging countries: does the level of income matter? Renew Sust Energ Rev 72:900–906
Machin A (2022) Climates of democracy: Skeptical, rational, and radical imaginaries. Wiley Interdiscip Rev Clim Chang 13(4):e774
Mainland I, Batey C (2018) The nature of the feast: commensality and the politics of consumption in Viking Age and Early Medieval Northern Europe. World Archaeol 50(5):781–803
Mann M (2012) The sources of social power: volume 1, a history of power from the beginning to AD 1760, vol 1. Cambridge University Press, Cambridge
Marino MD, Fargher LF, Meissner NJ, Johnson LRM, Blanton RE, Espinoza VYH (2020) Exchange systems in late postclassic mesoamerica: comparing open and restricted markets at Tlaxcallan, Mexico, and Santa Rita Corozal, Belize. Lat Am Antiq 31(4):780–799
Martín Gabaldón MM, Escalona Lüttig H, Güereca Durán R (2021) Impacto ambiental y paisaje en Nueva España durante el siglo XVI, vol 10. UNAM, Dirección General de Publicaciones y Fomento Editorial, Mexico City
Marx K (1857) Introduction to a contribution to the critique of political economy. Idem, Grundrisse, pp 81–114
Melville EG (2021) Environmental change and social change in the Valle del Mezquital, Mexico, 1521–1600. In: Wheatly H (ed) Agriculture, resource exploitation, and environmental change. Routledge, London, pp 69–98
Midlarsky MI (1998) Democracy and the environment: an empirical assessment. J Peace Res 35(3):341–361
Moore T, González-Álvarez D (2021) Societies against the chief? Re-examining the value of 'heterarchy' as a concept for studying European Iron Age societies. Power from below in premodern societies, pp 125–156
Mountjoy DC, Gliessman SR (1988) Traditional management of a hillside agroecosystem in Tlaxcala, Mexico: an ecologically based maintenance system. Am J Altern Agric 3(1):3–10
Murphy JT, Crumley CL, Isendahl C, Lucero LJ, Meunier J, Nijhuis S, Ostovar P, Reichel C, Scarborough VL, Sulas F, Thurston TL (2022) If the past teaches, what does the future learn? Ancient urban regions and the durable future. TU Delft, Delft
Neuman N (2024) Commensal attraction: eating together as a social tool. J Theory Soc Behav 54(4):556–570
Olsen AB (2014) Courtyard sites in western Norway. Central assembly places and judicial institutions in the Late Iron Age. In: Eriksen MH, Pedersen U, Rundberget B, Axelsen I, Berg HL (eds) Viking worlds: Things, spaces and movement, pp 43–55
Olson M (1965) Logic of collective action: Public goods and the theory of groups. Harvard University Press, Cambridge
Ostrom E (1990) Governing the commons: the evolution of institutions for collective action. Cambridge University Press, Cambridge
Ostrom E (2000) Collective action and the evolution of social norms. J Econ Perspect 14(3):137–158
Pauketat TR, Emerson TE (2007) Alternative civilizations: heterarchies, corporate polities, and orthodoxies. In DM Bondarenko & AA Nemirovskiy (eds) Alternativity in cultural history: heterarchy and homoarchy as evolutionary trajectories. Proceedings of the third international conference "hierarchy and power in the history of civilizations", June 18–21 2004, Moscow, pp 107–117
Pellegrini L, Gerlagh R (2006) Corruption and environmental policies: what are the implications for the enlarged EU? Eur Environ 16(3):139–154
Peregrine PN (2001) Cross-cultural comparative approaches in archaeology. Annu Rev Anthropol 30(1):1–18
Popkin SL (1979) The rational peasant. University of California, Berkeley

Postgate JN (2007a) The invisible hierarchy: Assyrian military and civilian administration in the eighth and seventh centuries BC. The land of assur and the yoke of assur: studies on Assyria 1971–2005, 331–360.

Postgate JN (2007b) Royal exercise of justice under the Assyrian Empire. The land of assur and the yoke of assur: studies on Assyria 1971–2005, 47–56.

Postgate JN (2007c) Some remarks on conditions in the Assyrian countryside. The land of assur and the yoke of assur: studies on Assyria 1971–2005, pp 27–45.

Postgate JN (2007d) The economic structure of the Assyrian empire. The land of assur and the yoke of assur: studies on Assyria 1971–2005, pp 71–99.

Raaflaub KA (2021) Caesar and genocide: confronting the dark side of Caesar's Gallic Wars. New England Class J 48(1):54–80

Redman CL, Grove JM, Kuby LH (2004) Integrating social science into the long-term ecological research (LTER) network: social dimensions of ecological change and ecological dimensions of social change. Ecosystems 7:161–171

Rivera-Núñez T, Fargher L, Nigh R (2020) Toward an historical agroecology: an academic approach in which time and space matter. Agroecol Sustain Food Syst 44(8):975–1011

Roach L (2013) Kingship and consent in Anglo-Saxon England, 871–978: assemblies and the state in the early middle ages, vol 92. Cambridge University Press

Rokkan S (1969) Models and methods in the comparative study of nation-building. Acta Sociol 12(2):53–73

Rosenzweig MS (2018) Assessing the politics of Neo-Assyrian Agriculture. Archeol Pap Am Anthropol Assoc 29(1):30–50

Rothstein B (2011) The quality of government: Corruption, social trust, and inequality in international perspective. University of Chicago Press, Chicago

Sabloff JA (2016) Archaeology matters: action archaeology in the modern world. Routledge

Sanmark A, Semple SJ (2008) Places of assembly: new discoveries in Sweden and England. Fornvännen: tidskrift for Svensk Antikvarisk Forskning 103(4)

Scarborough VL, Valdez F, Dunning NP (eds) (2003) Heterarchy, political economy, and the Ancient Maya: the three rivers region of the East-central Yucatan Peninsula. University of Arizona Press

Schmandt-Besserat D (2010) The token system of the ancient Near East: its role in counting, writing, the economy and cognition. The archaeology of measurement: comprehending heaven, earth and time in ancient societies, 27–34.

Scott JC (1990) Domination and the arts of resistance: hidden transcripts. Yale University

Scott JC (2009) The art of not being governed: An anarchist history of upland Southeast Asia. Yale University Press

Semple S, Sanmark A (2013) Assembly in North West Europe: collective concerns for early societies? Eur J Archaeol 16(3):518–542

Semple S, Sanmark A, Iversen F, Mehler N (2020) Negotiating the North: meeting-places in the Middle Ages in the North Sea zone. Taylor & Francis, p 370

Shahid SA, Zaman M, Heng L (2018) Soil salinity: historical perspectives and a world overview of the problem. In: Zaman M, Shahid SA, Heng L (eds) Guideline for salinity assessment, mitigation and adaptation using nuclear and related techniques. Springer, Cham, pp 43–53

Sherratt A (1989) V. Gordon Childe: archaeology and intellectual history. Past Present 125:151–185

Simonton M (2024) Politics and political institutions in Archaic and Classical Greek cities. In: Flohr, Zuiderhoek (eds) Companion to cities in the Greco-Roman World. Wiley, pp 221–235

Sinha A, Kathayat G, Weiss H, Li H, Cheng H, Reuter J, Schneider AW, Berkelhammer M, Adalı SF, Stott LD, Edwards RL (2019) Role of climate in the rise and fall of the Neo-Assyrian Empire. Sci Adv 5(11):eaax6656

Skinner AT, Semple S (2016) Assembly mounds in the Danelaw: place-name and archaeological evidence in the historic landscape. J North Atlantic 8(sp8):115–133

Slobodian Q (2018) Globalists: the end of empire and the birth of neoliberalism. Harvard University Press, Cambridge, MA

Smith P (1872) A smaller ancient history of the East: from the earliest times to the conquest by Alexander the great: including Egypt, Assyria, Babylonia, Media, Persia, Asia Minor, and Phoenicia. Harper

Smyth MP (2022) Urban organization and origins of complexity in the Puuc Region of Northern Yucatán. In: Framing complexity in formative Mesoamerica. University Press of Colorado, pp 170–200

Stasavage D (2020) The decline and rise of democracy: a global history from antiquity to today. Princeton University Press

Stephens L, Fuller D et al (2019) An archaeological assessment of global land use from 10,000 BP to 1850 CE. Science 365(6456):897–902

Thurston TL (2007) Rituals of rebellion: cultural narratives and metadiscourse of violent conflict in iron age and Medieval Denmark. J Confl Archaeol 3(1):269–295

Thurston TL (2009) Farming the margins: on the social causes and consequences of soil management strategies. In: Fisher CT, Hill B, Feinman G (eds) The archaeology of environmental change: socionatural legacies of degradation and resilience. University of Arizona, pp 106–134

Thurston TL (2010) Bitter arrows and generous gifts: what was a king in the European Iron Age? In: Price TD, Feinman GM (eds) Pathways to power new perspectives on the emergence of social inequality. Springer Scientific Publishing, New York, pp 193–254

Thurston TL (2016) Enduring nations and emergent states: rulership, subjecthood, and power in early Scandinavia. In: Fargher L, Heredia V (eds) Alternative pathways to complexity a collection of essays on architecture, economics, power, and cross-cultural analysis. University Press of Colorado, Boulder, pp 177–206

Thurston TL (2019) "Can you hear me now?" Heterarchy as an instrument and outcome of collective action in Iron Age and Medieval Europe. In: Historical ecologies, heterarchies and transtemporal landscapes. Routledge, pp 55–80

Thurston TL (2022) Reversals of fortune: shared governance, "democracy," and reiterated problem-solving. Front Polit Sci 4(870):773

Thurston TL, Fernández-Götz M (2021a) Power from below in the archaeological record: trends and trajectories. In: Thurston TL, Fernandez-Götz M (eds) Power from below in premodern societies: the dynamics of political complexity in the archaeological record. Cambridge University Press, Cambridge

Thurston TL, Fernández-Götz M (2021b) Restoring disorder: thoughts on the past and future of a politically and socially conscious archaeology. In: Thurston TL, Fernandez-Götz M (eds) Power from below in premodern societies: the dynamics of political complexity in the archaeological record. Cambridge University Press, Cambridge

Thurston TL, Pettersson C (2024) Old cities, 'new' agendas: Swedish cities across time. In: Hawken S, Barthel S, Isendahl C, Strickland K (eds) Long-term intergenerational perspectives on urban sustainability transitions special issue: urban studies. https://doi.org/10.1177/0042098023122021

Tilly C (1975) Western state-making and theories of political transformation. In: Tilly C (ed) The formation of national states in Western Europe. Princeton, Princeton University Press, pp 601–638

Trigger BG (2003) Understanding early civilizations: a comparative study. Cambridge University Press, Cambridge

Utting P (ed) (2015) Revisiting sustainable development. The United Nations Research Institute for Social Development (UNRISD), Geneva

van der Leeuw SE (ed) (1998) The archaeomedes project: understanding the natural and anthropogenic causes of land degradation and desertification in the Mediterranean Basin, research results. Office for Official Publications of the European Communities, Luxembourg

van der Leeuw SE (2003) Desertification, land degradation and land abandonment in the Rhône Valley, France. In: Barker G, Gilbertson D (eds) The archaeology of drylands: living at the margin. Routledge, pp 327–344

van der Leeuw SE, Redman CL (2002) Placing archaeology at the center of socio-natural studies. Am Antiq 67(4):597–605
van der Leeuw SE et al (2005) Climate, hydrology, land use, and environmental degradation in the lower Rhone Valley during the Roman period. Compt Rendus Geosci 337(1–2):9–27
Vidale M (2018) Heterarchic powers in the ancient Indus cities. J Asian Civiliz 41(2):1–46
Weber M (1947) The theory of economic and social organization. Translation A.M. Henderson and T. Parsons. Oxford University Press, New York
Weber H (2017) Politics of 'leaving no one behind': contesting the 2030 sustainable development goals agenda. Globalizations 14(3):399–341
Wernke SA (2013) Negotiated settlements: Andean communities and landscapes under Inka and Spanish colonialism. University Press of Florida, Gainesville
Whitley M (2018) Citizenship and commensality in archaic Crete: searching for the Andreion. Duplouy and Brock 2018:227–248
Wickham C (2017) Consensus and assemblies in the Romano-Germanic kingdoms: a comparative approach. Vorträge und Forschungen 82:389–426
Wilhelmina Q, Joost J, George E, Guido R (2010) Globalization vs. localization: global food challenges and local solutions. Int J Consum Stud 34(3):357–366
Wilken GC (1970) The ecology of gathering in a Mexican farming region. Econ Bot 24(3):286–295
Wilkinson TJ, Ur J, Wilkinson EB, Altaweel M (2005) Landscape and settlement in the Neo-Assyrian Empire. Bull Am Sch Orient Res 340(1):23–56
Wilson R (2020) Authoritarian environmental governance: Insights from the past century. In: Environmental governance in a populist/authoritarian era. Routledge, pp 14–23
Wolf E (1972) Ownership and political ecology. Anthropol Q 45(3):201–205
Wylie SA, Jalbert K, Dosemagen S, Ratto M (2014) Institutions for civic technoscience: How critical making is transforming environmental research. Inf Soc 30(2):116–126
You WH, Zhu HM, Yu K, Peng C (2015) Democracy, financial openness, and global carbon dioxide emissions: heterogeneity across existing emission levels. World Dev 66:189–207
Young AT (2015) From Caesar to Tacitus: changes in early Germanic governance circa 50 BC-50 AD. Public Choice 164:357–378

T. L. Thurston is a Professor of Anthropology and Environment and Sustainability, at the University at Buffalo. He is an environmental and historical archaeologist who combines material culture, the documentary record, and ecological data to understand the relationship between ordinary people and their leaders or rulers in protohistoric and historic societies, especially the interaction of livelihood strategies and political development in a cross-cultural perspective. With a focus on internal social disruption, he examines the differing experiences of rural and urban people, as well as the distinct strategies of producers in remote regions with difficult livelihoods versus those in agriculturally rich areas. In addition to contributing to the archaeological literature, he prioritizes connecting the past and the present across the broader social sciences, recently co-editing Power from Below in Premodern Societies: The Dynamics of Political Complexity in the Archaeological Record (Cambridge University Press) and co-authoring If the Past Teaches, What Does the Future Learn?: Ancient Urban Regions and the Durable Future (Delft), workshopped through the US National Science Foundation's National Socio-Environmental Synthesis Center (SESYNC), whose mission brings together research groups that merge the science of the natural world with the science of human behavior and decision-making.

Lane F. Fargher-Navarro is the Director of Research at The PAST Foundation and Adjunct Professor of Anthropology at the Ohio State University. He has a PhD in Anthropology from the University of Wisconsin-Madison and a postdoc in Historical Anthropology from Purdue University. His research interests include Mesoamerican archaeology, collective action and urbanism in pre-modern states, historical ecology, and STEM education. His research specializations include archaeological survey and household excavations, ethnographic methods, elemental chemostratigraphy of unconsolidated sedimental cores via XRF, and ceramic petrography. He has authored or coauthored two monographs and more than 60 research articles and book chapters, as well as co-edited two books. His research has been funded by the US National Science Foundation, Mexico's Consejo Nacional de Ciencia y Tecnología (Conacyt), Mexico's Secretaria de Educación Pública, Sigma Xi, and FAMSI, Inc. He is a member of the Academia Mexicana de Ciencias.

Climate Change Politics in a Polarized World

Reconstructing Ukraine's Post-War Future

Camille Wejnert-Depue

Abstract The Russian invasion of Ukraine has significantly damaged cities, buildings, and energy infrastructure, forced a significant portion of Ukraine's population to migrate out of the country, and led to a sharp decline in the country's GDP. The war, however, will end and for the Ukraine's post-war reconstruction to be successful, policymakers must recognize not only the opportunity to modernize and decarbonize Ukraine's buildings and energy sectors but also to do so in a way that advances energy security (due to a reduction in reliance on foreign-sourced fossil fuels and a domestically sourced green energy transition) and improves citizens' well-being (via a decrease in energy service costs and a reduction in air pollution). This study aims to inform future decisions through integrated assessment modeling and provide a potential pathway toward achieving net-zero economy-wide CO_2 emissions in Ukraine by 2050. While informing an economy-wide net-zero pathway, the report recognizes (a) improvements to human well-being through reductions in non-CO_2 emissions, a stable supply of energy services during extreme weather events, and reduced energy service costs, and (b) advancements in energy security via an increase in domestically sourced renewable energy consumption, as well as (c) which sector(s) should be a priority for decarbonization, as a way to guarantee economy-wide decarbonization.

Keywords Russian Invasion · Ukrainian post-war reconstruction · Energy security · Carbon-neutral economy · Decarbonize energy sectors · Integrated assessment modeling

---

C. Wejnert-Depue (✉)
Center for Global Sustainability, University of Maryland, College Park, MD, USA
e-mail: cwejnert@umd.edu

B. Wejnert (ed.), *Democracy in Retreat and the Climate Crisis*, Springer Studies on Populism, Identity Politics and Social Justice,
https://doi.org/10.1007/978-3-032-02881-5_6

Introduction

Russia began a full-scale invasion of Ukraine on February 24, 2022. The war has significantly damaged cities, buildings, and energy infrastructure, forced a significant portion of Ukraine's population to migrate out of the country, and led to a sharp decline in the country's economic GDP. Over the next 5 to 10 years, Ukraine will need to rebuild its cities, buildings, and energy infrastructure. This presents a significant opportunity to plan for a Ukrainian future, including European integration, modernization, and economy-wide decarbonization. Thus, Ukraine has developed a comprehensive reconstruction plan, prioritizing rebuilding the Ukrainian energy system using clean, sustainable, and resilient energy systems and fuels (Ukraine 2022). Goals within the plan include shifting away from fossil fuel use in the buildings sector (mainly for heating and cooling), introducing more energy-efficient and negative emissions technologies (e.g., electric heat pumps), and moving toward a net-zero CO_2 emissions goal by 2050 to successfully integrate into the EU (an update from the country's original decarbonization goal of 2060) (Ukraine 2022). In addition, the plan states that the Ukrainian government needs to advance overall energy security while improving citizens' well-being (Ukraine 2022).

Thus, for Ukraine's post-war reconstruction to be successful, policymakers must recognize not only the opportunity to modernize and decarbonize Ukraine's buildings and energy sectors but also to do so in a way that advances energy security (due to a reduction in reliance on foreign-sourced fossil fuels and a domestically sourced green energy transition) and improves citizens' well-being (via a decrease in energy service costs and a reduction in air pollution). This study and report are designed to help inform those future decisions via integrated assessment modeling and provide a potential pathway toward net-zero economy-wide CO_2 emissions in Ukraine by 2050. While informing an economy-wide net-zero pathway, the report recognizes (a) improvements to human well-being through reductions in non-CO_2 emissions, a stable supply of energy services during extreme weather events, and reduced energy service costs, and (b) advancements in energy security via an increase in domestically sourced renewable energy consumption, as well as (c) which sector(s) should be a priority for decarbonization, as a way to guarantee economy-wide decarbonization.

Ukraine's Fossil Fuel History

Ukraine has had a rich history with fossil fuels. In the past, the country was a significant supplier of coal, and although coal reserves have diminished since the late 1990s, Ukraine remained the world's ninth largest steel producer in 2021 (O'Reilly 2023). In addition, the country consumes large quantities of natural gas, particularly for residential building energy use (Cahill and Palti-Guzman 2023; Kuznetsov et al. 2020). The following subsections of this literature review address Ukraine's history and current situation regarding fossil fuel supply and consumption, as well as Russian-Ukrainian conflicts based on fossil fuel consumption and supply, which point toward transitioning away from imported fossil fuels (primarily Russian natural gas) to domestically sourced renewables for energy consumption.

Ukraine and Coal

Before the rise of and continuing during the Soviet Union, Ukraine's territory was a significant area of coal supply. In fact, in 1913, the Donbas region (in what later became known as Ukraine) produced 87% of the coal and 74% of the pig iron supplied to the Russian Empire (Graziosi 1995). By 1976, Ukraine's territory was the third-largest coal-producing region in Europe, producing 218 million metric tonnes (Vorutnikov 2015). After the collapse of the Soviet Union in 1991, Ukraine gained its independence, which resulted in Russia losing territorial control over the coal-rich Donbas region. However, in 2014, Russia launched a pro-Russian campaign for Russian separatists in the Donbas region to regain control of the area (Masters 2023). Continued geopolitical offensives against Ukraine in this region damaged coal reserve locations, flooded mines, and decreased coal production (Fadek 2022). Coal production in Ukraine in 2013 was reduced to 83 million metric tonnes and further reduced in 2016 to 42 million metric tonnes (Fadek 2022).

Ukraine and Natural Gas

As coal production (and therefore supply and consumption) fell due to becoming an unreliable energy source, the country searched for a new, more sustainable energy source. Geographically, Ukraine shares its Eastern border with Russia, one of Europe's largest natural gas suppliers. Thus, Ukraine saw an opportunity to bargain, establishing a deal to trade its geographic positioning as a prime natural gas transit corridor to supply more conveniently to the rest of Europe as an opportunity to get discounted gas prices from Russia's state-owned energy company, Gazprom. At one point, Ukraine received discounts on natural gas as significant as $100 off Russia's benchmark gas price per thousand cubic meters in return for a quicker transit line to the rest of Europe (Roth 2015). The reduced cost of Russian natural gas led to large amounts of gas consumption in residential households for space and hot water heating (Cahill and Palti-Guzman 2023; Kuznetsov et al. 2020). For example, natural gas consumed for residential heating purposes accounted for 40% of Ukraine's gas consumption in 2020 and 52% of Ukraine's gas consumption in 2021 (Naftogaz 2021).

Russia-Ukraine Conflicts

Gas Wars

Russia has continually used Ukraine's reliance on natural gas to engage in a geopolitical battle with the country (Prokip 2020). Even with discounted prices, Ukraine had historically been unable to afford Russian natural gas, resulting in significant financial debts for the Ukrainian gas company Naftogaz (Prokip 2020). However, in 2004, a deal was made, and Ukraine's gas debts owed to Russia were reduced in exchange for a 5-year promise of Russian gas transit to Europe via Ukraine. Unfortunately, in May 2005, the security of Russian natural gas came into question. Gazprom had promised Naftogaz 7.8 billion cubic meters of gas reserves due to an extra need for household heating during the more frigid–than-usual winter period of 2004 (October 2004 to March 2005), which Naftogaz claimed had not been made available for use. Gazprom eventually blamed the lack of gas reserves on transit issues and negotiated with Naftogaz to supply 2.55 billion cubic meters of gas at no cost (Prokip 2020). However, this event led to initial questions regarding the insecure nature of Russian-supplied natural gas for Ukrainian household heating consumption.

A similar dispute between Ukraine and Russia occurred in 2006 when Gazprom accused Ukraine of stealing supplies from pipelines going to other European countries through Ukraine's territory. This dispute led to Russia cutting its gas supply to Ukraine for more than 24 h in January 2006, reducing the available gas supply to Europe (Lowe and Polityuk 2009). A third gas cut-off occurred in 2008 and 2009 when Gazprom accused Ukraine of having additional debts associated with gas payments. These debt-payment disputes led Russia to cut off the gas supply to Ukraine again for multiple days in late December and early-to-mid-January (Lowe and Polityuk 2009). Lastly, Ukrainian concerns about the security of Russian-supplied gas peaked after their discount on Russian gas ended after the 2014 Crimea conflict and after additional accusations of Ukraine owing Russia money for its gas (Polityuk and Soldatkin 2014). After the annexation of Crimea, Russia priced gas at $385 per thousand cubic meters, much higher than the original price for the fuel (Polityuk and Soldatkin 2014), fortifying all major Ukrainian concerns about the long-term stability of using this fuel source for energy consumption.

A Full-Scale Invasion of Ukraine

The current Russian invasion of Ukraine has also amplified concerns about the dependence on natural gas, mainly due to the risks of damage to natural gas pipelines and boilers. Many natural gas reserves are in the occupied east of the country (European Union 2018), such as the Yuzvitska block, a natural gas reserve in the Donetsk region, with an area of 7886 km^2, 70.8 trillion cubic feet of natural gas reserves, and a production rate of 960 million cubic feet per day. Studies show that the field could deliver up to 10 billion cubic meters of natural gas annually, one-third of Ukraine's annual natural gas consumption (Spravdi 2021). The full-scale invasion of Ukraine furthers concerns about Ukraine's reliance on Russian-supplied natural gas, particularly as Russia seeks to control areas where gas reserves for extreme weather (e.g., winter storms) are most abundant.

Risks Associated with Fossil Fuel History

Well-being Risks of Ukrainian Coal Use

The first set of associated concerns is attributed to coal use in Ukraine. These concerns mainly fall under the category of citizens' well-being, as using coal for building energy services (e.g., residential heating) can contribute to air pollution. Ukrainian coal plants expose an estimated 8.7 million people to exceedances of World Health Organization air-quality guidelines (Myllyvirta and Gierens 2021). In 2019, 2700 deaths of Ukrainian citizens were attributed to air pollution from coal plants (Myllyvirta and Gierens 2021). Thus, if Ukraine continues supplying and consuming coal for energy services, the well-being of citizens may be drastically impacted.

Well-being Risks of Reliance on Russian Natural Gas

Russian-supplied gas cut-offs, increased prices for Russian-supplied gas, and geopolitical attacks such as the invasion of Crimea (partially seen as an attempt to capture the Black Sea's offshore oil and gas reserves) have led to significant citizen well-being concerns. For example, during the Russian invasion of Ukraine in 2022, missile attacks targeted energy infrastructure to destroy the heat supply for Ukrainian residents during winter. Previous Russian gas cut-offs in the winter have also had similar impacts on human well-being, leaving many residents without heat for days. Furthermore, looking beyond health concerns, energy affordability concerns are also associated with Russian-supplied gas price spikes, as seen in recent Ukrainian history. With Russia demanding such high gas prices, gas-based heating could become unaffordable for Ukrainian citizens, potentially leading to sizeable residential energy burdens (shares of annual income spent on household energy services).

Energy Security Risks of Ukrainian Reliance on Russian Natural Gas

Regarding energy security concerns associated with Ukraine's reliance on Russian natural gas, the lack of stability in supply is the most significant issue. With historical trends of Russian gas cut-offs in Ukraine, unreasonably high gas price demands from Russia, and the geopolitical tension between Russia and Ukraine currently at its highest, the stability and guarantee of secure supplies of Russian natural gas, particularly for residential heating, are in question. Thus, the resulting energy security concerns associated with Russian natural gas consumption point toward expanding Ukraine's use of domestic renewable energy sources.

Research on the Energy Security in Ukraine's Post-War Future

Methodology

Research Introduction

While Ukraine undergoes post-war reconstruction, the country has an opportunity to improve and guarantee energy security and well-being for its returning citizens while also achieving economy-wide decarbonization to help the country achieve EU member state status. However, the country's NDC currently sets a decarbonization goal by 2060. To join the EU, Ukraine's NDC goal of net-zero will need to be advanced from 2060 to 2050. Thus, this study is based on the following sets of research questions:

Research Question Set 1

(1) What would Ukraine's energy supply mix look like for residential buildings without fossil fuel consumption for energy services? (2) Do energy security and well-being improve with a clean energy transition in residential buildings, and are CO_2 emissions reduced in buildings and economy-wide?

Research Question Set 2

(1) Can Ukraine reach economy-wide decarbonization by 2050, and if so, how? (2) If economy-wide decarbonization is achieved, which sector(s) are decarbonized the most to reach net-zero emissions?

Both sets of research questions consider positive implications such as (a) having an adequate amount of resource reserves to heat Ukrainian households during the winter, (b) strengthening energy security via reducing (particularly Russian-supplied) natural gas consumption, increasing consumption of domestically supplied renewable energy, and becoming an EU member state, (c) reducing air pollution (non-CO_2 emissions) associated with fossil fuel combustion, and (d) reducing CO_2 emissions associated with fossil fuel combustion. In addition, these research questions also seek to understand negative implications, such as an increase in buildings' energy service costs for consumers associated with a transition to domestically supplied renewable energy sources and a goal of economy-wide CO_2 emissions by 2050. One crucial factor will be the resulting residential heating method in the Fossil Fuel Phaseout and Decarbonization Scenarios. Currently, in Ukraine, 40% of the population (approximately 5.5 million households) is served by District Heating (DH) (Kuznetsov et al. 2020). DH is a centralized heat distribution system where combustion occurs at a central plant in the city, which then distributes heat via underground pipelines to buildings (IEA 2021). With DH able to use low-carbon technologies such as heat pumps more efficiently than traditional heating systems, it could be a critical technology to reduce gas consumption and CO_2 emissions in the building sector (IEA 2021).

Scenario Development

Reference Scenario: This scenario is like the reference scenario in GCAM 6.0 global. The scenario does not include (a) new or accelerated penetration of renewable energy sources or (b) new or added constraints on coal for energy use or natural gas for residential heating use. The reference scenario for GCAM-Ukraine also includes updated post-war assumptions for GDP and population based on research from the Ukrainian Statistical Department (Ukrstat) and the Institute for Economics and Forecasting of the National Academy of Science of Ukraine (IEF NASU). With no advancements in energy efficiency or constraints on fuel use or CO_2 emissions, the reference scenario is a baseline for projections of energy supply and consumption, emissions, and energy costs based on a business-as-usual (BAU) world (See Table 1)

No Coal in Residential Energy (No Coal Scenario): The no coal in residential energy scenario includes the post-war updates for GDP and population that were included in the reference scenario, and a detailed buildings and District Heating (DH) sector, to include building efficiency advancements Ukraine expects in the next few years (e.g., the inclusion of heat pumps[1] and improvements in building insulation). In addition, the scenario includes a constraint on the use of coal, in which no coal will be used in residential energy services by 2030 in Ukraine. This

[1] For heat pumps, technology costs are given based on Danish Energy Agency (DEA) technology cost data (DEA 2021. Energy Statistics 2020. Danish Energy Agency.)

scenario was created based on the current goals of European countries to phase out coal consumption for energy use by 2030 (Wacket and Alkousaa 2022). With coal being one of the leading fuels for energy supply in Ukraine (a share of 39% in the reference scenario in 2015), and with buildings and DH sector energy efficiency improvements included, the scenario will be an essential predecessor to the economy-wide Decarbonization Scenario (Table 1). The scenario was modeled by editing share-weight values for coal in residential energy, which were smoothed over time until consumption reached '0' in 2030.

No Gas in Residential Energy (No Gas Scenario): The no natural gas in residential energy scenario includes post-war GDP and population projections from the reference scenario and the detailed buildings and DH sectors mentioned in the no coal in residential energy scenario. In addition, the no natural gas in the residential sector scenario includes a constraint on the consumption of natural gas for residential energy services. The constraint leads to no natural gas consumption in Ukraine for residential buildings by 2035. The goal of no gas consumption by 2035 comes from the EU's plan to phase out fossil fuel heating by 2035 (Garcia 2023). With Ukraine's historically significant natural gas consumption for residential services, this scenario will also be an essential predecessor to the economy-wide Decarbonization Scenario (Table 1). The scenario was modeled by editing share-weight values for gas in residential energy, which were smoothed over time until consumption reached '0' in 2035.

Fossil Fuel Phaseout in Residential Energy (Fossil Fuel Phaseout Scenario): The no fossil fuel consumption for residential energy scenario was created as a combination of the no coal in the residential sector and no natural gas in the residential sector scenarios. With the no gas and no coal in residential energy scenarios combined, the no fossil fuel consumption for residential energy scenario points toward decreasing emissions in the buildings sector rather than forcing (or constraining) economy-wide decisions to reduce emissions. This scenario is a *necessary precursor* to show the impact reducing fossil fuel consumption in buildings alone has on future CO_2 emission trends in Ukraine (Table 1). The scenario was modeled by combining the updated share-weight values for coal and gas in residential energy, which were smoothed over time until consumption reached '0' in 2030 and 2035, corresponding to each fuel type.

Economy-Wide Decarbonization (Decarbonization Scenario): Following Ukraine's desire to be an EU member state, decarbonizing the energy sector will be crucial to reaching the bloc's net-zero emissions goal by 2050. This scenario will include using heat pumps in buildings, industry, and DH. This scenario will also have reduced use of coal and gas, as discussed in the two prior scenarios. Finally, the scenario will include negative-emission technologies (NETs) associated with carbon capture and storage (CCS), such as direct air capture (DAC) (Table 1). The scenario was modeled by applying a constraint on CO2 emissions beginning in 2025, where emissions gradually decline to reach net-zero in 2050 (see Fig. 13).

Table 1 GCAM-Ukraine scenario development

Scenarios	Description	Implications	Model Development
Reference (BAU)	• Business-as-usual • No changes in energy efficiency or constraints on fuel type or CO2 emissions.	• No forced (modeled) improvements in energy security or well-being, no forced (modeled) reductions in CO2 emissions.	• Updated XML files for post-war population and GDP for Ukraine (protections based on Ukrstat statistics).
No Coal in Residential Energy (No Coal Scenario)	• No Coal in Residential Energy (Residential Heating and Residential Others) by 2030 (based on Ukraine goal of coal (phaseout in residential heating by 2030). • Detailed building & DH sectors. • No constraints on CO2 emissions.	• Reduction in CO2 emissions (if fuel consumption shifts away from all fossil fuels simultaneously). • Improvements in energy security (if fuel consumption shifts away from coal and gas simultaneously). • Improvements in well-being (reduced air pollutants and non-CO2 emissions, reduced residential energy costs).	• Updated XML files for post-war population and GDP for Ukraine (projections based on Ukrstat statistics). • Updated Shell Conductance to represent post-war impacts on shell efficiency. • New share-weight XML for Coal is Residential Heating and Others set to 0 by 2030 but smoothed over time. • New XML files for detailed buildings and DH sectors.
No Gas in Residential Energy (No Gas Scenario)	• No Gas in Residential Energy by 2035 (EU plan to phase OUI fossil fuel residential heating). • Detailed building & DH sectors. • No constraints on CO2 emissions.	• Reduction in CO2 emissions (if fuel consumption shifts away from all fossil fuels simultaneously). • Improvements in energy security (reduction in reliance on Russian-sourced gas, increase in reliance on domestic renewable resources). • Improvements in well-being (guaranteed energy supply during Winter months, reduced residential heating costs).	• Updated XML files for post-war population and GDP for Ukraine (protections based on Ukrstat statistics). • Updated Shell Conductance to represent post-war impacts on shell efficiency. • New share-weight XML for Gas in Residential heating set to 0 by 2035 but smoothed over time. • New XML files for detailed buildings and DH sectors to indicate energy efficiency.

(continued)

Table 1 (continued)

Scenarios	Description	Implications	Model Development
Fossil Fuel Phaseout in Residential Energy (Fossil Fuel Phaseout Scenario)	• No Gas in Residential Energy by 2035 (EU plan to phase fossil fuel residential heating). • No Coal in Residential Energy by 2030 (based on Ukraine's goal of coal phaseout in residential heating by 2030). • Detailed buildings & DH sectors. • No CO2 emissions constraints.	• Combination of implications from No Gas and No Coal scenarios. • Reduction in CO2 emissions, increase in renewable energy share. • Improvements in energy security (reduction in reliance on Russian-sourced gas. increase in reliance on domestic renewable resources). • Improvements in well-being (reduced air pollution, guaranteed energy supply during Winter months, reduced residential heating costs).	• Used same GDP and population assumption as reference • Used new share-weight XML for Gas m Residential heating set to 0 by 2035 but smoothed over time. • Used new share-weight XML for Coal in Residential Heating and Others set to 0 by 2030 but smoothed over time.
Economy-wide Decarbonization (Decarbonization Scenario)	• Net-zero CO2 Emissions by 2050 constraint. • Updated Ukrainian NDC target from 2060 to ensure EU integration (EU net-zero CO2 target by 2050). • Detailed buildings & DH sectors.	• Net-zero CO2 emissions by 2050. • Improvements in energy security (reduction in fossil fuels, including Russian-sourced gas, increase in reliance on domestic renewable resources). • Improvements in well-being (reduction in air pollutants and non-CO2 emissions). • Consequences for well-being (potentially higher energy costs).	• Used same GDP and population assumption as reference. • New XML file for CO2 constraint by 2050.

Model Methodology

The Global Change Analysis Model (GCAM) and GCAM-Ukraine

The model used for this analysis *(GCAM-Ukraine)* is derived from the Global Change Analysis Model (GCAM). GCAM is an integrated assessment model (IAM) that links five major systems—energy, water, land, climate, and the economy—through an integrated assessment modeling framework. Composed of 32 energy-economy regions, 384 land regions, and 235 water basins, the model simulates the complex interactions, synergies, and trade-offs between these systems at global and regional scales at 5-year time steps through 2100 (Calvin et al. 2019). The energy component of GCAM captures both the supply and demand for energy across end-use sectors. Within each sector, competition for differing fuel types and energy sources occurs based on the combined price for extraction, refinement, and generation, as well as the capital and overnight costs for energy supply systems. Based on the original global model of GCAM, PNNL developed the GCAM-Ukraine model, in which Ukraine is added as a 33rd energy-economy region, allowing for increased country-level representation of Ukraine in GCAM. Ukraine is responsible for about 80% of energy consumption in the "Europe_Eastern" region, so all assumptions from the region apply to Ukraine.

Important Modeling Considerations and Assumptions

It is essential to consider that in 2014, Russia annexed Crimea and occupied part of eastern Ukraine (known as Donbas). These areas were densely populated, and Donbas was a center of heavy industry with rich iron ore and coal deposits. As a result of the Russian aggression, the Ukrainian government lost control over 2/3 of the coal mines. Between 2014 and 2022, Russia effectively controlled 8% of the territory, where about six million people, or 14% of the population, lived before 2014. After 2014, the Ukrainian Statistical Service (Ukrstat) only reported data on government-controlled areas (GCA). For example, total primary energy supply (TPES) in Ukraine in 2015 was 14% lower than in 2014, as Donbas and Crimea were omitted (Ukrstat 2022). However, with Ukraine's assertion that all occupied territories will be returned to the country, Ukraine's population and GDP values include Donbas, Crimea, and all currently Russian-occupied territories due to the war.

Conclusions

Model Results

This section includes results for three primary scenarios considered in this study (Reference (BAU), Residential Fossil Fuel Phaseout, and Decarbonization).[2] Results are organized into six main categories impacted the most by clean energy transitions and decarbonization pathways: Total Primary Energy Supply (TPES), Electricity Generation, Total Final Consumption (TFC), Buildings, District Heating (DH), and Emissions.

Total Primary Energy Supply (TPES)

Overall, the results for TPES show that in both the Residential Fossil Fuel Phaseout and Decarbonization Scenario, the amount of energy supplied for energy demand decreases (Fig. 1). Although TPES only slightly reduced by 5.9% in 2050 in the Fossil Fuel Phaseout Scenario compared to the Reference Scenario, energy supply significantly decreased by 16.9% in 2050 for the Decarbonization Scenario compared to the Reference Scenario (Fig. 1). In addition, renewable energy supply (Geothermal, Biofuels, Hydro, Wind, Solar, and Nuclear) increases in 2050 by 4%

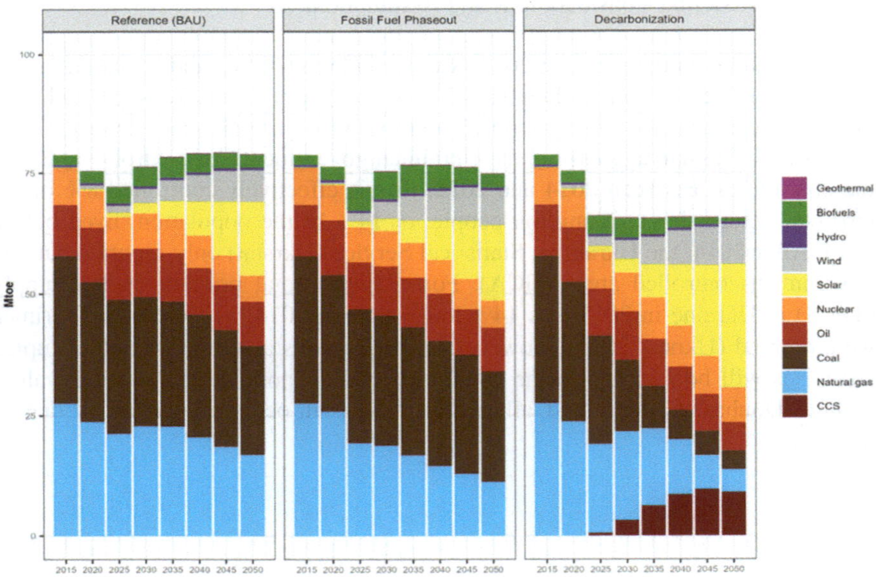

Fig. 1 Total primary energy supply by fuel, GCAM-Ukraine (2015–2050)

[2] Since the no gas and no coal in residential energy scenarios are combined in the residential Fossil Fuel Phaseout Scenario, results from these two scenarios are not shown.

in the Fossil Fuel Phaseout Scenario and by 38% in the Decarbonization Scenario compared to the Reference Scenario. Lastly, solar energy is a crucial energy source in the Decarbonization scenario in 2050, contributing to 39% of all energy supplied (Fig. 1). CCS is also becoming more prominent, contributing to 14% of all energy supplied in the Decarbonization Scenario in 2050 (Fig. 1).

Electricity Generation

Results from GCAM-Ukraine for the Fossil Fuel Phaseout and Decarbonization Scenarios indicate that electricity generation will be critical in both a clean energy transition and an economy-wide decarbonization pathway (Fig. 2). This conclusion can mainly be drawn from the fact that demand for electricity generation in 2050 is 4% larger in the Fossil Fuel Phaseout Scenario and 28% larger in the Decarbonization Scenario compared to the Reference Scenario (Fig. 2). In particular, electricity generated by renewable energy sources (Geothermal, Hydro, Wind, Solar, Nuclear, and Biofuels) will be crucial in 2050, increasing the share of electricity generation from renewables in the Fossil Fuel Phaseout Scenario by 9% and increasing the share of electricity generation from renewables in the Decarbonization Scenario by 50% compared to the Reference Scenario (Fig. 2). In the Decarbonization Scenario in 2050, solar accounts for 59% of electricity generation, with Wind and Nuclear being the second-largest share combined, accounting for 36% of electricity generation (Fig. 2).

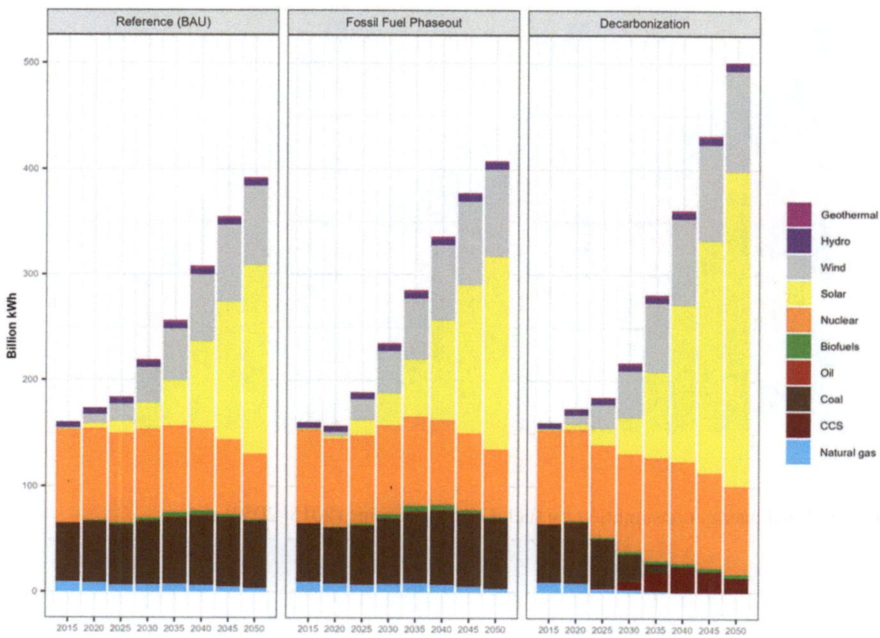

Fig. 2 Electricity generation by source, GCAM-Ukraine (2015–2050)

Total Final Consumption (TFC)

Results for final energy consumption by fuel for GCAM-Ukraine show that energy consumption decreases in the Fossil Fuel Phaseout and Decarbonization Scenarios compared to the Reference Scenario (Fig. 3). Low-carbon-emitting fuels such as biofuels and hydrogen increase in consumption in the Fossil Fuel Phaseout and Decarbonization Scenarios compared to the Reference Scenario, and District Heating (DH) remains a significant supply of energy consumption (third and fourth most prominent share of consumption in Fossil Fuel Phaseout and Decarbonization Scenarios). In addition, electricity becomes a pivotal fuel to achieve economy-wide decarbonization, contributing to a 60% share of total consumption in 2050 in the Decarbonization Scenario.

Looking deeper into final consumption for Ukraine by sector (Fig. 4), industry is the largest energy-consuming sector, accounting for more than ~50% of final consumption in all modeled scenarios. However, it is essential to recognize that buildings also contribute to a large share of energy consumption in Ukraine by sector, accounting for ~30% of all energy consumption in each scenario, with residential buildings accounting for ~20% of all energy consumption in each scenario (Fig. 4).

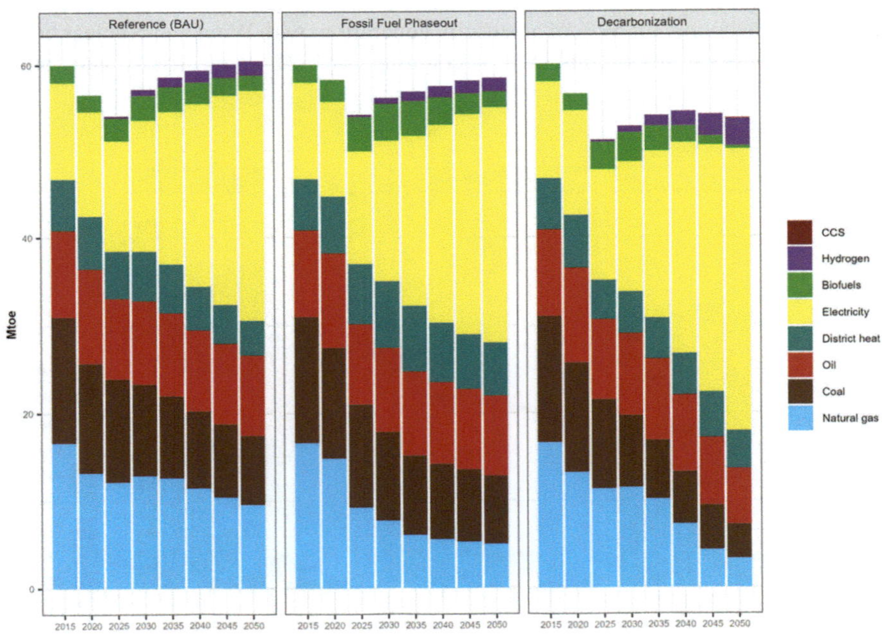

Fig. 3 Final energy consumption by fuel, GCAM-Ukraine (2015–2050)

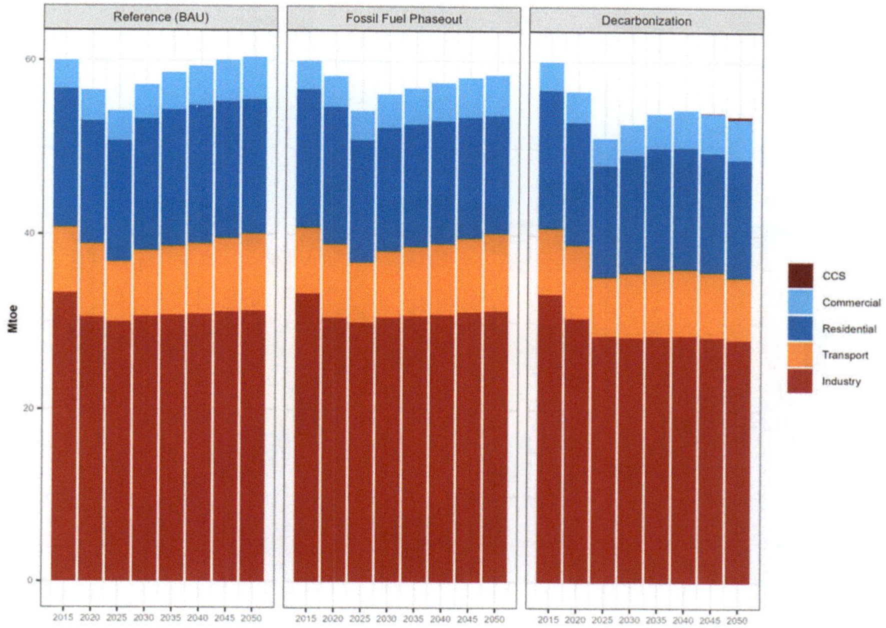

Fig. 4 Final energy consumption by sector, GCAM-Ukraine (2015–2050)

Buildings

With buildings being the second-largest contributor to final energy consumption and a significant source of fossil fuel consumption in Ukraine, consecutive results highlight critical takeaways regarding the buildings sector. In general, the model results of the building sector illustrate that with energy efficiency improvements and low-carbon fuels for buildings and DH, the demand for building energy decreases as you move through each scenario compared to the Reference scenario (Figs. 5 and 6). For instance, in the Fossil Fuel Phaseout Scenario, it is interesting to note that fossil fuel consumption for residential building energy reduces from 2025 onward until reaching '0' in 2035 (due to the constraint imposed on GCAM-Ukraine) and is replaced by energy-efficient DH (Fig. 6). In contrast, in the Decarbonization Scenario, fossil fuel consumption by residential buildings never ultimately reaches '0', and the increase in the use of DH is more spread out over time (Fig. 6). Nonetheless, DH remains a crucial source of energy consumption in the Decarbonization Scenario (Figs. 5 and 6), contributing to 19% and 20% share of residential building demand in 2050 in the Fossil Fuel Phaseout and Decarbonization Scenarios, compared to 5% of residential building energy demand in 2050 in the Reference Scenario (Fig. 6). Lastly, to decarbonize buildings in Ukraine, electricity will be a crucial source of fuel for consumption, especially with the introduction of electric heat

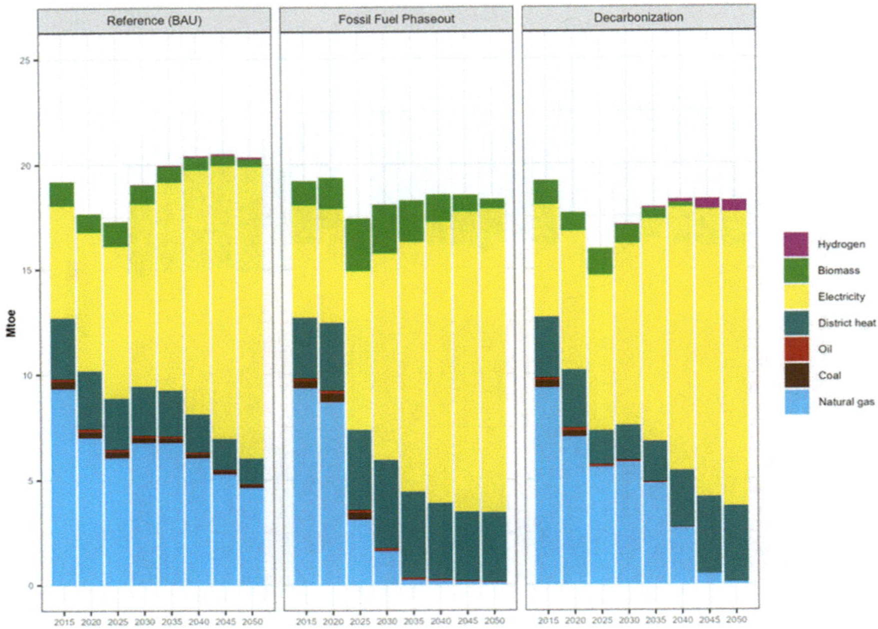

Fig. 5 Buildings sector final energy consumption by fuel, GCAM-Ukraine (2015–2050)

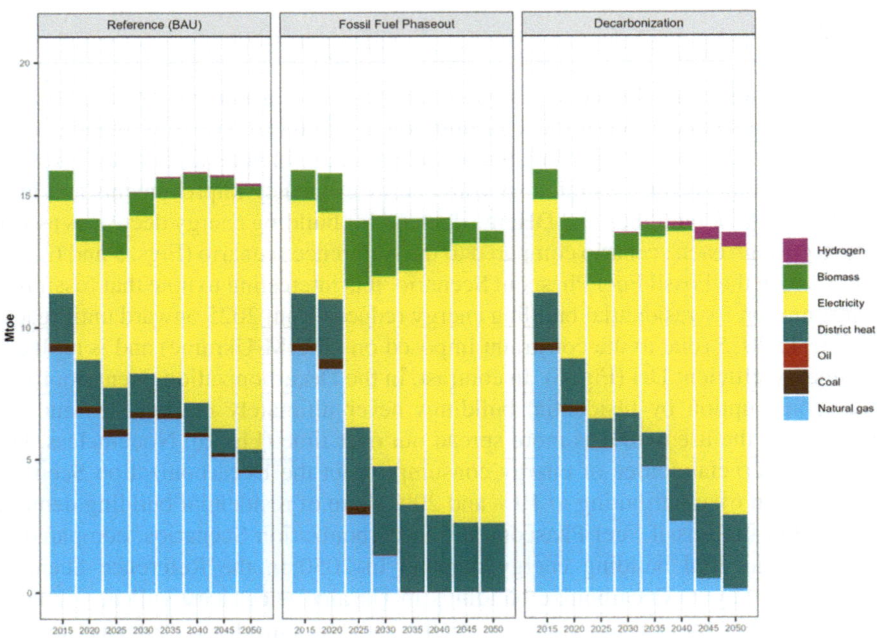

Fig. 6 Final energy consumption for residential buildings by fuel, GCAM-Ukraine (2015–2050)

pumps for heating, contributing to 76% of final building energy consumption in 2050 in the Decarbonization Scenario (Fig. 5).

As fossil fuel consumption (primarily natural gas) decreases in buildings (Figs. 5 and 6), it is essential to recognize potential impacts on energy security and citizens' well-being. Positive effects on energy security and citizens' well-being can be seen via the reduction in gas consumption for building services, as domestically supplied low-carbon fuel consumption increases. This reduction in gas consumption can help Ukraine depart from its reliance on Russian-supplied gas, improving energy security by strengthening the stability of fuel supply during extreme weather events (e.g., winter storms) and consequently improving citizens' well-being by guaranteeing energy services during these extreme weather events.

Although there are positive impacts of a reduction in the use of fossil fuels for residential building energy, relevant literature identifies that in some cases, decarbonizing buildings can be highly costly for consumers as electricity prices increase, creating an unwanted result of residential consumer energy poverty (EC 2009; Frerk and MacLean 2017; Mahone et al. 2019). Higher energy costs associated with residential energy services due to a clean energy transition or a decarbonization pathway can negatively impact citizens' well-being by increasing the share of household income dedicated to household energy services and decreasing the ability to spend income on other needs (food, medical bills, education costs, etc.) (EC 2009; Frerk and MacLean 2017; Mahone et al. 2019). According to model results, residential consumers would not have a more significant energy burden (share of annual income spent on energy services) in the Fossil Fuel Phaseout and Decarbonization Scenarios than in the Reference Scenario. Fuel prices for buildings in 2050 in the Fossil Fuel Phaseout and Decarbonization Scenarios are 0.37% and 18.4% lower than the Reference Scenario in 2050 (Fig. 7).[3]

In addition to fuel prices to buildings reducing through all scenarios, building energy service costs reduce through all scenarios as well, with a 1.7% decrease in residential service costs in 2050 in the Fossil Fuel Phaseout Scenario and a 1.9% decrease in residential service costs in 2050 in the Decarbonization Scenario compared to the Reference Scenario (Fig. 8).

District Heating (DH)

Model results in the DH sector indicate a major switch toward electric heat pumps in both the Fossil Fuel Phaseout and Decarbonization Scenarios (Fig. 9). In the Fossil Fuel Phaseout Scenario, ~50% of DH to all buildings still comes from fossil fuels (Fig. 9). This result is expected, as in the Fossil Fuel Phaseout Scenario, fossil fuel consumption is only constrained for residential buildings, meaning that DH's remaining fossil fuel consumption is for commercial (public) buildings. Lastly, it is essential to note that the share of the supply of DH from electric heat pumps in the

---

[3] The lower fuel prices to buildings in the Decarbonization Scenario are mainly due to the negative costs of biomass starting in 2040 (due to financial incentives for using biomass).

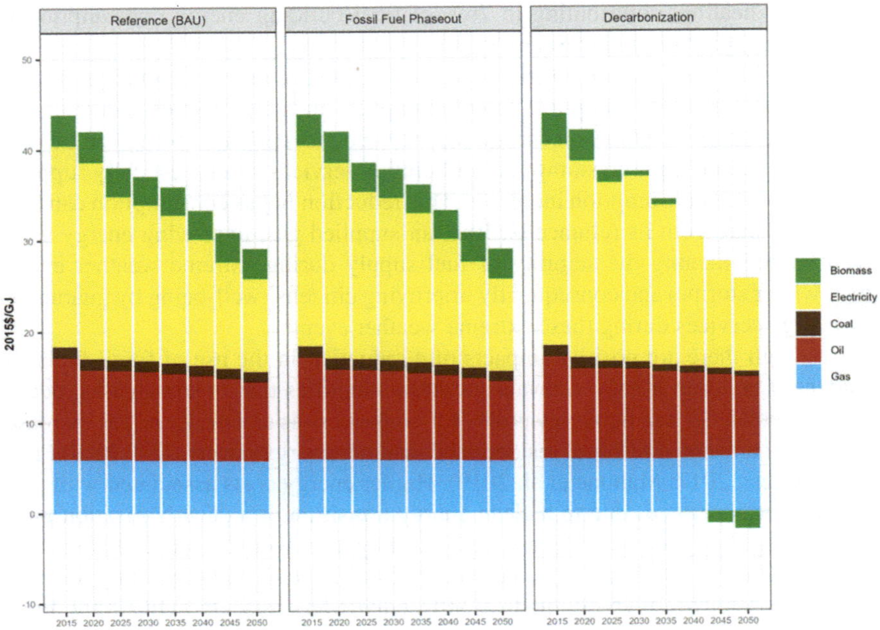

Fig. 7 Fuel prices for buildings' energy use, GCAM-Ukraine (2015–2050)

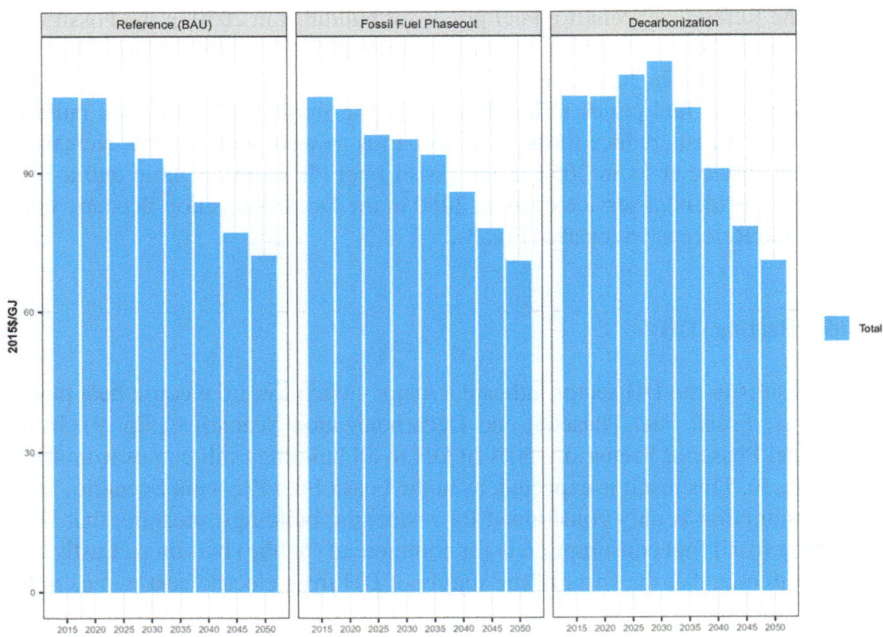

Fig. 8 Building service costs, GCAM-Ukraine (2015–2050)

Decarbonization Scenario is dominated by electric heat pumps starting in 2045, reaching a 99% share of DH supply in 2050 (Fig. 9).

Emissions

CO_2 emissions decrease in all three scenarios, with net-zero emissions achieved in 2050 in the Decarbonization Scenario (Fig. 10). In the Fossil Fuel Phaseout Scenario, CO_2 emissions reductions are mainly seen in the buildings sector, where the total share of emissions from buildings in 2050 is 1% compared to a 13% share of CO_2 emissions in 2015 (Fig. 10). In the Decarbonization Scenario, the buildings sector also significantly reduces emissions, making up the most negligible share of emissions (0.04% of total emissions) compared to any other sector in 2050 (Fig. 10).

Regarding the implication of citizens' well-being, we can see that non-CO2 emissions remain stable in the Fossil Fuel Phaseout Scenario compared to the Reference Scenario, but decrease in the Decarbonization Scenario (Fig. 11). Thus, it is concluded that a decarbonization pathway will also reduce harmful non-CO2 emissions and air pollutants (e.g., Black Carbon, NH3, SO2, NOx, etc.) and improve citizens' well-being.

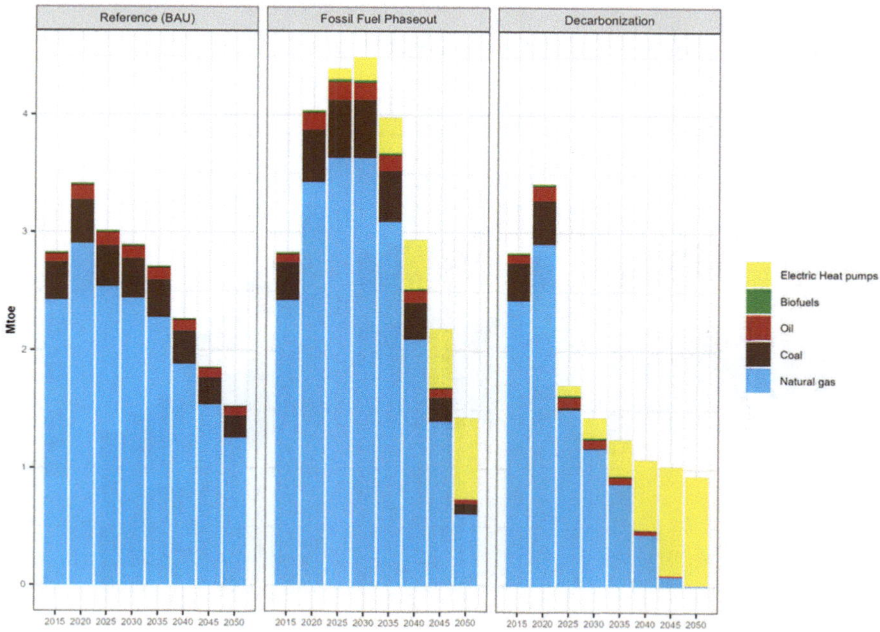

Fig. 9 Sources of District Heating (DH), GCAM-Ukraine (2015–2050)

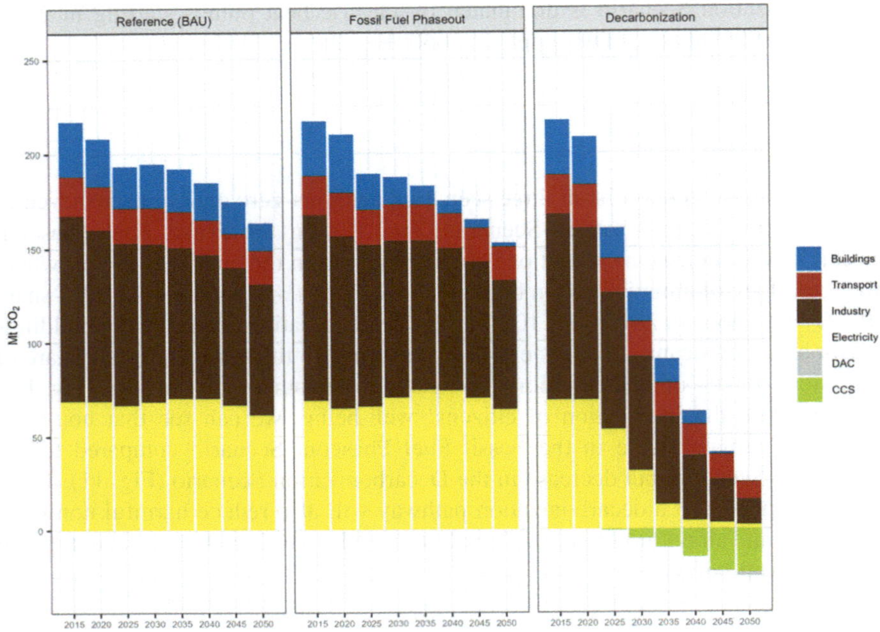

Fig. 10 CO_2 emissions by sector, GCAM-Ukraine (2015–2050)

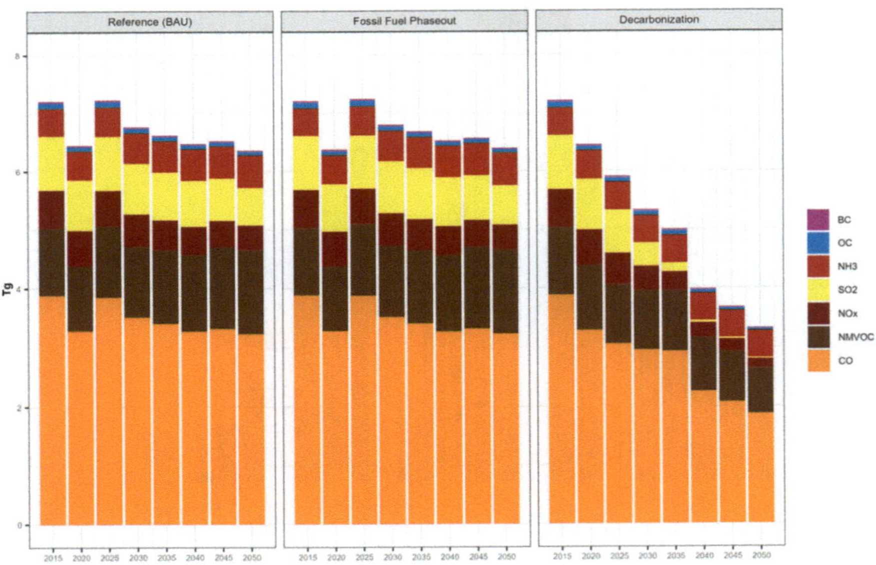

Fig. 11 Non-CO_2 emissions by pollutant, GCAM-Ukraine (2015–2050)

Policy-Based Recommendations from Modeled Results

Based on the results provided using GCAM-Ukraine to model both the Fossil Fuel Phaseout in Residential Energy and the Decarbonization Scenarios, there are three main takeaways. ***First***, rebuilding and decarbonizing Ukraine will require a significant shift to modern, low-carbon options, such as electrification with domestically sourced renewable energy and fuel-switching in the buildings sector. Thus, policies that provide renewable energy share goals in buildings and energy efficiency improvements via building energy codes should be strongly considered to support this clean energy transition in building energy consumption. ***Second***, DH (primarily sourced via electric heat pumps) will remain an essential heating source in Ukraine's resilient and decarbonized future and could help bolster energy security, improve human well-being, and help Ukraine reach net-zero CO_2 emissions by 2050. Keeping DH as a primary heat source will require policies aimed at increasing total DH network connections in Ukraine, improving DH energy efficiency (limiting heat losses), and advancing DH electrification via financial incentives for installing heat pumps, all as a way to reinforce its impact on achieving net-zero emissions in the buildings sector. ***Third***, economy-wide decarbonization is not only achievable by 2050 but will also provide (i) a means for Ukraine's integration into the EU and advance the country's energy security, (ii) improve citizens' well-being by guaranteeing energy services in extreme weather and reducing air pollution (non-CO_2 emissions), and (iii) benefit the entire planet by keeping global mean temperatures below 2 °C in 2050 (Fig. 12).

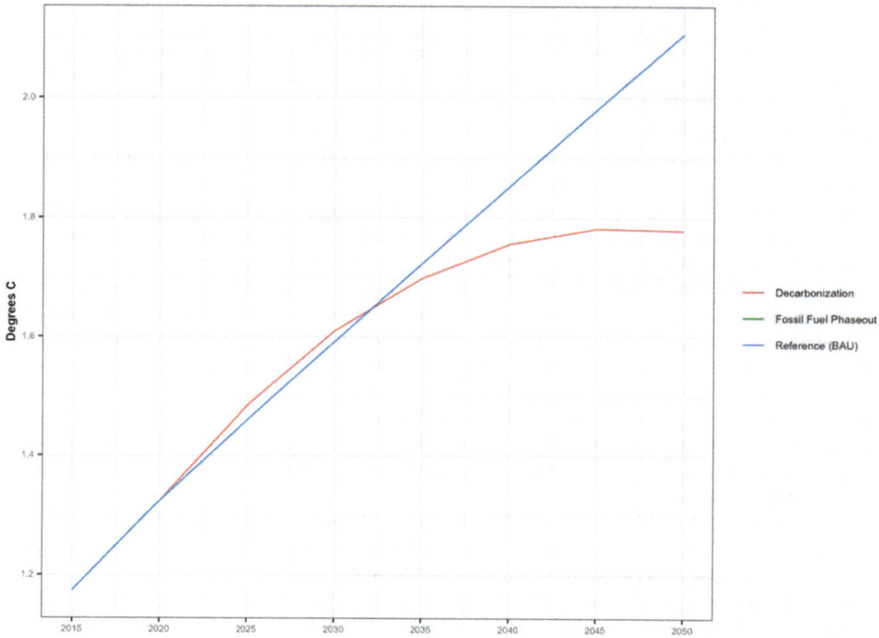

Fig. 12 Global mean temperatures, GCAM-Ukraine (2015–2050)

Appendix (Additional Figures)

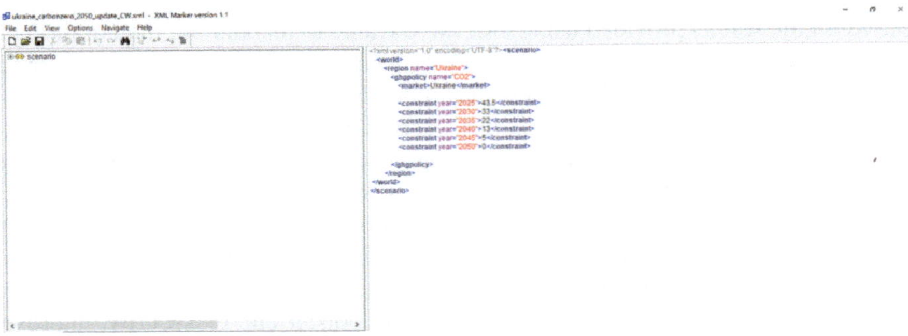

Fig. 13 Net-zero CO_2 emissions constraint, GCAM-Ukraine (2025–2050)

References

Cahill B, Palti-Guzman L (2023) The role of gas in Ukraine's energy future. Center for Strategic and International Studies. https://csis-website-prod.s3.amazonaws.com/s3fs-public/publication/230113_Cahill_Ukraine_Energy.pdf?VersionId=kDKcGINYrCRR0iV6rmXs3YBKxX.PSD52

Calvin K, Patel P, Clarke L, Asrar G, Bond-Lamberty B, Cui RY, Di Vittorio A, Dorheim K, Edmonds J, Hartin C (2019) GCAM v5. 1: representing the linkages between energy, water, land, climate, and economic systems. Geosci Model Dev 12:677–698

DEA (2021) Energy Statistics 2020. Danish Energy Agency

EC (2009) Tackling fuel poverty in Europe - recommendations for policymakers. Europen Union. https://www.finlombarda.it/c/document_library/get_file?p_l_id=1313844&folderId=1327936&name=DLFE-6278.pdf%20

European Union (2018) Heat Roadmap Europe 4. Aalborg University

Fadek T (2022) How war has devastated Ukraine's coal industry. https://www.cnn.com/2022/02/05/europe/gallery/ukraine-coal-miners/index.html

Frerk M, MacLean K (2017) Heat decarbonisation: potential impacts on social equity and fuel poverty, National Energy Action. Neat Energy Action (UK). https://www.nea.org.uk/publications/heat-decarbonisation-potential-impacts-on-social-equity-and-fuel-poverty/

Garcia DE (2023) With fossil fuel boilers on the way out, heat pumps and green hydrogen jockey for position. El Pais. https://english.elpais.com/international/2023-04-20/with-fossil-fuel-boilers-on-the-way-out-heat-pumps-and-green-hydrogen-jockey-for-position.html

Graziosi A (1995) At the roots of Soviet industrial relations and practices. Piatakov's Donbass in 1921. https://www.jstor.org/stable/20170947

IEA (2021) District heating. International Energy Agency

Kuznetsov A, Evans M, Cherviachenko E, Kholod N, Delgado A (2020) White paper on transforming district heating in Ukraine: assessment and recommendations. Pacific Northwest National Lab (PNNL), Richland, WA

Lowe C, Polityuk P (2009) TIMELINE: Gas crises between Russia and Ukraine. Reuters. https://www.reuters.com/article/us-russia-ukraine-gas-timeline-sb/timeline-gas-crises-between-russia-and-ukraine-idUSTRE50A1A720090111

Mahone A, Li C, Subin Z, Sontag M, Mantegna G, Karolides A, German A, Morris P (2019) Residential building electrification in California Energy & Environmental Economics. https://

www.ethree.com/wp-content/uploads/2019/07/CA_Res_Building_Electrification_Final_Presentation.pdf
Masters J (2023) Ukraine: conflict at the crossroads of Europe and Russia. Council on Foreign Relations. https://www.cfr.org/backgrounder/ukraine-conflict-crossroads-europe-and-russia
Myllyvirta L, Gierens R (2021) Health impacts of coal power plant emissions in Ukraine. Center for Research on Energy and Clean Air, https://energyandcleanair.org/wp/wp-content/uploads/2021/09/Ukraine-coal-health-impacts.pdf
Naftogaz (2021) Annual Report 2021. https://www.naftogaz.com/short/0ac0b2d1
O'Reilly F (2023) How Ukraine's battered steel industry galvanized its war effort. The New York Times. https://www.nytimes.com/2023/03/29/world/europe/ukraine-steel-plants-war.html#:~:text=Ukraine's%20steel%20production%2C%20which%20was,makers%20association%20said%20in%20January
Polityuk P, Soldatkin V (2014) Ukraine prepares for gas cuts as Russia stands firm. Reuters. https://www.reuters.com/article/us-ukraine-crisis-naftogaz/ukraine-prepares-for-gas-cuts-as-russia-stands-firm-idUSKBN0EO20020140613
Prokip A (2020) A new era of gas wars between Ukraine and Russia. Wilson Center. https://www.wilsoncenter.org/blog-post/new-era-gas-wars-between-ukraine-and-russia
Roth A (2015) Gazprom halts natural gas deliveries to Ukraine. The New York Times. https://www.nytimes.com/2015/07/02/business/international/gazprom-ukraine-natural-gas-deliveries-russia.html
Spravdi UA (2021) Yuzivka: between fakes and the truth of Ukraine's gas independence. Center for Strategic Communication, Ministry of Culture and Information Policy of Ukraine. https://spravdi.gov.ua/en/yuzivka-between-fakes-and-the-truth-of-ukraines-gas-independence/
Ukraine (2022) Ukraine's national recovery plan. National Recovery Council. https://uploads--ssl.webflow.com/621f88db25fbf24758792dd8/62c166751fcf41105380a733_NRC%20Ukraine%27s%20Recovery%20Plan%20blueprint_ENG.pdf
Ukrstat (2022) Ukraine Energy Balance 2020, in: Ukraine, S.S.S.o. (Ed.)
Vorutnikov V (2015) Ukrainian coal: an industry divided. Coal Age. https://web.archive.org/web/20171014083502/http://www.coalage.com/features/4409-ukrainian-coal-an-industry-divided.html#.WeHMQ3bP2ux
Wacket M, Alkousaa R (2022) Germany's cabinet approves accelerated coal exit by 2030 in western state. Reuters, https://www.reuters.com/business/energy/germanys-cabinet-approves-accelerated-coal-exit-by-2030-western-state-2022-11-02/

Camille Wejnert-Depue, is a Ph.D. student at the University of Maryland's School of Public Policy. His research is conducted at the Center for Global Sustainability (CGS), where he works on modeling global and domestic decarbonization pathways and policies, utilizing the Global Change Analysis Model (GCAM) as a methodology for analysis. His innovative research interests lie within the intersection of integrated assessment modeling, climate pathways and scenarios, and human wellbeing. He is originally from Ithaca, NY, and he moved to the DMV for my undergraduate and master's studies at American University in Washington, D.C., in 2016.

Gender, Dictatorship, and Environment: The Military Government Response to Environmental Disasters in Burma (Myanmar)

Soe Win

Abstract Burma faces heightened vulnerability to environmental disasters as a result of widespread deforestation and the accelerating effects of climate change. However, ongoing internal conflicts—particularly civil war and political unrest—have led the ruling military regime to deprioritize environmental concerns, exacerbating the consequences of ecological degradation. Women and children are disproportionately affected by these disasters, as intersecting factors such as entrenched patriarchy, traditional gender roles, economic instability, and political repression compound their vulnerability. This paper critically examines the role of military dictatorship in shaping disaster preparedness and response and explores the broader humanitarian crises emerging in the context of environmental collapse and authoritarian governance in Burma.

Keywords Dictatorship · Environmental disasters · Deforestation · Humanitarian crisis

Introduction: Burma (Myanmar): Geography, Climate Vulnerability, and the Challenges of Authoritarian Governance

Burma, officially known as Myanmar, is a Southeast Asian nation bordered by Thailand and Laos to the east, China to the northeast, and Bangladesh and India to the northwest. To the south and west, it is flanked by international waters—the Andaman Sea and the Bay of Bengal, respectively. With a total land area of approximately 676,578 square kilometers (261,228 square miles), Burma is marginally

S. Win (✉)
Department of Women and Gender Studies, SUNY Brockport, State University of New York, Brockport, NY, USA
e-mail: soewin@buffalo.edu

B. Wejnert (ed.), *Democracy in Retreat and the Climate Crisis*, Springer Studies on Populism, Identity Politics and Social Justice,
https://doi.org/10.1007/978-3-032-02881-5_7

smaller than the US state of Texas. As of July 2021, the country's population stood at roughly 57.1 million, encompassing more than 135 recognized ethnic groups (Myanmar National Portal; The World Factbook 2022) (see Fig. 1).

Burma's topography is notably diverse, consisting of mountainous regions, rolling hills, and expansive plains. The central lowland areas—such as the Mandalay Region in upper Burma and the Yangon Region in lower Burma—are relatively flat but are surrounded by mountain ranges and river valleys extending toward neighboring countries. The nation experiences a tropical to subtropical monsoon climate, characterized by three primary seasons: a rainy monsoon season (June to September), a dry summer season, and a cooler winter season. The monsoon season brings intense rainfall, making the country particularly susceptible to environmental hazards such as flooding, landslides, and cyclones—especially in coastal and deltaic regions like the Irrawaddy Delta, where the river branches into multiple channels before emptying into the Andaman Sea. Conversely, the dry season often results in drought conditions in upper Burma due to extreme heat (The World Factbook 2022).

This rich and varied geography presents both opportunities and challenges. While the land is fertile and abundant in natural resources, making it conducive to agriculture and economic development, it is also highly vulnerable to environmental degradation and natural disasters—particularly in the context of rapid urbanization, deforestation, and unsustainable resource extraction.

Since gaining independence from British colonial rule through the Burma Independence Act of 1947, the country has been predominantly governed by military regimes, with the exception of a brief civilian-led administration between 2011 and 2021. These extended periods of military rule have been marked by political instability, weak governance, ongoing civil conflict with ethnic armed groups, and a general neglect of environmental management and humanitarian needs. The state's authoritarian governance has failed to adequately address the consequences of ecological destruction and environmental disasters, contributing to a worsening crisis. In recent years, Burma has faced a series of severe natural disasters—such as floods, storms, and cyclones—exacerbated by deforestation, unchecked urban expansion, excessive logging, and mining (Climate Change Knowledge Portal 2021; Kim 2024; The World Factbook 2022).

Given these complex and interlinked challenges, this paper begins by reexamining the theoretical foundations of democracy and autocracy to contextualize Burma's authoritarian political trajectory. It then proceeds to analyze the military regime's responses to environmental disasters and associated humanitarian crises, highlighting the governance failures that have intensified the country's vulnerability to climate-related threats.

Fig. 1 Map of Burma (Myanmar) adopted from The World Factbook on 6/14/24

Democracy vs. Autocracy

Recently, due to the Russian invasion of Ukraine and the increase in military governments starting coups around the world, many scholars have been conducting increased research on autocracy (authoritarianism). They have been comparing democracy and autocracy based on which system better supports economic development, civil liberty, and citizens' well-being, as well as environmental protection. It is, therefore, helpful to reexamine the definitions of these frequently used terms.

While democracy is a familiar term, and many countries proclaim themselves as democratic, it is more complicated than some believe. Democracy has no single definition. Democracy is interpreted in different ways by various scholars.

In the book "Diffusion of Democracy," Barbara Wejnert points out that democracy has a lengthy history and has meant different things for different people at different times and places. However, Wejnert argues that democracy should guarantee every adult citizen the right to vote, as well as provide basic rights, freedom and opportunities, gender and political equality, and human development.

In Dahl's book, "On Democracy," he argues that democracy offers many advantages over a non-democratic society. He expands the tenets of democracy to be a system that prevents government from cruel acts, such as jailing, arresting, and prosecuting political participants from the opposing party. It guarantees basic human rights, political rights, rights to vote, which of these rights are written in laws as a constitution. Democracy also provides general freedom, such as opportunities, freedom of expression and assembly, self-determination, human development, political equity, and many more. In his argument, autocratic government doesn't offer these benefits to its citizens (Dahl 1998).

According to the Freedom House Report in 2022, when one defines democracy in its ideal form, the government system must protect all citizens under the rule of law and respect human rights. The power of the government must be checked internally and externally (e.g., courts, independent press, and civil society). It requires all parties to compete fairly to govern the people, and the citizens to vote fairly and freely. All people, regardless of their age, gender, religion, or background, must enjoy basic human rights.

However, democracy is not only in an ideal form. In a practical form, it is a tool that can be used in a governmental system to demonstrate fairness, justice, freedom, and liberty to its citizens. The US Department of State's Bureau of International Information Programs interprets democracy as being ruled and vested by the people, which offers freedom, justice, and equality. It is important to note that while not all democracies in the world are perfect, they do offer freedoms and institutions that include everyone for better lives.

Summarily, one could argue that in a democratic society, the government should protect citizens, guarantee rights for women and minorities, value self-determination, encourage development, and provide freedom for everyone. Although there are many interpretations of democracy, one thing that remains largely uniform is the belief that democracy offers and promotes basic human rights, including for women and minorities.

In contrast to democracy, autocracy is defined as "politically exclusive rule" by scholars Ashley Anderson, Jason Brownlee, and Killian Clarke. They describe autocracy as an oppressive form of politics, where one leader makes all the rules and decisions, and their opponents are excluded from any decision-making. In other words, one ruler makes all the decisions and controls the state, and citizens who live in those countries are restrained from determining the nation's law. Today, it is presented mostly as a form of dictatorship[1]National Geographic). A modern dictatorship usually involves military coups, rejecting democratic principles, prosecuting opponents, limiting civil liberties, canceling or denying elections, and suspending the constitution (Levitsky and Ziblatt 2018). These autocrats work as a team or a group of networks known as "Autocracy, Inc.," a term coined by Anne Applebaum. In her argument, Autocracy Inc. is a group of autocrats who collaborate and support each other through financial, technological, military, and disinformation. They also assist each other to maintain power, for example, Russian's government supporting aide to North Korea, or the Chinese government providing military aide to Burma. Most importantly, they attack democracy and spread their propaganda known as autocracy promotion. They share similar ideology such as limiting media or restricting civil rights (Applebaum, 2024). Countries like China and Russia have heavily invested in military equipment and financial aid to their neighboring countries to expand their power on the international stage, which in turn challenges democracy (Way 2015).

By definition, democracy and autocracy differ in that democracy values and promotes fairness, freedom, justice, and liberty, while autocracy tends to restrict civil liberties, limit freedoms and rights, and oppress political opponents. According to Barbara Wejnert, democracy is a preferred system due to its rules and laws designed to protect citizens from human rights violations. Additionally, she finds that democratic societies offer and promote a higher standard of living, happiness, liberty, and freedom. Women's empowerment is also found to be greater in democratic countries than in autocratic ones. In democracies, women are more empowered to participate in politics, the workforce, education, and decision-making, which helps reduce gender-based violence, poverty, and social development. Healthcare for women and children is also better in democracies than in autocracies (Wejnert 2014).

Unfortunately, the Freedom House Report in 2022 maintains that democracy has declined in many countries, even those that self-define as democratic. For instance, India, the largest democracy in the world, has experienced a decline in functioning based on democratic principles. The country has dropped from "free" to "partly free" status in 2021 due to the limited freedom of speech and the increase in Hindu nationalist ideology against other minorities, as well as the attacks on the Muslim population. Under the Narendra Modi administration, there has been a dramatic decline in democracy due to the crackdown on peaceful protesters opposing a discriminatory citizenship law and the arrest of journalists for criticizing the government.

---

[1] I will use the terms "autocracy" and "dictatorship" interchangeably to mean a ruler who uses violence to stage a coup to control power and limit freedom and justice in their country.

In addition, the United States of America, a free country that values democracy and human rights, also has suffered from human rights violations under President Donald Trump. In 2018, Freedom House reported that America saw declines in democracy and political rights following Trump's election and an increase in the violation of basic ethical standards, including the failure of protection for minorities, women, refugees, immigrants, and certain religious groups. In 2022, Freedom House reported that the United States continued to see a decline in democracy due to President Trump's refusal to accept the loss of his election and the storming of the Capitol Building.

There are many other self-professed democracies throughout the world that struggle with issues related to refugees and immigration, which seem to menace their claims of democracy. They state that they are democracies but continue to be brutal and exercise violence against minorities and refugee populations. Although each country has its own authority on how to handle immigration and refugee issues, they must do so without violating their own constitutions. For example, the separation of migrant children from their families in the United States, Australia's imprisonment of seaborne migrants on the remote island of Nauru, and the detention of migrants by the Libyan military are violations of human rights, as well as violations of their own laws and constitutions. In addition, ethnic cleansing and the imprisonment of the opponents' political party seem to be the biggest issues in 2018 that led to a decline of freedom and democracy around the world (Freedom House Report 2018).

Overall, more than sixty countries experience limited freedoms and a decline of democracy due to factors such as violence, politicians' promotion of hate speech, military coups, the abuse of power, and other human rights violations. More than 38% of the global population lives in 'not free' countries (see Fig. 2). This figure illustrates the countries that see a decline in democracy, manifested in a decline of freedom of justice, expression, and rules of law. These countries include many in highly developed regions, such as the United Kingdom, Israel, and Spain.

One fact remains consistent between democratic and autocratic societies. Violence against women remains pronounced in both democratic and autocratic countries. According to the United Nations (UN), women face many forms of violence, including sexual violence, femicide, and domestic violence, despite the UN's Convention on the Elimination of all Forms of Violence Against Women. The Freedom House has stated that while women enjoy a wider range of freedom and justice in recent years, they continue to suffer from various forms of discrimination, including freedom of speech and gender policy that infringe upon women's rights (e.g., abortion law), and the exclusion of women from political participation (Freedom House Report 2022). When women do participate in politics or public offices, they often encounter a backlash under gender ideology (Han and Heldman 2017). Since politics is gendered in many parts of the world, female politicians often face violence and discrimination. For instance, a study by the Inter-Parliamentary Union (2016) covering 42 parliaments found that 81% of women experienced psychological violence, 44.4% received threats such as death or rape, and 65.5% faced sexual humiliation.

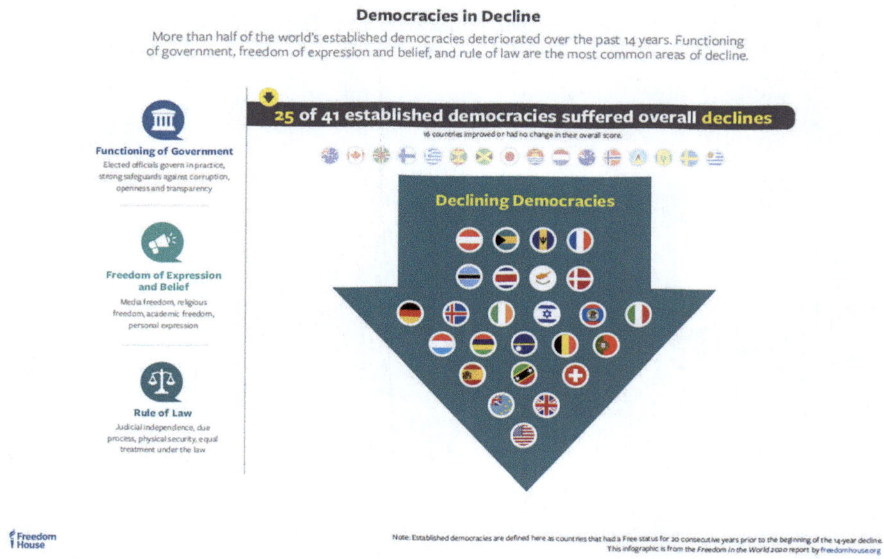

Fig. 2 Countries that saw the decline of democracy in 2021, adapted from the Freedom House Report on 7/29/22

Sadly, Burma also made the list in the Freedom House Report going from partly free to not free in 2022 (see Fig. 3). Burma has been largely ruled by a military government since gaining independence from the British. Since then, it has also been engaged in conflicts with various ethnic armed groups seeking self-autonomy. Due to the dictatorship, issues such as women's rights, LGBTQ rights, environmental protection, and disaster response programs have been neglected in the political agenda. Although the country experienced a brief transition to democracy from 2010 to 2020, it was short-lived. This is due to the military staging another coup in 2021.

This coup abruptly ended a 10-year democracy in Burma that shocked the international community. However, fearing a loss of their rights under the military dictatorship, this coup sparked a significant response by women. They have organized and participated in the movement against the coup that has become one of the largest protests in Burmese history. Freedom House Report stated that "democracy today is beleaguered but not defeated." It has endured destructive situations with resistance in recent years, and it remains an indication of justice, enjoyment, and freedom for many people around the world, including those fighting for freedom and justice in Burma. When one looks to the longer history of Burma, it is possible to see how this longing for democracy, and willingness to fight for it, has been established and nurtured for many centuries. Burma has often dealt with shifts between varying degrees of colonialism, democracy, and military dictatorship. The more people experienced the potentiality of democracy, the more people were

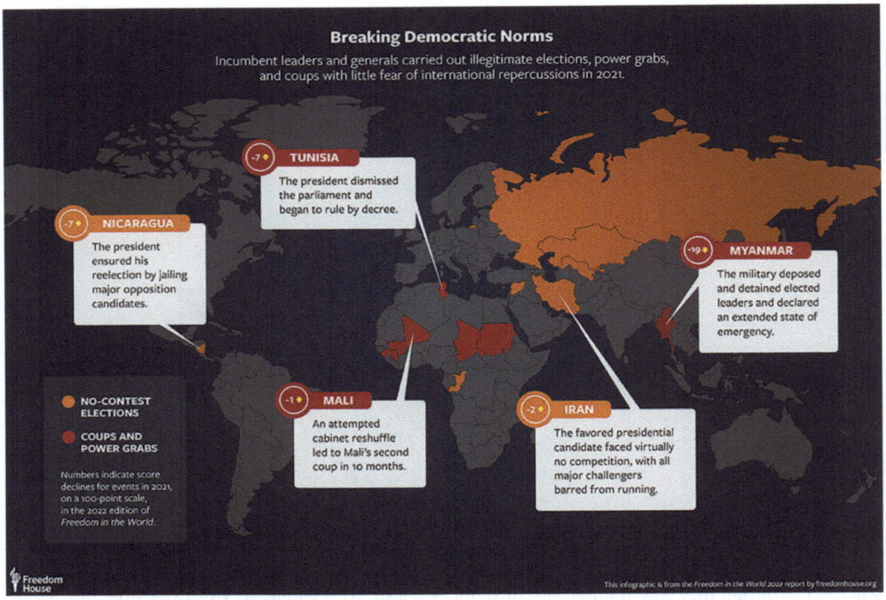

Fig. 3 Countries that break the democratic norms in 2021, adapted from the Freedom House Report

willing to fight for it. The next section will examine Burma's dictatorship, its environmental protection policies, and its response to environmental disasters.

The Military Responds to Environmental Disasters

In recent years, more scholars have been studying whether environmental protections are better under democracy or autocracy. The results are mixed. Some argue that environmental protections are less prevalent in a democratic society. Scholars acknowledge that in free market economies, where large corporate interests lie in profits, one finds extensive exploitation of natural resources. Further, it is argued that environmental degradation happens for the benefit of elites and not every citizen in a state of democracy (Li and Reuveny 2006). In a democracy, governments must listen to the public opinions and use that information to determine laws and policies. That is not required in autocratic countries where decision-making is centralized around one person (or a small group). In these instances, decisions can be made faster, such as in China's climate policies that aim to reduce carbon dioxide emissions, which were implemented by the authoritarian regime (Stein 2022). Moreover, some scholars argue that democracies may be the preferred political system for many citizens since they support environmental protection. However, they do not necessarily reduce CO_2 emissions that contribute to climate change (Chesler et al. 2023). For instance, many powerful democratic countries, such as the United

States and Germany, continue to produce high levels of CO_2 emissions that are responsible for climate change.

However, more recent scholars disagree with the above-presented arguments and point out that democracy promotes environmental enhancement and growth. These arguments are based on three main factors. First, people are better informed about environmental problems in a democracy. Second, citizens play an important role in environmental protection, and governments are more likely to listen to them. Environmental activists are more likely to speak up against pollution, promote land protection, and address other environmental issues without any fear of prosecution. These activists are more likely to put pressure on governments that help to implement policies that promote and protect the environment and address climate change. Lastly, democracy has lower deforestation rates, natural resources are most likely to be protected, and it is more likely to promote environmental quality. In autocracy, a few elites own the majority of resources and lands and are less likely to care for the environment (Stein 2022; Li and Reuveny 2006; Sinha et al. 2023).

This paper is in alignment with these findings using Burma as an example. I argue that military governments (using dictatorship) do not promote environmental protection, growth, and awareness by depleting natural resources, not promoting environmentally friendly policy, suppressing citizens' voices about environmental issues, and having a slow response to environmental disasters when they happen.

As briefly described above, Burma is vulnerable to disasters due to climate change and human activities, such as the depletion of natural resources. Deforestation is the largest problem faced by Burma today. Burma is the seventh most deforested country in the world. Deforestation is caused by several factors, such as agriculture, both legal and illegal logging, and infrastructure development (AFoCO 2021; Thu 2019). Burma's economy is mainly based on agriculture, and in many regions, deforestation is used for traditional farming. Lack of development and agricultural technologies means that many regions continue to practice traditional farming, where thousands of trees are cut down and burned for agriculture. In addition, trees are cut down for fuel as well as to make charcoal due to a lack of electricity, especially in rural areas (Kim 2024). Exploitation of timber and both legal and illegal logging are common in the country due to the demand from global trading, urban development, and infrastructure building (AFoCO 2021). Deforestation also endangers wildlife habitats and is a major cause of climate change. Further, women perform household duties by being responsible for collecting water and wood for cooking. Due to deforestation, women are required to travel long distances in search of firewood. Some woods do not burn well and produce a lot of smoke. Inhaling the smoke from cooking increases the risk of respiratory diseases, such as coughs or lung cancer, in women (Momsen 2010). However, people have no choice but to be involved in impeding climate action, mainly for survival.

In addition, civil war and conflict escalate the situation by destroying civilization, infrastructure, and land access. Deforestation has been rapid since the first military coup in 1962. There are two sides to this story. On the one hand, the military government promotes logging and timber for export and trading. Rapid forest depletion was found in regions under military control (Thu 2019). Natural resource extraction increased after the coup due to international sanctions, and this is the only way for the military to finance themselves. The government depends on the natural resources, such as logging timber (Lo and Hoy 2023). Moreover, deforestation increased during the recent coup in 2021, which corresponds to the military demolition of villages, infrastructure, farmlands, and nearby forests (Kim 2024). For instance, there is evidence of the military burning villages and towns during recent civil conflicts (Reuters 2021). On the other hand, it is believed that ethnic armed groups depend on timber and wildlife trading to support themselves during the conflicts. The ethnic minorities also depend on natural resources such as gold, timber, and wildlife. Due to the conflict, illegal logging and wildlife smuggling increased in ethnic regions and many rural areas. Many villagers engage in wildlife trading, logging, and traditional farming due to poverty and economic instability caused by the conflicts. With limited employment opportunities, they feel compelled to participate in these activities (Nikkei Asia, 2023).

Although there are several policies that protect the forest (e.g., Myanmar Forest Policy in 1995, Environmental Conservation Law in 2012, and Forest Law in 2018), the military government does not enforce them. When the country was a democracy, the coalition government (National League of Democracy & the military government) recognized the need to protect the forest and work diligently to limit the potential of climate change. They adopted the Myanmar Climate Change Strategy (2018–2030) that aims to promote climate resilience and low-carbon emissions for future sustainability. However, the goals of this policy were uncertain after the coup. As mentioned earlier, due to the ongoing civil war, environmental protection is not the military's top priority, let alone enforcing related policies. However, if the military had not initiated the coup and seized power, this policy would have promoted inclusivity by involving marginalized groups, including women, men, and ethnic communities, in climate resilience and low-carbon emission development. It would also promote a green economy for a climate-friendly as well as a sustainable future for the younger generation. The policy also aimed to reduce rural poverty and expand environmental programs to address the aftermath of natural disasters. (Lo and Hoy 2023; The Myanmar Climate Change Strategy 2019). This demonstrates that democracies are generally more likely to prioritize environmental issues compared to autocracies.

Additionally, the military government does not value citizens' concerns regarding environmental issues. Lack of freedom of speech makes it impossible for many activists to speak up regarding environmental and climate-related issues. Human Rights Watch stated the military government had arrested many environmental activists in Burma for criticizing it. Due to a coup in 2021, many young activists were prosecuted and fled the country. This could also lead to many consequences in the future as these climate activists are involved in advocacy (Lo and Hoy 2023). Their advocacy is essential for environmental protection and to limit climate change. Such advocates help to increase knowledge about climate change and protection.

Furthermore, the military governments' slow response to environmental issues shows their true disdain for their own people. For instance, the Indian Meteorological Department in New Delhi noticed the threatening storm coming from the Bay of

Bengal and notified the Burmese government on April 28, 2008. In Bangladesh, the government announced a state of emergency, evacuated the people, and closed schools and many businesses. However, the Burmese military government took no action and assured the people in the affected area that this storm was weak and the chance of landfall was unlikely. Unfortunately, Nargis made a landfall. It became one of the most devastating and deadly natural disasters in Burma's recorded history, claiming nearly 140,000 lives and causing approximately $4 billion in damage. Many international organizations believe the death toll may have been higher, as the military government stopped counting after the number reached 140,000. After one hour and even after one day of the cyclone, the military government initially refused to receive international humanitarian aid, as well as denied visas for foreign aid workers. As a result, the local community, such as churches and Buddhist monks, took the lead in helping people in affected areas (Human Rights Watch 2009). The military government was met with severe criticism, faced chaos, and government officials did not know how to respond to this disaster. Human Rights Watch reported that the government's neglect and ignorance of Cyclone Nargis was the most brutal approach the military could have taken toward its people. While they eventually allowed aid help, their initial denial of the cyclone and their slow response demonstrated their disdain for the humanity of their people.

More recently, Burma has seen more frequent cyclones and flooding. Cyclone Mocha made a landfall in May 2023 in Rakhine State, where the areas affected are the same as those experiencing conflict between the military government, the Rakhine ethnic armed group (AA), and the Rohingya population. The cyclone brought flooding that also impacted Rohingya refugees, a population that was already devastated by persecution from the military government. Once again, the government's response to this disaster was slow, and some witnesses stated that they did not offer any assistance to the Rohingya population (Human Rights Watch 2023). Instead, the government restricted and limited access to humanitarian aid to the marginalized population. Therefore, many international donors withdrew their assistance (Kim 2024).

In September 2024, Burma was once again devastated by severe flooding. According to the United Nations, the floods destroyed more than 70 towns, resulting in 330 deaths by drowning and leaving 100 people still missing. The damage was estimated at around $one million. The floods also destroyed crops, farmland, and livestock. Once again, the military's response was slow, forcing local communities and volunteers to step in and assist flood victims. Due to the ongoing civil war, humanitarian aid was delayed in reaching the areas most in need. Citizens remain vulnerable and helpless during and after environmental disasters under military rule, demonstrating the actions of a dictatorship that violates basic human rights.

The aftermath of environmental disasters is understudied in Burma, especially since there is a lack of research on gender and disaster. However, based on available data, women, children, and ethnic minorities are mostly affected by climate burden (Kim 2024). Burma is a male-dominated patriarchal society, which continues to hold traditional values such as men as the breadwinners of the family and women as the caretakers (Myanmar Asia Barometer 2019). Due to this ideology, women continue to experience gender-based violence, discrimination, and exclusion in socioeconomic and political spheres. The military government neglects women's development and empowerment by limiting access to education, healthcare, and political participation. Women are unable to speak up against rape and domestic violence. Due to high rates of poverty and political instability in the country, women may drop out of school to help family members (Belak 2000). Therefore, women are more likely to take care of their family members first before taking care of themselves in the aftermath of a disaster. They may experience increased food shortages when they are displaced in the aftermath of a disaster, and women may feed their husband and children first before eating the leftovers. Food insecurity is often linked to the aftermath of a disaster or civil war. Female-headed households are more vulnerable, since many of the disaster-related policies tend to focus less on women's issues. For instance, women may be left out without male relatives during the distribution of aid due to their inability or power to fight for assistance. Since women perform the majority of the household chores, limited food supplies and clean water also lead them to diseases and gender-related problems such as gender-based violence (Khakimova 2011; United Nations 2024a, b). Women may develop mental health issues, but these issues are rarely discussed or addressed due to the stigma attached to mental health. Unfortunately, climate change concerns, environmental disasters, and women's rights issues have not been the first priority for the military government since they are currently facing internal conflicts with ethnic armed groups.

The information presented, using Burma as an example, demonstrates that dictatorships often show little concern for environmental destruction and the depletion of natural resources, and they limit the freedom of environmental activists. After disasters, these regimes are slow to respond to humanitarian crises, conceal accurate information from the public, restrict access to foreign aid, and even arrest and imprison environmental activists who criticize their actions. Sadly, many Burmese citizens face this harsh reality, oppressed by the military government. Their fundamental rights, such as access to safe housing, clean water, sanitation, and the ability to live without fear, are being stripped away.

Conclusion

The history and politics of Burma are complicated. Burma has a history of colonialism, independence, military dictatorship, and democracy. The country has struggled to maintain democracy since its independence from the British. It has been mostly ruled by the military. However, by the 2010s, the country gradually transitioned itself to democracy. However, this reality was destroyed by the military government staging another coup in 2021. These coups escalate the ethnic tensions and civil wars. As a result, their citizens are facing poverty and economic hardship and internal sanctions. Ethnic people are also persecuted and forced to flee the country due to the ongoing civil war. Due to these internal conflicts and issues, climate change and environmental disasters have not been their priority. The military government often ignores or disengages from environmental protections, conceals disaster warnings from meteorologists, limits access to humanitarian aid, and/or delays response in the aftermath of a disaster. When activists speak up against the disaster's response or articulate their concerns, they are faced with prosecution or jailing. Presently the situation is more dire. Since I completed the second round of edits on this article, I was not able to incorporate research on the recent 2025 Myanmar earthquake. Burma was struck by a powerful 7.7-magnitude earthquake in central Burma, which killed thousands and left many people missing. The quake devastated several cities, including Mandalay, and severely damaged infrastructure. According to United Nations reports, women and children are particularly vulnerable in this disaster, with many unable to access basic needs such as food, shelter, and medical care. Women and girls face heightened risks of gender-based violence and lack access to essential reproductive health services. Unfortunately, as noted in the article, the military government ignored international aid efforts and responded slowly, worsening the humanitarian crisis. Therefore, Burma serves as a case study to demonstrate that under military or autocratic rule, leaders are less inclined to honor the values of freedom, justice, and liberty. While all individuals are negatively impacted in these situations, women, children, and ethnic minorities are particularly vulnerable due to gender roles, displacement, unstable infrastructure, and civil war. Unfortunately, they are paying the price for their government.

References

Applebaum A (2024) *Autocracy, Inc: the dictators who want to run the world*, 1st edn. Knopf Doubleday Publishing Group

Asian Forest Cooperation Organization (AFoCO) (2021, December 1) Forests and forestry in Myanmar: an overview of forestry laws, policies, and strategies. Retrieved from https://afoco-sec.org/newsroom/news/forestry-news/forests-and-forestry-in-myanmar-an-overview-of-forestry-laws-policies-and-strategies/

Belak B (2000) Double Jeopardy: abuse of ethnic women's human rights in Burma. Cult Surv Q 24(3):24

Bauchner S (May 18, 2023). Cyclone Mocha Devastates Myanmar's Rohingya. Human Rights Watch. Retrieved from https://www.hrw.org/news/2023/05/18/cyclone-mocha-devastates-myanmars-rohingya

Chesler A, Javeline D, Peh K, Scogin S (2023) Is democracy the answer to intractable climate change? Glob Environ Polit 23(4):201–216. https://doi.org/10.1162/glep_a_00710

Climate Change Knowledge Portal (2021) Retrieved from https://climateknowledgeportal.worldbank.org/country/myanmar

Dahl RA (1998) On democracy. Yale University Press

Freedom House Report (2018–2022). Retrieved from https://freedomhouse.org/reports

Han LC, Heldman C (2017) Women, power, and politics: the fight for gender equality in unite states. Oxford University Press

Human Rights Watch (2009) The lesson of cyclone Nargis. Retrieved from https://www.hrw.org/news/2009/05/03/lessons-cyclone-nargis

Inter-Parliamentary Union (2016). Sexism, harassment, and violence against women parliamentarians. Retrieved from https://www.ipu.org/resources/publications/issue-briefs/2016-10/sexism-harassment-and-violence-against-women-parliamentarians

Khakimova S (2011) The nutrition and health of women and children in the aftermath of natural disasters. Res Polit Soc

Kim K (2024, June 9) Concurrent challenges of conflict and climate change in Myanmar Georgetown J Int Aff Retrieved from https://gjia.georgetown.edu/2024/06/09/concurrent-challenges-of-conflict-and-climate-change-in-myanmar/#:~:text=Myanmar%20is%20one%20of%20the,low%2Dlying%20and%20coastal%20regions

Kipgen N (2011) Political change in Burma: transition from democracy to military dictatorship (1948-62). Econ Polit Wkly 46(20):48–55

Levitsky S, Ziblatt D (2018) How democracies die. Broadway Books. Crown Publishing, New York

Li Q, Reuveny R (2006) Democracy and environmental degradation. Int Stud Q 50(4):935–956

Lo A, Hoy ST (2023, November 9) Conflicts intensify climate change risks in Myanmar. East Asia Forum. Retrieved from https://eastasiaforum.org/2023/11/09/conflicts-intensify-climate-change-risks-in-myanmar/

Momsen J (2010) Gender and development, 2nd edn. Routledge

Myanmar Asia Barometer (2019) Retrieved from https://www.asianbarometer.org/content?nid=20

Myanmar Climate Change Strategy (2018–2023) Retrieved from https://myanmar.un.org/sites/default/files/2019-11/MyanmarClimateChangeStrategy_2019.pdf

Myanmar National Portal. Retrieved from https://myanmar.gov.mm/

National Geographic. Retrieved from https://education.nationalgeographic.org/resource/autocracy/

Nikkei Asia (2023, September 14) Myanmar's forests fall victim to conflict. Retrieved from https://asia.nikkei.com/Life-Arts/Life/Myanmar-s-forests-fall-victim-to-conflict

Rueters (June 16, 2021) Myanmar Village Burned after Fighting; Residents Blame Security Forces. Retrieved from https://www.reuters.com/world/asia-pacific/myanmar-village-burned-after-fighting-residents-blame-security-forces-2021-06-16/

Sinha A, Sedai AK, Kumar A, Nepal R (2023) Are autocracies bad for the environment? Global evidence from two centuries of data. Energy J *44*(2):47–78

Stein JV (2022) Democracy, autocracy, and everything in between: how domestic institutions affect environmental protection. Br J Polit Sci 52(1):339–357

The World Factbook (2022) Washington, DC: Central Intelligence Agency. Retrieved from https://www.cia.gov/the-world-factbook/countries/burma/

Thu, W. M. (2019, September 30) The deforestation dilemma in Myanmar. Asia Dialogue. https://theasiadialogue.com/2019/09/30/the-deforestation-dilemma-in-myanmar/

United Nations (2024a, July 2) Myanmar: new report urges robust support for women, girls, and LGBT people in post-coup Myanmar. Retrieved from https://www.ohchr.org/en/press-releases/2024/07/myanmar-new-report-urges-robust-support-women-girls-and-lgbt-people-post#:~:text=%E2%80%9CIn%20cities%2C%20villages%2C%20and,in%20violent%20and%20chaotic%20environments

United Nations (2024b) Myanmar: flood situation report # 3, 27 September 2024. Retrieved from https://myanmar.un.org/en/279958-myanmar-flood-situation-report-3-27-september-2024

Way LA (2015) The limits of autocracy promotion: the case of Russia in the "near abroad". Eur J Polit Res 54(4):691–706. https://doi.org/10.1111/1475-6765.12092

Wejnert B (2014) Diffusion of democracy: the past and future of global democracy. Cambridge University Press

Soe Win is a scholar and activist with a Ph.D. in Global Gender and Sexuality Studies from the University at Buffalo. Her research focuses on women's rights, gender-based violence, minority rights, LGBTQ rights, global gender inequality, and the intersections of gender, democracy, and politics. Her dissertation explored the women's movement in Burma, analyzing the role of women in social and political struggles under an authoritarian regime.

Beyond academia, Soe is deeply engaged in advocacy and direct support work. She currently serves as a care manager, working with vulnerable populations, including survivors of domestic violence and sexual assault, homeless individuals, mental health patients, refugees, and LGBTQ individuals. She is a Teaching Assistant Professor at the State University of New York: Brockport, where she offers courses such as *Introduction to Women and Gender Studies, Gender in Global Perspective, Feminist Theory, Research Methods,* and *Race, Class, & Gender*. She is currently working on her book, *The Karen Women: Reclaiming History, Feminism, and the Struggle for Justice,* which examines the historical and contemporary struggles of Karen women through a feminist lens.

The Bosphorus Bottleneck: Navigating the Tension Between Development, Political Conflicts, and Climate Crisis

Senay Imre

Abstract Exploring how local governance actors and the central government work together and/or against one another to address the climate crisis is an important area to explore. Political will and governing ideology and development models are important areas of exploration to understand challenges in addressing the climate crisis. What stands in the way of designing a city management policy with long-term sustainable goals? What role does politics play? What type of political model works better to address the climate crisis? What are areas of contention and what are areas of collaboration? When and where does political contention get ahead of climate concerns? This paper will try to answer these questions by looking into the example of climate approaches in Turkey at the local and national level. Through a close analysis of the example of Turkey's most developed urban mega city of Istanbul, this paper will scrutinize climate crisis policies and rhetoric taken by two different political leaders in the local and central government. By looking into the example of the proposed Canal Istanbul project as a case study, this paper will lay out the complexities of addressing the climate crisis under different political ideologies and governance models. Lastly, the chapter will discuss transnational climate justice solidarity networks and city diplomacy efforts as important mechanisms and efforts to counter political challenges.

Keywords Climate crisis · Democracy · Political polarization · Turkey · Sustainable development

S. Imre (✉)
Global Gender Studies, University at Buffalo, SUNY, Buffalo, NY, USA
e-mail: senayimr@buffalo.edu

B. Wejnert (ed.), *Democracy in Retreat and the Climate Crisis*, Springer Studies on Populism, Identity Politics and Social Justice,
https://doi.org/10.1007/978-3-032-02881-5_8

Introduction

In February 2023, Turkey was struck by a devastating 7.8-magnitude earthquake that caused widespread destruction and highlighted significant concerns about environmental management and government preparedness. This major seismic event not only resulted in tragic loss of life and property but also exposed critical weaknesses in the country's disaster response capabilities. Compounding the tragedy, urban development in earthquake-prone areas, often not grounded in the realities of seismic risks, has led to particularly devastating outcomes. The disaster comes amid a backdrop of increasing environmental challenges, including climate change and natural resource depletion, which have further strained Turkey's ability to manage and mitigate such crises. The situation underscores the urgent need for more robust government planning—at all levels of government with active participation of civil society - and proactive measures to address both environmental vulnerabilities and enhance resilience against future disasters.

Global climate change without a doubt has become the number one concern for many citizens in the world. According to the United Nations, 3.6 billion people live in areas that are highly vulnerable to climate change, which represents nearly half of the world's population (United Nations Development Programme 2024). According to a 2022 Pew Research Center survey, citizens in advanced economies are especially concerned about climate change. A median of 75% across 19 countries in North America, Europe and the Asia-Pacific region label global climate change a major threat (Poushter et al. 2024). Those living in areas that have high vulnerability to climate change due to population size, limited resources and geographical disadvantages feel the utmost danger and anxiety. The increasing scale, frequency and intensity of catastrophic disasters that affect people's day-to-day life is on the rise, creating "radical uncertainties" followed by "eco anxiety" that endanger wellbeing for all, now and in the future.

The last five years showed the world that a rapidly warming planet poses an existential threat to all life on earth. From earthquakes, hurricanes to forest fires, from water scarcity to flooding, from excessive heat to shrinking natural resources, the world has woken up to a new world order that is no longer run by people but by the climate crisis itself. The insistence on continuing a world order based on economic growth, urbanization, population increase and excessive consumption that comes with a high cost of social and environmental consequences is no longer possible nor desirable. This human-induced crisis is calling for an urgent, long-term intervention to regulate human-environment interaction at the local, national, and global levels.

Urban environments and cities emerge as the main areas of concern. According to research shared by the World Bank, around 78% of energy and 75% of natural resources are consumed in urban areas, while cities contribute to more than 70% of greenhouse gas emissions (Dasgupta et al. 2024). These numbers call for urgent intervention in city management as it is evident that the impact of climate change could be more devastating in urban cities than anywhere else. Experts call for sustainable urbanization efforts and underline the important role of local governance in addressing the climate crisis. However, even though the climate crisis affects everyone equally, it would be a false assumption to believe the climate crisis would be a unifying common national problem across the political spectrum. Exploring how local governance actors and the central government work together and/or against one another to address the climate crisis is an important area to explore. Political will and governing ideology and development models are important areas of exploration to understand challenges in addressing the climate crisis. What stands in the way of designing a city-management policy with long-term sustainable goals? What role does politics play? What type of political model works better to address the climate crisis? When and where does political contention get ahead of climate concerns? This paper will try to answer these questions by looking into the example of climate approaches in Turkey at the local and national level. Through a close analysis of the example of Turkey's most developed urban mega city of Istanbul, this paper will scrutinize climate crisis policies and rhetoric taken by two different political leaders in the local and central government. By looking into the example of the proposed Canal Istanbul project as a case study, I will lay out the complexities of addressing the climate crisis under different political ideologies and governance models. Lastly, the paper will discuss transnational climate justice solidarity networks and city diplomacy efforts as important mechanisms to counter political challenges.

Literature: Democracy, Sustainable Development and Climate Crisis

"Ecological democracy" has been advocating for the relationship between democratic political systems and sustainability for a long time. However, efforts to implement ecological democracy have been hindered by an increase in democratic backsliding. Additionally, the role, impact and sufficiency of democracy has also been questioned more than ever especially in relation to tackling the climate crisis. Andrea Westell's *Exploring the tensions: The relationship between democracy and sustainable development* puts forth the worrying increase in dissatisfaction with current democratic practices (Westell 2023). Westell shares that the 2021 report by Freedom House underlines that 18 countries in Europe and Eurasia saw a decline in their democracy scores, six experienced an improvement, and five remained unchanged (Westell 2023, pg3). The report highlights that this marks the 17th consecutive year of overall decline, with the number of countries classified as 'democratic' by their criteria reaching the lowest level in the history of the report (Westell 2023, p.3). Separately, Westell cites a survey from the World Values Survey (2017–2021) that shows people continue to place a high value on the 'importance of democracy' question (Westell 2023, p.3). Westell's statistics suggest that democracy as a concept is not the issue. Democracy itself is neither unpopular nor inherently problematic; rather, concerns focus on its implementation and the issues associated with unpopular leadership (Westell 2023, p.3). For example, Westell cites widening inequality and a growing tendency toward authoritarian leaders as a key driver of dissatisfaction with democracy (Westell 2023, p.3) and warns that if liberal democracies fail to tackle climate change, biodiversity loss, and extreme inequality, they could likely jeopardize the foundations of future democracies. A couple of ways this could happen is through diminished social cohesion, overly technocratic decision making, or an increase in authoritarianism (Westell 2023, p.1). Byford Tsang and Juan Pablo Osornio emphasize in their work *Autocracy vs Democracy: A Climate Edition* that current research indicates that the level of democracy experienced by the average global citizen today has returned to the levels seen in 1989, due to the rise of authoritarian and nativist politics over the past decade (Tsang and Osornio 2024). They analyze the patterns in highly centralized government models where they notice that "enlightened despots" have the tendency to commit to long-term interventions, policies, and programs as they serve for an extended time and do not carry the same election concern as democratic leaders (Tsang and Osornio 2024). They underline that these long-term interventions are more common in highly centralized government models and most of the time these programs and policies are developed without any citizen consent or civil society participation but through the participation of specific groups of people that are close to the government (Tsang and Osornio 2024).

Their research also underlines that the same autocratic countries also show an increase in the level of carbon emissions since 2010 up to now (Tsang and Osornio 2024). They share that while 35% of carbon emissions in G20 countries came from autocracies in 2010, this number reached 52% in 2023 and is predicted to go up to 58% by 2030 (Tsang and Osornio 2024). The increase in carbon emission levels in autocratic governments is an alarming development for Tsang and Osornio. While they underline democracy's deep commitment to human dignity, rights, collective well-being of our planet and the importance of individual voices, a free press, freedom of speech, and freedom of assembly as necessary democratic values to influence decision making processes, they also think there is value in offering economic incentives such as access to global clean tech markets and the prospect of inbound investment on green industries to influence both democratic and non-democratic governments in the fight for the climate crisis (Tsang and Osornio 2024).

In his book *Democracy and Sustainable Development*, Dan Banik, akin to Westell's viewpoint, also emphasizes that democracy alone is not enough to ensure sustainable outcomes (Banik 2022, p. 243). Banik also observes that democratic governments often prioritize short-term policies aimed at securing re-election rather than committing to long-term objectives and projects. The drive for re-election shifts attention away from sustainable, long-term planning, as electoral pressures tend to promote and shape populist policies that focus on immediate gains (Banik 2022, p. 243). A recent example of this phenomena would be from one of the most democratic states in the USA, New York. NY Governor Kathy Hochul's last-minute, surprise decision to halt New York City's congestion pricing program that was planned to start on June 30, 2024, with the intent to reduce traffic congestion, improve air quality, and fund the public transit system by charging $15 per car to enter Lower Manhattan. Many saw this move as a politically motivated pre-election decision taken against a long term-environmentally friendly policy. Banik's book underlines that there is a growing amount of literature that argues that "democracies are better able than non-democracies to mobilize considerable public action when harm is imminent and visible rather than when threats are more long-term and currently less visible" (Banik 2022, p.244) which is true, and that "democracies guarantee that citizens can articulate an alternative vision of development, which may often differ in substantial ways from the visions of their leaders" (Banik 2022, p.244) which is also true (Banik 2022, p.244), however, this doesn't always turn in favor of people's mobilization on the ground, especially when short-termism wins over, as experienced in the case of NYC.

To add to Banik's argument, Westell also identifies "short-termism" as major tension standing between democratic systems and sustainable development. Westell lists the tension between liberal democracies and sustainable development in an effort to show that a change in approach and thinking is needed to address climate change issues. She shares that liberal democracies are "characterized by short-term thinking and policy design focused on individual freedom, that they value competition among ideas and political parties to ensure pluralism, they lack a unified vision of the 'good life' or how to achieve it, and they emphasize representative government with relatively limited public and stakeholder involvement" (Westell 2023, p. 6). These democracies operate within defined political boundaries and legally defined citizenship, view economic growth as a measure of success and a means to improve citizens' quality of life, and do not systematically account for environmental constraints (Westell 2023, p. 6). On the other hand, Westell argues that sustainable development acknowledges that its drivers and impacts span political boundaries, time, and levels of governance, affecting not only people in other jurisdictions but also future generations and the natural environment (Westell 2023, p. 6). It highlights "the integration of economic, environmental, and social considerations to enhance individual and societal well-being, recognizes environmental limits to economic and social activities, and supports integrated and precautionary policies to address complex and uncertain impacts, utilizing multi-criteria decision-making tools" (Westell 2023, p. 6). Westell contends that maintaining democracy will be increasingly difficult in a world marked by inequality, resource limitations, and high temperatures. She suggests that prioritizing democratic values such as justice, legitimacy, and transparency in democratic processes and protections can make the pursuit of sustainable development fairer, more justifiable, and more broadly accepted (Westell 2023, p. 11).

These articles offer a strong foundation for understanding governance models and the political motivations behind various policies. They also shed light on the contrast between long-term authoritarian or restrictive democratic models and short-term democratic narratives, a topic this paper will explore through the Turkish example. Over the past 20 years under President Recep Tayyip Erdoğan's leadership, his government has navigated a shifting landscape between democratic, restrictive democratic, and authoritarian practices. I refer to this fluctuating approach as 'meandering democracy,' a phenomenon strategically employed during significant political moments to appeal to both national and international audiences. This approach allows the government to adjust its stance depending on the issue at hand—sometimes highlighting democratic principles to garner support from Western allies, while at other times adopting more restrictive or authoritarian measures to consolidate domestic power. This tendency becomes especially pronounced in discussions surrounding the climate crisis, where policy shifts are used to signal alignment with global environmental standards or reinforce control over internal narratives, depending on the broader political context. The following sections will explore the political strategies and shifting dynamics surrounding the decision-making process of the proposed Canal Istanbul project, highlighting how the tension between local and central governments obstructs the implementation of long-term, sustainable solutions for public well-being. Additionally, it will emphasize the importance of political cooperation in decision making as a critical factor for effectively addressing the climate crisis.

Research Methods

This study employs a mixed-methods approach, utilizing secondary qualitative and quantitative data to analyze the case of a proposed large infrastructure project in Istanbul, Turkey within the broader context of climate change, sustainability, and development, with a focus on the 2020–2024 timeframe. Istanbul serves as a focused case study, providing a localized lens through which broader national and international sustainability challenges can be examined.

An extensive literature review was conducted, drawing from international, national, and local research papers on climate change, sustainability, and development. This review provided context for the study within existing academic research and helped identify relevant theoretical frameworks. Sources included books, peer-reviewed journal articles, conference papers, and research reports to explore themes and issues at the intersection of climate change, sustainability, and development. These sources served as lenses to analyze climate policies in Istanbul within global discussions on urban sustainability.

Qualitative data were analyzed from official local and national governmental sources, such as Istanbul Metropolitan Municipality's Climate Change Action Plan and Sustainability Report and Ministry of Environment, Urbanization and Climate's climate change mitigation strategy and action plan, Ministry of Transportation and Infrastructure policy documents, public statements, and published reports specific to Istanbul. These sources offer insights into the official narratives, policies, and responses surrounding climate change and sustainability initiatives within the city. The article, however, heavily relies on quantitative data collected by analyzing book, journal and news articles from both pro-government and anti-government newspapers. This dataset captures a range of perspectives on public discourse surrounding climate change and development in Istanbul, reflecting diverse opinions and representations of sustainability issues. A content analysis quantified the frequency and context of specific themes, such as urban policy impact, climate resilience efforts, and local government actions. By comparing coverage across political orientations, this analysis highlighted public sentiment, policy critiques, and the influence of media biases on narratives surrounding sustainability in Istanbul. Given the use of publicly available secondary data, this study did not involve direct interaction with human subjects. Particular care was taken to interpret data impartially, especially in analyzing media sources with varying political biases, to ensure that findings reflect an unbiased view of sustainability issues in Istanbul.

Istanbul, Turkey: A Case Study

Turkey is a middle-income country, with a large manufacturing base, and a growing population that is much younger than neighboring countries in Europe and Central Asia. The country is home to several major cities, but the biggest by far is Istanbul, the country's economic, media, and cultural hub.

Istanbul boasts more than 16 million residents and stands as a testament to the complexities of modern urban life. This vibrant metropolis has not only experienced rapid growth but has also navigated significant challenges and transformations over the past decade. The city's evolution is shaped by a unique terrain,a city that straddles two continents, bifurcated by the Bosphorus Strait which connects the Black Sea to the Sea of Marmaraand a combination of ambitious infrastructure projects, shifting political dynamics, and pressing social issues, reflecting a rich and intricate narrative of development, environmental concerns, and democratic struggles.

At the heart of Istanbul's recent transformation are its economic and infrastructure advancements. Strategically located as a bridge between Europe, Asia, the Middle East, and Africa, Istanbul has cemented its role as a major global transit hub. This status has been bolstered by a series of high-profile infrastructure projects initiated under President Recep Tayyip Erdogan's administration. The construction of the massive new Istanbul Airport, large-scale city hospitals, a third bridge across the Bosphorus Strait, and extensive railway tunnels are all part of President Erdogan's vision to enhance the city's global connectivity and economic appeal. These "mega projects" have not only advanced Istanbul's infrastructure but have also positioned it as a key player on the international stage, highlighting the city's growing significance in global transit and commerce.

Social and political issues also further complicate Istanbul's narrative. The city's population grew immensely following large-scale migration from Anatolia's provinces, and in recent years, the city has seen a significant influx of refugees due to conflicts in Syria, Afghanistan, Ukraine, and other regions, leading to complex social dynamics. The legal ambiguity and socioeconomic challenges faced by refugees and immigrants have created a vulnerable population. Gender inequality and rising urban poverty are additional pressing issues, with women, children, and the elderly bearing the brunt of these challenges. The deepening poverty within both host and refugee communities highlights the urgent need for comprehensive social policies that address these disparities.

Economic growth and social challenges in Istanbul are accompanied by severe environmental issues. As Turkey's largest emitter of greenhouse gasses, the city is highly susceptible to the impacts of climate change, including rising temperatures, shifting precipitation patterns, drought, sea level rise, and intense rainfall (Istanbul Policy Center 2022, p. 19). The expanding population, coupled with a push for rapid urbanization and economic development, is placing immense pressure on the city's resources and infrastructure. This leads to increased energy and water use, higher waste production, a reduction in green spaces, and a growing need for housing (Istanbul Policy Center 2022, p. 19). Additionally, Istanbul faces significant seismic risks due to its proximity to the Eastern Anatolian fault line. Over its more than 2000 years of human history, the city has experienced numerous massive earthquakes, including a 1999 quake that killed more than 17,000 people. A 2023 earthquake in central Anatolia killed more than 50,000 people, and displaced thousands towards main cities Istanbul, the capital Ankara, and Izmir on the Aegean seeking refuge.

Amidst these economic advancements and social setbacks, the most significant change has occurred in Turkey's political landscape which was also triggered by the public's environmental concern to a government project.

A small sit-in protest in 2013 against the planned redevelopment of Gezi Park, a public park in Istanbul, quickly grew into a broader movement against the policies and governance of the ruling Justice and Development Party (AKP) and President Erdogan. The park, located in the heart of Istanbul, was one of the few green spaces in a densely urbanized area and was seen as an important public space for recreation and community activities. These developments, along with a failed coup attempt in 2016, led to significant changes in governance. One major shift was the transition from a parliamentary democracy to an executive presidential system in 2018, resulting in a more centralized governance model and effectively reversing the decentralization efforts of previous decades. This centralization has profoundly impacted local governance, particularly in major cities like Istanbul and Ankara.

While the centralized model functioned more effectively when local municipalities were governed by the same party as the central government, issues became more apparent when these municipalities changed hands. The election of Ekrem İmamoğlu in Istanbul and Mansur Yavaş in Ankara in 2019, representing the opposition Republican People's Party (CHP), has heightened tensions between local and central authorities, underscoring the challenges of balancing local autonomy with central control. This issue has been more pronounced in Istanbul than in Ankara and Istanbul mayor İmamoğlu's administration has encountered numerous obstacles, highlighting the friction between local governance and centralized authority. The fact that President Erdoğan has a special connection to Istanbul, having been born and raised there, and he began his political career as the city's mayor in 1994 plays an important role. Additionally, President Erdoğan has a strong appreciation for Istanbul's historical prominence, particularly its significance and heritage as the former seat of Ottoman Empire.

The tension between President Erdogan and Mayor İmamoğlu is an important example that shows the limitations of city management and planning under a highly centralized government model. For example, Mayor İmamoğlu accused the central government of obstructing crucial urban projects, such as metro expansion, bus fleet replacement, and the redesign of the city's central public square, Taksim. The Mayor shared that there were delays in approving international financial deals, including Eurobond sales, which have further complicated the municipality's efforts (Bloomberg 2022). All of these developments encouraged İmamoğlu to seek international funding from European countries after Turkey's state-run banks showed reluctance to provide routine loans, including for essential expenditures like salaries (Bloomberg 2022). The municipality has since secured financing from foreign commercial banks, the Black Sea Trade and Development Bank, and the European Bank for Reconstruction and Development. (Bloomberg 2022) Urban planner Murat Guvenc and other experts argue that these political conflicts are exacerbating Istanbul's ecological challenges, including the looming risk of earthquakes, environmental threats, and urban poverty. (Bloomberg 2022) These experts argue that a more cooperative relationship between central and local governments is necessary for effective urban planning in order to address climate related issues Bloomberg 2022). One of the main examples where this tension becomes very visible is the debate over President Erdogan's mega project "Canal Istanbul."

President Erdogan proposed an ambitious project to construct an artificial waterway parallel to the Bosphorus Strait, primarily to alleviate congestion in the heavily trafficked Bosphorus, provide an alternative path for more hazardous cargoes, and reduce the risk to residents and buildings along the highly populated shores (Time 2020). However, debates between President Erdogan's party (AKP) and the opposition party (CHP) revolve around the potential environmental impacts and benefits of the project. Ekrem İmamoğlu, the CHP-run Istanbul Municipality Mayor has emerged as a leading opponent of the President's large-scale urban projects, supported by a coalition of experts and environmentalists who have raised significant concerns about the potential environmental damage the Canal Istanbul Project could cause to Istanbul and the greater Marmara Region (Pehlivanli and Eslen-Ziya 2023, p. 45) (Fig. 1).

Discussion: "The Canal or Istanbul: Which Holds Greater Significance?"

While the Turkish government has committed to key international UN-led environmental and climate agreements—such as the United Nations Framework Convention on Climate Change (UNFCCC), the Paris Agreement (2015), the Montreal Protocol (1987), and the United Nations Convention to Combat Desertification (UNCCD)—its development policies, which heavily prioritize urbanization and industrialization through mega-projects, rapid population growth, and increasing consumption, present

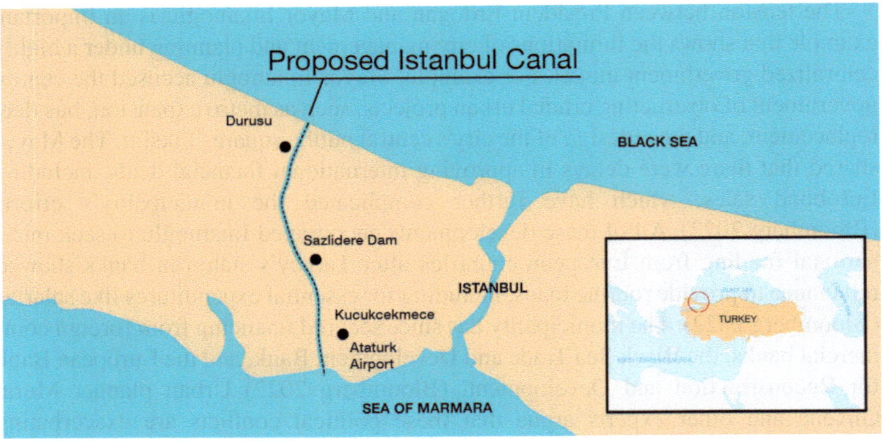

Fig. 1 Adapted from Canal İstanbul: Erdoğan's canal and the trench of protest, by K. Kirişci 2021, Foreign Policy

significant challenges. Additionally, political divergence between local and national agencies disrupts effective coordination on climate initiatives, further contributing to Turkey's lag upon other European countries in climate action.

Turkey ratified the Paris Agreement in October 2021, committing to its goals of limiting global warming to well below 2° C above pre-industrial levels, with efforts to limit the temperature increase to 1.5° C (Istanbul Policy Center 2022, p.4). Turkey also updated its first Nationally Determined Contribution (NDC) to the UNFCCC and the Paris Agreement in April 2023, pledging to reduce greenhouse gas (GHG) emissions by 41% by 2030 (targeting 695 Mt. CO2 equivalent) compared to a business-as-usual (BAU) scenario based on 2012 levels. This updated NDC includes economy-wide mitigation and adaptation measures with the goal of peaking emissions by 2038 and achieving net-zero emissions by 2053. The updated target reflects a significantly more ambitious approach, aiming for a science-based, equitable trajectory (Republic of Türkiye Updated First Nationally Determined Contribution, 2023, p. 1).

While this commitment appears promising on paper, a closer examination reveals contradictions between the government's stated goals and its ongoing development projects and political stance, particularly regarding opposition-led local governments. For example, credible climate organizations, like Climate Action Tracker, rate Turkey's climate efforts as "critically insufficient" (Türkiye. climateactiontracker.org/countries/turkey). Beyond political and coordination challenges, Turkey faces intensified reliance on imported oil, coal, and natural gas due to rising energy demand, and plans to expand domestic coal and gas production to reduce import dependence. (Türkiye. climateactiontracker.org/countries/turkey.) Additionally, insufficient investment and incentives for renewable energy further hinder Turkey's climate progress. The effectiveness and implementation of these commitments have been subject to ongoing discussion and evaluation, particularly considering projects

	Canal Istanbul	Strait of Istanbul
Length (km)	45	31
Min. width (m)	275	698 (minimum)
Depth (m)	20.75	35

Comparison titles are given in the first column, and the Canal İstanbul and SoI data are given in the second and third columns.

Fig. 2 Comparison of Canal Istanbul and SoI. Source: Mersin et al. (2024)

like the Canal Istanbul which has become one of Turkey's most ambitious and controversial infrastructure ventures in recent years due to its potential environmental consequences.

First proposed by President Erdogan in 2011, the artificial waterway is designed to link the Black Sea with the Sea of Marmara and onto the Mediterranean, offering an alternative route to the Bosphorus Strait for international shipping (Istanbul Policy Center 2022, p. 14). The canal is projected to cost approximately $15 billion and will feature a depth of about 25 kilometers and width ranging from 250 meters to 1 kilometer to accommodate large vessels (TrtWorld 2018). The depth of the canal will be nearly 25 km (16 miles) and the width will change from 250 meters to 1 km allowing maneuvering space for the ships (TrtWorld 2018). While President Erdogan and his cabinet presents the canal as a symbol of Turkey's progress under President Erdogan's leadership, leveraging emotional appeal and historical references to bolster their position, Istanbul Mayor İmamoğlu as part of the opposition portrays the project as solely economic, benefiting only the current administration fueling skepticism and mistrust towards the ruling party. Initially proposed by Erdoğan in 2011, when his party also held the Istanbul mayoralty, Canal Istanbul faced growing opposition in 2019 after İmamoğlu made resistance to the project a centerpiece of his mayoral campaign.(Fig. 2).

President Erdoğan and the Ruling Justice and Development Party (AKP): The Government's Voice and Vision

President Erdoğan often presents his mega projects in Istanbul as central to Turkey's economic development. He frames these projects as a manifestation of his deep affection for the city, portraying himself as a visionary leader, representing a continuation of Istanbul's historical legacy. Proponents of the project view Canal Istanbul as crucial for the city's future transformation, opening a new page in the history of Turkey's development. A professor at Istanbul Commerce University and former AK Party parliament member Dr. Mustafa Ilicali, for instance, argues that the project represents a significant achievement with the potential to reshape

Istanbul. Ilicali underlines that "With this project, we will have undertaken the largest project in the world in the last decade" (TrtWorld 2018). He also contends that the canal will reduce the risks associated with large cargo ships and tankers navigating the Bosphorus, which currently pose threats to the city's historic sites (TrtWorld 2018). Ilicali adds that it will significantly contribute to the urban transformation of Istanbul and enhance the city's resilience to natural disasters such as earthquakes (TrtWorld 2018). But how these are going to be achieved by building a canal is not clearly explained.

Supporters of Canal Istanbul view the project as a transformative development, taking Istanbul's urban development within the last decade to the next level. They assert that the canal will improve accessibility and facilitate mobility, thus promoting urban and economic development. As the world relies more on global shipping and with the increased size of super tankers carrying raw materials and shipping containers, Canal Istanbul is promoted as a means to meet the needs of consumers and governments of Black Sea littoral states and benefit the Turkish economy (TrtWorld 2018). Additionally, the Turkish government views the project as a strategic initiative to prevent environmental disasters in the Bosphorus, an essential maritime corridor (TrtWorld 2018) (Fig. 3).

Fig. 3 Adapted from *Realities about Canal Istanbul video* by Turkish Ministry of Transportation and Infrastructure 2021

Mayor İmamoğlu and the Republican People's Party (CHP): Leading the Opposition Voice

Mayor İmamoğlu argues that the project could harm the ecology of not only Istanbul, but the entire Marmara Region and he has mobilized a coalition of experts and environmentalists to oppose the Canal Istanbul project, highlighting the potential environmental risks. Additionally, he filed a formal legal objection against the Canal Istanbul Project in February 2020 (Hincks 2020). He has criticized and questioned the government's inconsistent cost estimates, with Erdogan placing the project's cost at 75 billion Lira (approximately $12.4 billion), while some economists predict it could be twice as expensive (Hincks 2020). İmamoğlu argues that the canal is unnecessary, asserting that existing underground pipelines are a more efficient means of transporting hydrocarbons (Hincks, 2020). He also contends that rather than alleviating congestion in the Bosphorus, the canal would merely replicate existing hazards and create an island of eight million people, increasing the city's vulnerability to earthquakes (Hincks 2020). İmamoğlu has expressed concerns that the construction of the canal could increase earthquake risks, especially since Istanbul is located near a major fault line. Seismologists have warned that a significant earthquake could result in up to 30,000 casualties (Hincks 2020). İmamoğlu cites warnings from Turkish geologist Dr. Naci Gorur, who predicts that extensive excavations and the creation of islands from the excavated earth could worsen the region's seismic risks (Hincks 2020).

However, President Erdoğan asserts that the plans include building projects on both sides of the planned canal to accommodate around 500,000 people and incorporate earthquake precautions (Sozcu 2023). By contrast, İmamoğlu argues that creating a new development for 16 million people living in Istanbul would increase the city's vulnerability to earthquakes (Hincks 2020). İmamoğlu faced investigation due to his opposition to the Canal Istanbul project in 2020, with allegations that his resistance constitutes separatism against a state initiative (Euronews 2020) (Fig. 4).

Environmental Experts' Perspectives: Weighing In on Key Issues

Environmental experts have raised serious concerns about the project's potential impact. Critics argue that the canal will redraw the map of one of Europe's biggest cities, turning Istanbul's western side into an island, resulting in severe ecological damage. (TrtWorld 2018). They express concerns that the Canal may exacerbate deforestation in Istanbul's northern areas, which are crucial for maintaining the city's air quality (TrtWorld 2018). Afik Burak Atlar of Istanbul's Chamber of Urban Planners says that "Istanbul's northern forests "are the city's lungs" (TrtWorld 2018). Cemal Saydam, a professor of environmental engineering at Hacettepe University in Ankara comments that "there are two flows in the Bosphorus. At the bottom of the

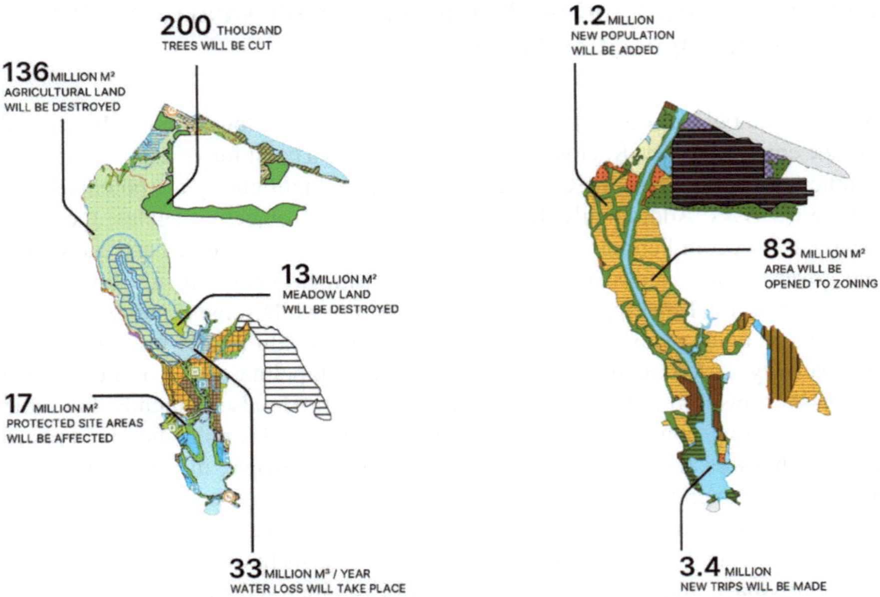

Fig. 4 Adapted from *Canal Istanbul Workshop Report* by Istanbul Metropolitan Municipality 2020

Bosporus there is a [denser] flow going north, from the Mediterranean to the Black Sea, and at the surface another flow coming south." The Union of Chambers of Turkish Engineers and Architects (TMMOB) warns that "The canal will demolish two basins that provide nearly a third of Istanbul's fresh water" (TrtWorld 2018).

The Environmental Impact Assessment approved in January 2020 revealed that the canal would disrupt a 25-year-old dam near Sazlibosna village, which is crucial to an ecosystem providing nearly 30% of Istanbul's water supply (Hincks 2020). Hydrologists have warned that the canal could alter the depths of the connected seas and disrupt Bosphorus currents, which currently balance the cold freshwater of the Black Sea with the warmer, saltier waters of the Sea of Marmara (Hincks 2020). Prominent oceanographer Cemal Saydam has cautioned that the canal might deplete the Black Sea and introduce polluted water into the Sea of Marmara and the Mediterranean, negatively affecting marine life and disrupting the delicate balance of sea currents and temperatures (Hincks 2020). The introduction of saltier Mediterranean water into the less saline Black Sea could alter sea currents and temperatures and would potentially harm marine life. Saydam's research indicates that such changes could reduce oxygen levels in the Sea of Marmara and adversely impact the Black Sea's ecological balance (Hincks, 2020). Water management expert Akgun Ilhan has also highlighted the risk of all groundwater to be contaminated by salty water as an irreversible process (TrtWorld 2018).

The project also received criticism and backlash from women's groups in Turkey. Women's groups argue that the project will cause the relocation of thousands of residents and the destruction of neighborhoods, which disproportionately affects women, particularly those from lower-income backgrounds. The social and economic impacts on these communities, including the disruption of community ties and access to essential services will be a major one, which will exacerbate social inequalities, including gender-based disparities ("Kanal İStanbul'a Itaat Yok, Isyan Var! – Sezen ÖZkan").

Conclusion: What Is Next?

The devastating earthquake that struck southeastern Turkey and northwestern Syria, with a magnitude of 7.8 on February 6, 2023, caused extensive damage in multiple cities, including Gaziantep, Kahramanmaraş, and Şanlıurfa, leading to widespread destruction of buildings and infrastructure. The disaster resulted in thousands of casualties and left many people homeless. The recovery process is ongoing and challenging, requiring substantial resources and international support. A year later on February 13, 2024, a month before local elections—including for Istanbul where Ekrem İmamoğlu was seeking re-election—the Istanbul 11th Administrative Court canceled the development plan of Turkish government's Canal Istanbul Project, a plan to build an artificial waterway parallel to the Bosphorus (Duvar English 2024). The court finalized the lawsuit filed by the İmamoğlu-governed Istanbul Metropolitan Municipality regarding the zoning plan of Canal Istanbul Project andcancelled the zoning plan amendment made by the Environment, Urbanization, and Climate Change Ministry regarding the Yenişehir Reserve Building Area (Kanal Istanbul Project) Phase 1. Announcing the court's ruling, İmamoğlu underlined that a "big mistake" was made in the Canal Istanbul Project and argued that, "This decision is not final as it is an administrative court decision. The requirements of the decision were asked to be fulfilled within 30 days. It is a judgment proving a great mistake has been made in the Canal Istanbul Project" (Duvar English 2024).On the other hand, the timing of the decision was questioned, with critics suggesting it was driven more by political motives tied to the upcoming election than by a genuine reflection of an effective and just legal system grounded in sound, rational principles.

The ongoing power struggle between the opposition-led Istanbul Municipality and the country's ruling party along with the difficulty of implementing local decisions without central government cooperation stands as the main challenge for climate crisis in the context of Turkey. İmamoğlu, along with climate activists, non-governmental organizations, and the chamber of architects strongly opposes the Canal due to its potential to cause significant environmental damage to Istanbul and the Marmara region of Turkey. Yet, the central government still signals that they are committed to the project. The project's potential to disrupt ecosystems, exacerbate deforestation, and alter marine environments raises high concerns. Domestically, the Canal has become a contentious political issue. The weakening of civil society and grassroots organizing since 2016 has limited the scope for activists to influence this proposal, leaving only a handful of academics and civil society members to voice concerns and advocate for more sustainable practices. There has been a heavy reliance on political action and activism from the opposition party and İmamoğlu himself. It is unclear if İmamoğlu's opposition to the project was a driver in earning votes from constituents concerned about protecting the environment, or if opposition to the Canal was a political talking point, namely, to criticize the ruling party. Meanwhile, the political tensions between the local and central authorities have diminished tangible efforts to combat climate change. If these issues are not addressed effectively, Istanbul—and the Bosphorus bottleneck—could become increasingly vulnerable, worsening the challenges faced by its communities and lagging the global trend towards smart, resilient cities.

On the other hand, the necessary changes in city diplomacy are still lacking in Turkey as the current government has pushed for further centralization, especially in city governance. In such cases, cities such as Istanbul that are governed by opposition parties struggle to access the required resources for an effective governance structure. To overcome this limitation, they must seek new channels of cooperation other than from the national authorities. For instance, İmamoğlu explored new possible forms of cooperation with European counterparts who have effectively implemented climate policies within their city governance structures. İmamoğlu's new strategy is to establish closer relations and an extended network with its European partners, especially in the field of climate crisis and solutions. Mayor İmamoğlu also actively participated in global climate related summits, including the United Nations Climate Conference, European Climate Summit, C40 cities and "Mayors' Summit for Ecological Transition." İmamoğlu, along with a group of climate experts who share his concerns, is increasing engagement with transnational climate justice networks, particularly through "city diplomacy" efforts in Europe, aiming to integrate Istanbul into broader European climate dialogues and exchange best practices on urban sustainability. For example, C40 Cities is a network of the world's megacities committed to addressing climate change and driving urban action that leads to a healthier and more sustainable future. Mayor İmamoğlu, has been actively involved with C40 Cities as part of his broader efforts to tackle climate change and improve sustainability in Istanbul. He participated in the C40 Cities Climate Leadership Summit in 2023, held in Buenos Aires and engaged in discussions with other global city leaders about innovative solutions and commitments to address climate challenges. Therefore, Istanbul has been involved in various C40 initiatives focused on sustainability, such as improving urban mobility, reducing greenhouse gas emissions, and enhancing resilience to climate impacts (C40 2024). As part of these efforts, the Istanbul Metropolitan Municipality emphasizes their Istanbul Climate Change Action Plan, a key component of their Istanbul 2050 Vision, as a major effort to tackle the climate crisis through sustainable urbanization. They assert that this plan represents the country's most comprehensive and human-centered approach, initiating a full-scale climate change mobilization (Istanbul Metropolitan Municipality 2022). The plan aims to achieve a carbon-neutral city resilient to climate impacts by 2050, with reduction targets set as follows:

- 52% absolute reduction by 2030
- 89% absolute reduction by 2040

- 100% absolute reduction by 2050 (Istanbul Metropolitan Municipality 2022).

The report identifies Canal Istanbul as a significant threat to meeting these 2050 goals. It argues that the project will disrupt the ecological balance, damage centuries-old cultural heritage, and increase disaster risks. The report claims that the canal will undermine urban rights and negatively impact the quality of life for people and the environment. In essence, the local government identifies the project as not only an urban but also a regional and global climate threat, potentially undermining progress in combating climate change (Istanbul Metropolitan Municipality 2022).

While these efforts appear promising and suggest long-term, sustainable governance models, the extent of İmamoğlu's influence and the effectiveness of his strategy for managing the tension between local and central government through his European connections remains uncertain. By aligning with European democratic values, İmamoğlu's strategy to put pressure on President Erdogan's government on climate related issues might backfire. At the same time, his initiatives bring back Turkey's complex relationship with Europe into focus under President Erdoğan's leadership. This could revive political discussions about "Europeanization" and the influence of Western pressures on the central government.

References

Banik D (2022) Democracy and sustainable development. Anthropocene Sci 1(2):233–245. https://doi.org/10.1007/s44177-022-00019-z

Bloomberg (2022) Turkey's biggest city falls victim to the Cold War with Erdogan. Bloomberg. https://www.bloomberg.com

C40 Cities (2024, June 25). A global network of mayors taking urgent climate action. C40 Cities https://www.c40.org/

Canal Istanbul Workshop (2020) Kanal Istanbul Calistayi [Digital report]. https://calistay.ibb.istanbul/wp-content/uploads/2020/07/KanalIstanbulCalistayi_Dijital_EN.pdf

Dasgupta S, Lall S, Wheeler D (2024, March 16) Cutting global carbon emissions: Where do cities stand? World Bank Blogs. https://blogs.worldbank.org/en/sustainablecities/cutting-global-carbon-emissions-where-do-cities-stand

Duvar English (2024, February). Turkish court cancels zoning plans of Kanal Istanbul. Duvar English. https://www.duvarenglish.com

Euronews (2020, November 16) İmamoğlu'dan Kanal İstanbul soruşturmasına ilişkin açıklama: Benim için devlet projesi değil. Euronews. https://tr.euronews.com

Hincks, Joseph. "The Multibillion Dollar Canal Carving a Rift Through Erdogan's Turkey." TIME, 14 Feb. 2020, https://time.com/5783560/canal-istanbul-erdogan-imamoglu-infrastructure/

Istanbul Metropolitan Municipality (2020) Canal Istanbul workshop report (p. [page number]). BİMTAŞ. https://calistay.ibb.istanbul/wp-content/uploads/2020/07/KanalIstanbulCalistayi_Dijital_EN.pdf

Istanbul Metropolitan Municipality (2022) Istanbul climate report. https://cevre.ibb.istanbul/wp-content/uploads/2023/04/2022-iklim-izleme-raporu-eng-1.pdf

Istanbul Policy Center (2022) Overcoming global problems through local cooperation: The case of the Istanbul Metropolitan Municipality. https://ipc.sabanciuniv.edu/Content/Images/CKeditorImages/20220707-12073921.pdf

Kirişci K (2021, April 28) Canal İstanbul: Erdoğan's canal and the trench of protest. Foreign Policy. https://foreignpolicy.com/2021/04/28/canal-kanal-istanbul-erdogan-channel-trench-protest/

Mersin K, Bayırhan İ, Yıldırım M (2024) Possible effects of the Kanal Istanbul Waterway Project on ship emissions: challenges and prospects. J Coast Res 40(1):210–222. https://doi.org/10.2112/JCOASTRES-D-22-00094.1

Pehlivanli E, Eslen-Ziya H (2023) A political game?: The dispute over the Kanal Istanbul project. Turk Stud 25(1). https://doi.org/10.1080/14683849.2023.2262090

Poushter F, Fagan M, Gubbala S (2024, April 14) Climate change remains the top global threat across 19-country survey. Pew Research Center. https://www.pewresearch.org/global/2022/08/31/climate-change-remains-top-global-threat-across-19-country-survey/

Republic of Türkiye (2023) Updated first nationally determined contribution. UNFCCC. https://unfccc.int/node/627744

Sozcu Gazetesi. (2023, August). President Erdoğan'ın Kanal İstanbul inadı! Deprem sorusuna ilginç yanıt. Sozcu. https://www.sozcu.com.tr

TIME (2020, February 14) The multibillion-dollar canal carving a rift through Erdogan's Turkey. TIME. https://time.com

TrtWorld (2018, October 16) Turkey's Istanbul canal project explained. TrtWorld. https://www.trtworld.com

Tsang BT, Osornio JP (2024, March) Autocracy vs. democracy: climate edition. Carnegie Endowment for International Peace. https://carnegieendowment.org/research/2024/03/autocracy-vs-democracy-climate-edition?lang=en

Turkish Ministry of Transportation and Infrastructure. 2021. Realities about Canal Istanbul. https://www.youtube.com/watch?v=0KmvYFMx214

United Nations Development Programme (2024) Climate finance. UNDP. Retrieved September 6, 2024, from https://www.undp.org/romecentre/focus-areas/climate-finance#:~:text=It%20is%20estimated%20that%203.6,4%25%20of%20global%20carbon%20emissions

Westell A (2023, January) Exploring the tensions: the relationship between democracy and sustainable development. FDSD. https://www.fdsd.org/wp-content/uploads/2023/03/The-Relationship-between-Democracy-and-Sustainable-Development104461.pdf

Senay Imre is an academic, researcher, policy advisor, and strategic leader at the forefront of gender equality and women's empowerment, driving sustainable and transformative change both locally and globally. Her work bridges academic research and real-world impact, with a focus on the intersection of gender and sustainable development. She is a PhD candidate in Global Gender Studies at the State University of New York (SUNY) at Buffalo, where she also serves as an adjunct professor, teaching across topics of women's studies, international relations, and political science. Her research interests examine women's economic participation, social and cultural movements, and the disproportionate impact of climate change on women and marginalized communities. Through this lens, she works to reshape dominant narratives and develop inclusive, actionable policy solutions. Senay holds a Master's degree in International Relations with a concentration in Development from The New School for Social Research, and a Bachelor's degree in International Relations from Marymount Manhattan College in New York. She has conducted field research in Turkey, India, and China, analyzing the legal, political, and cultural structures that perpetuate gender-based violence and restrict women's full participation in society. In her role as a strategic advisor to municipal leaders, NGOs, and international stakeholders, she has led initiatives that promote women's political, economic, and social empowerment. Her policy work has informed global dialogues at the United Nations General Assembly and the Commission on the Status of Women, where she has represented the Federation of Women's Associations of Turkey. A Ford Foundation Fellow and Clinton Global Initiative Fellow, Senay is deeply committed to ensuring that gender equity remains central to both scholarly inquiry and policy-making.

The Impact of Climate Change on Women's Rights in Sri Lanka: A Critical Review

Muthukuda Arachchige Dona Shiroma Jeeva Shirajanie

Abstract This study examines how climate change affects women and their human rights in Sri Lanka. The research study aims to identify the problems Sri Lankan women encounter due to climate change, how their human rights—civil rights including socio-economic rights are violated as victims of climate change and evaluate the adequacy of the current legal frameworks concerning the protection of the rights of women victims of climate change effects. To reach the said goal this research endeavors to appraise the following. Climate change impacts differ by gender and specific challenges that women encounter due to climate change; existing international legal instruments address the intersection of climate change issues and gender equity; the adequacy of national laws in Sri Lanka to protect the rights of women from the adverse effects of climate change; role of women in climate change mitigation and adaptation with the legal recognition; and suggestions for improving the protection of the rights of women those who encounter the adverse climate change effects by incorporating the relevant international standards. The methodology includes a rigorous review of the existing literature in the subject area and content analysis. This study aims to deliver actionable insights for policymakers and legal professionals to improve legal (human rights) protections and assistance for women who suffer climate change effects (the findings and the observations).

Keywords Climate change · Gender equality · Legal frameworks · Women's rights · Sri Lanka

Muthukuda Arachchige Dona Shiroma Jeeva Shirajanie (✉)
Department of Public and International Law, Faculty of Law,
University of Colombo, Colombo, Sri Lanka
e-mail: jeeva@law.cmb.ac.lk

B. Wejnert (ed.), *Democracy in Retreat and the Climate Crisis*, Springer Studies on Populism, Identity Politics and Social Justice,
https://doi.org/10.1007/978-3-032-02881-5_9

Introduction

Climate change offers significant and far-reaching global challenges, with its effects and impacts falling unequally on certain vulnerable populations—particularly women, children, the elderly, persons with disabilities, and trans persons (Niriella 2025). This study specifically investigates the ways in which climate-induced consequences infringe upon the civil rights of women in Sri Lanka and critically evaluates whether the country's existing legal system is adequately equipped to safeguard those rights under escalating climate stress. The study delves into the capacity of prevailing national legislation to address the gender-specific risks and disadvantages that arise from climate-related disruptions. It seeks to identify the distinctive hardships encountered by women as a result of climate change (environmental changes) and evaluate the strength and inclusiveness of legal protections, emphasizing the principle of non-discrimination. By conducting an in-depth analysis of available academic and policy literature, alongside a detailed examination of legal content, the study intends to offer informed and practical recommendations to improve legal structures for better protecting and empowering women in Sri Lanka in the face of climate change.

Climate Change: Gender-Specific Implications and Legal Challenges for Women

Climate change establishes an inevitable and pervasive worldwide phenomenon, exerting profound and multifaceted impacts across diverse dimensions of human existence, including but not limited to public health, economic security, and social cohesion (Niriella 2025). However, the consequences of this environmental crisis are not homogeneously distributed; rather, they manifest differently depending on a range of interconnecting determining factors—most notably, gender. In this context, women—especially those living in developing countries like Sri Lanka—bear an uneven portion of climate-related burdens. This inequality/discrepancy is largely attributable to ingrained gender roles, socio-economic susceptibility, unequal access to critical resources, and existing discriminatory cultural norms. This section of the study aims to rigorously scrutinize the deleterious consequences of climate change within the Sri Lankan context, focusing specifically on the gendered dimensions of environmental adversity. It highlights the differentiated impacts experienced by women and identifies the distinctive risks, vulnerabilities, and legal challenges that emerge as a result. Ultimately, the analysis seeks to underscore the urgent need for gender-responsive legal and policy frameworks capable of mitigating these inequities in the face of ongoing climatic disruptions.

Climate Change Adverse Effects

As of the beginning of 2024, Sri Lanka's estimated population was approximately 23.1 million, according to United Nations data. The population has a slight annual growth rate of around 0.61%, with projections suggesting it may remain stable for the next decade. The gender distribution at the beginning of 2024 was males—10,756,287 and females—11,044,777. This results in a sex ratio of about 0.974, meaning there were approximately 974 males for every 1000 females.

Sri Lanka, as a developing island nation, is particularly vulnerable to the adverse impacts of climate change, affecting its environment, economy, and population. In essence, climate change issues detrimentally impact the three primary pillars of sustainable development—environmental protection, social development, and economic development in Sri Lanka. The following discussion addresses the adverse effects of climate change in Sri Lanka.

(i) **Sea-Level Rise and Coastal Erosion**: Rising sea levels threaten coastal communities and ecosystems, with the potential to submerge low-lying areas, erode beaches, and increase saline intrusion into freshwater resources. This affects agriculture, drinking water supplies, and rich biodiversity in the country. The coastline (coastal belt) of Sri Lanka is experiencing considerable erosion due to rising sea levels. Studies show that approximately 40% of the western and southern coasts are affected by coastal erosion. For instance, in the city of Negombo, it is estimated that the coastline receded by about 0.5 meters annually between 2010 and 2020, impacting fisheries and tourism (Ranasinghe et al. 2022).

(ii) **Extreme Weather Events**: Sri Lanka has experienced more frequent and intense weather events (Department of Meteorology, Sri Lanka, 2020; Disaster Manangement Centre of Sri Lanka, (n.d.); Ministry of Heath Sri Lanka, 2016), such as floods, droughts, and cyclones. Flooding often impacts urban areas and agricultural regions, while droughts harm water supply and food production, affecting both rural and urban populations. According to the Disaster Management Centre of Sri Lanka (DMC), the 2017 floods and landslides affected over 630,000 people across the 15 districts, caused 213 deaths and led to a financial loss exceeding USD 197 million. Droughts have similarly impacted large populations, with the 2016–2017 drought affecting approximately 900,000 people across multiple districts (DMC, 2017). The severe drought from 2016 to 2017 impacted nearly one million people and led to a 35% reduction in rice production in the dry zone regions. The drought led to shortages of potable water, especially in the Northern, North Central, and Eastern provinces, putting additional strain on communities that rely on agriculture as their primary source of income (World Bank, 2018). Cyclones and storms have become more intense, with storm surges significantly damaging coastal communities. Cyclone Burevi, which hit Sri Lanka in 2020, affected over 70,000 people across multiple districts. The storm caused considerable damage to homes, infrastructure, and crops, with the government allocating

LKR 10 billion for immediate disaster relief and infrastructure repair in affected areas (National Disaster Relief Services Centre 2020). Increasing temperatures and heatwaves have become more frequent and intense in Sri Lanka, especially affecting urban areas. In 2019, a sudden spike in temperatures in Colombo reached over 37 °C, leading to a surge in heat-related illnesses, including heat exhaustion and dehydration, particularly among the elderly and children. The Meteorological Department of Sri Lanka recorded an average increase of approximately 0.01 °C annually, contributing to increased heat stress and associated health risks (Department of Meteorology, Sri Lanka, 2020). Sudden weather changes disrupt Sri Lanka's agricultural cycle, which (heavily) relies on predictable weather patterns. The combined impact of floods and droughts has caused significant crop losses, with a 10% drop in tea and rice production (Performance of the Export Sector of Sri Lanka, 2016) in recent years due to extreme weather fluctuations. In 2016 alone, drought-related losses in agriculture were estimated at USD 120 million, affecting (both) national food security and export revenue (Central Bank of Sri Lanka, 2016).

(iii) **Temperature and Rainfall Variability**: Fluctuations in rainfall patterns and rising temperatures have disrupted traditional agricultural cycles, impacting rice production and cash crops like tea, rubber, coconut, and spices. Unpredictable weather also threatens food security and rural livelihoods. Data from the Meteorological Department of Sri Lanka show that average temperatures in Sri Lanka have increased by about 0.6 °C over the past 50 years. The variability in rainfall has disrupted agriculture, particularly in the rice and tea sectors. For example, a 10% decrease in tea yield was observed during a drought in 2016, reducing Sri Lanka's export earnings by USD 80 million that year (Sri Lanka Tea Board, 2017).

(iv) **Loss of Biodiversity**: Climate change is pushing endangered Sri Lanka's rich biodiversity, with ecosystems like rainforests, coral reefs, and mangroves at risk. Species adapted to specific climate conditions may face extinction while changing temperatures and precipitation patterns affect habitats and breeding cycles. The World Conservation Union (IUCN) highlights that climate change causes a severe threat to Sri Lanka's biodiversity. For instance, coral bleaching events in 2016 resulted in the death of around 60% of the corals in some areas, notably the Hikkaduwa Marine National Park. This loss affects marine biodiversity and local economies reliant on tourism (IUCN Sri Lanka 2018).

(v) **Health Risks**: Rising temperatures and extreme weather patterns cause an increase in vector-borne diseases, such as dengue fever, and heat-related illnesses. Flooding can also contaminate water sources, posing risks of waterborne diseases. Dengue cases have risen substantially in recent years, with climate variability as a contributing factor. According to the Ministry of Health, Sri Lanka, there were over 186,000 cases of dengue fever in 2017, the highest recorded to date. This surge has been linked to changes in rainfall patterns, which create breeding grounds for mosquitoes (Ministry of Health 2017).

(vi) **Economic Impacts**: Sectors like agriculture, fisheries, and tourism—critical to Sri Lanka's economy—are severely affected by climate change. Losses in crop yield, fisheries production, and tourism revenues threaten national income and local livelihoods. Dengue cases have risen substantially in recent years, with climate variability as a contributing factor. According to the Ministry of Health, Sri Lanka, there were over 186,000 cases of dengue fever in 2017, the highest recorded to date. This surge has been linked to changes in rainfall patterns, which create breeding grounds for mosquitoes (Ministry of Health 2017).

(vii) **Water Scarcity**: Changes in rainfall patterns, coupled with higher evaporation rates, reduce freshwater availability. This exacerbates water scarcity in drier regions, impacting agriculture, drinking water, and sanitation. The Mahaweli Authority notes that water availability for irrigation has decreased in the country's Dry Zone, impacting rice production. During the 2016–2017 drought, water levels in reservoirs fell by over 40%, resulting in a 35% reduction in agricultural output in certain areas (Mahaweli Authority, 2017).

Challenges on Women Due to Climate Change Adverse Effects

Sri Lankan women face unique and disproportionate challenges due to the adverse impacts of climate change which finally leads to violation of their (human) rights. These impacts span economic, health, social, and safety dimensions, with significant repercussions for women, especially those in rural and coastal communities. Here are some specific examples and statistics that illustrate how Sri Lankan women are particularly vulnerable to climate-related effects and finally violate their rights:

(i) **Increased Workload and Economic Strain**: In rural Sri Lankan communities, where agriculture (Sri Lanka Central Bank Report, 2016) is a primary livelihood, women often bear the brunt of reduced crop yields due to climate events. The severe drought in 2016–2017, which affected nearly 1 million people, had severely impacted women in farming households. This drought led to a 35% decrease in rice production, resulting in food insecurity and income loss, and pushing many women into informal, low-paying labor to support their families. Studies indicate that women in these areas spend up to 40% more (time) on agricultural and household tasks during droughts (United Nations Development Programme 2018).

(ii) **Health Risks**: Climate-induced health issues, like the increase in waterborne diseases, have particularly affected women, who are typical primary caregivers in families. Flooding events, such as those in 2017, led to widespread water contamination, which increased incidences of waterborne diseases like dysentery and cholera. Women, especially pregnant women and young mothers were disproportionately affected due to their role in collecting water and caring for sick family members. During the 2017 floods, approximately 53% of those hospitalized with waterborne illnesses were women (Ministry of Health, Sri Lanka 2017). As discussed, earlier research in Sri Lanka does show that the impacts of climate change are felt disproportionately by women, especially in vulnerable communities. However, studies specifically linking climate change effects to menstrual health in Sri Lanka are limited. Broadly, climate impacts—such as extreme weather events, water scarcity, and increased temperatures—can worsen health conditions for women, affecting their ability to manage menstrual hygiene due to reduced water availability and heightened risks in accessing healthcare services in rural areas. These challenges are exacerbated by existing cultural taboos surrounding menstruation, which hinder discussions on menstrual health needs, making climate-resilient healthcare and support for menstrual hygiene management a pressing issue. Efforts by organizations like UNFPA Sri Lanka focus on addressing menstrual health and endometriosis through public awareness and reducing stigma around reproductive health. These initiatives, however, emphasize general menstrual health rather than climate-specific impacts, which suggests a gap in research specifically addressing climate-related menstrual health challenges in Sri Lanka.

(iii) **Vulnerability to Displacement and Loss of Safety**: Women in coastal areas are highly vulnerable to sea-level rise and coastal erosion, which displace families and disrupt livelihoods. For instance, following Cyclone Burevi in 2020, thousands of people, predominantly women and children, were displaced. Displaced women faced limited access to sanitation facilities and a heightened risk of gender-based violence in overcrowded temporary shelters. The National Disaster Relief Services Centre reported that women constituted 65% of those who sought shelter in temporary camps during and after Cyclone Burevi (National Disaster Relief Services Centre 2020).

(iv) **Loss of Livelihoods for Coastal Women**: Many Sri Lankan women rely on fisheries and coastal tourism for their livelihoods. With increased coastal erosion and coral bleaching—approximately 60% of corals in areas like Hikkaduwa were affected in 2016—female fish processors and tourism service providers experienced severe income loss. This economic hardship forced many women into alternative, often informal, labor, exacerbating their financial instability and social vulnerability (IUCN Sri Lanka, 2018).

(v) **Water Scarcity and Increased Domestic Burden**: Women are primarily responsible for water collection in many rural Sri Lankan households, and climate-related droughts have made this task significantly more challenging. During the 2016–2017 drought, women in affected areas spent an average of 4 to 5 h daily fetching water from distant sources, which limited their ability to engage in other income-generating activities or education. This increased workload often leads to physical exhaustion and health issues, such as musculoskeletal pain and fatigue (World Bank, 2018).

Specific Rights of Women Violated Due to the Climate Change Adverse Effects

The adverse effects of climate change in Sri Lanka significantly impact women's rights, leading to potential violations of several fundamental rights enshrined in the Constitution of Sri Lanka. Here are some specific rights that may be violated (as I argue) due to the challenges faced by women:

(i) **Right to Equality (Article 12)**: Women may face discrimination in access to resources, livelihood opportunities, and decision-making processes. For instance, during climate-related disasters like floods or droughts, women often receive less assistance and support compared to men, violating their right to equal treatment and opportunities.

(ii) **Right to Life (Article 11)**: The impacts of climate change, such as increased health risks from waterborne diseases and displacement due to natural disasters, can threaten the lives of women. The lack of adequate safety measures in temporary shelters and insufficient healthcare can result in increased mortality rates among vulnerable populations, infringing on the right to life.

(iii) **Right to Health (Article 13)**: Women's health is adversely affected by climate change, particularly through increased exposure to waterborne diseases and the burdens of care during health crises. This violates their right to attain the highest standards (quality) of physical and mental health, as defined in the Constitution.

(iv) **Right to Privacy (Article 14(1)(g))**: Displacement due to climate events often forces women into temporary shelters where privacy is compromised. Such conditions can increase their vulnerability to gender-based violence and exploitation, infringing upon their right to privacy and security.

(v) **Right to Education (Article 27)**: The increased workload on women due to climate-induced hardships—such as fetching water or engaging in informal labor—can limit their ability to pursue education. This lack of educational opportunity can perpetuate cycles of poverty and gender inequality, violating their right to education.[1]

---

[1]Article 27 of the Constitution of Sri Lanka is part of the *Directive Principles of State Policy and Fundamental Duties* under Chap. VI. This article outlines the responsibilities and guiding principles of the State, although these are not enforceable by courts. Article 27 emphasizes the following key objectives:

1. Promotion of National Welfare: It encourages the State to pursue social, economic, and cultural development aimed at improving the welfare of its citizens.
2. Promotion of Social Justice and Economic Equity: The article calls for reducing social and economic inequalities and fostering a fair distribution of wealth.
3. Provision of Fundamental Rights and Freedoms: The State is directed to ensure that basic rights, such as adequate shelter, food, and clothing, are accessible to all citizens.
4. Protection of the Environment: It emphasizes the importance of preserving the environment and conserving natural resources.

(vi) **Right to Participate in Public Life (Article 27)**: Climate change often marginalizes women in decision-making processes related to disaster management and resource allocation. Their exclusion from these processes infringes on their right to participate in public life and governance, limiting their influence in shaping policies that affect their lives.

(vii) **Right to Property (Article 28)**: Women, particularly in rural areas, may lose access to land and property due to climate change impacts like erosion and flooding. Displacement and loss of livelihoods can undermine their economic independence and property rights, leading to further marginalization.[2]

These rights violations highlight the need for gender-sensitive approaches to climate action in Sri Lanka, ensuring that women's voices are heard, and their rights are protected in the face of climate change challenges. Therefore, it should be pertinent to consider the international framework on climate change.

Nexus Between Climate Change-Related International Instruments and Rights of Women

Climate change embodies a formidable transnational challenge with extensive complications, disproportionately affecting the sections of the population who are burdened with structural vulnerabilities. Among those sections' women are said to be the most susceptible. Recognizing the inherent link between climate issues and gender-based differences, the global community has progressively recognized the necessity of integrating gender perspectives into universal climate governance (Niriella 2025). As a result, several binding and non-binding international

---

5. Promotion of National Unity and Integrity: It encourages fostering unity among diverse ethnic, linguistic, and religious communities.

Article 27 serves as a moral and policy guide for governance in Sri Lanka, outlining the State's commitment to ensuring balanced social, economic, and environmental development, with a focus on public welfare and justice.

[2] Article 28 of the Constitution of Sri Lanka, found within Chap. VI under the *Directive Principles of State Policy and Fundamental Duties*, sets out the fundamental duties of all Sri Lankan citizens. It mandates that every citizen has the duty to:

1. Preserve and protect public property and to combat misuse and waste of public resources.
2. Promote harmony and unity among all ethnic, religious, and linguistic groups in Sri Lanka.
3. Work conscientiously in their chosen occupations, while upholding the dignity of labor.
4. Protect nature and conserve its riches.

Though Article 28 is not legally enforceable by itself, it serves as a guiding framework, emphasizing the ethical and moral responsibilities expected from citizens, complementing the rights and freedoms provided by the Constitution. This article reflects the spirit of civic responsibility and aims to foster a society where citizens actively contribute to national unity, environmental protection, and social welfare.

instruments and mechanisms have started to incorporate provisions aimed at addressing and protecting the rights and interests of women within the broader context of climate action. This section undertakes a critical evaluation of the efficiency of such frameworks in upholding and promoting women's rights, with special reference to initial agreements and institutional procedures. Particularly, women in developing countries such as Sri Lanka are disproportionately dependent upon natural ecosystems and agricultural resources for existence and earnings. Despite this dependence, they come across intrinsic limitations in accessing financial capital, educational opportunities, and participatory decision-making platforms, thereby heightening their exposure to the detrimental effects of climate-induced phenomena such as crop failure, water shortages, and forced migration. Accordingly, this analysis will examine chief international climate agreements that not only advance environmental purposes but also set normative standards aimed at encouraging gender equity and the legal protection of women in the face of a changing climate (Niriella 2025).

United Nations Framework Convention on Climate Change (UNFCCC)

UNFCCC has, over time, progressively recognized the unequal and complex impacts of climate change on women, particularly in vulnerable and marginalized communities. It has concurrently recognized the need of addressing gender-specific susceptibilities and upholding the rights of women as a critical component of effective and equitable climate action. The efficacy of the UNFCCC in this regard can be examined through several interrelated dimensions.

One of the most noteworthy progressions under the UNFCCC background is the elevation of gender-balanced representation and inclusive participation within climate negotiations and decision-making forums. Efforts have been made to escalate the existence of women in national delegations to the Conference of the Parties (COP) and other subsidiary bodies, as well as to enhance their roles in leadership and negotiation positions. These initiatives aim to ensure that women's perspectives, experiences, and needs are adequately reflected in global climate policy. The Lima Work Programme on Gender (2014) and the enhanced Gender Action Plan (2019) have further institutionalized commitments to gender balance and women's participation across all levels of climate governance.

Align with its gender mainstreaming objectives, the UNFCCC facilitates a wide range of capacity-building initiatives designed to empower women as informed and active agents of change within their communities. These programs incorporate specialized training modules, technical workshops, and access to educational resources, all intended to enhance women's skills in climate change adaptation, mitigation strategies, and environmental governance. Such efforts are essential in equipping women with the skills and knowledge required to engage meaningfully in decision-making processes, both locally and universally.

The UNFCCC has taken significant steps in introducing Gender-Responsive Climate Finance. A notable example is the **Green Climate Fund (GCF)**, which has adopted a comprehensive **Gender Policy and Action Plan** to ensure that gender equality is integrated into the entire project cycle—from design and planning to implementation and monitoring. This framework mandates gender assessments and action plans for all funded projects, aiming to guarantee **equitable access to climate finance and benefits for women** and to lessen any unintentional gender-related harm. By embedding gender responsiveness into climate finance mechanisms, the UNFCCC seeks to **rectify structural inequalities** and improve the effectiveness of climate interventions.

Recognizing the importance of informed policymaking, the UNFCCC encourages the systematic collection and analysis of sex-disaggregated data as well as gender-sensitive indicators. This data is instrumental in revealing the differentiated impacts of climate change on women and men and is critical for designing and implementing targeted policies that address the unique needs, capacities, and vulnerabilities of women and girls. The availability and use of such data contribute to more transparent, accountable, and equitable climate governance.

A growing number of programs and projects implemented under the UNFCCC framework now explicitly incorporate gender perspectives into their design and execution, particularly in the context of adaptation and resilience-building activities. These projects commonly focus on enhancing women's access to critical resources such as land, water, clean energy, and climate-resilient technologies. Moreover, they aim to promote information sharing, sustainable livelihood development, and access to early warning systems, which are vital to strengthening community resilience to climate-induced shocks and stresses. By addressing the socio-economic and cultural barriers that hinder women's full participation, these projects promote inclusive, community-led adaptation approaches.

The adoption and implementation of the Gender Action Plan (GAP) represent a landmark achievement in institutionalizing gender equality within the UNFCCC. The GAP outlines concrete activities and priority areas to advance the integration of gender considerations across all aspects of climate policy and action. Its primary objectives include ensuring women's full, effective, and equal participation in climate decision-making, enhancing coherence between gender and climate policies, and improving accountability and resource allocation. The GAP also serves as a guiding document for Parties, UN entities, and civil society actors in aligning their climate efforts with gender equality principles.

Civil society organizations have played a pivotal role in promoting and operationalizing gender equality within the climate regime. The Women's Environment & Development Organization (WEDO) has been particularly influential in advocating for the integration of gender perspectives into international environmental and climate policies. Through its advocacy and grassroots initiatives, WEDO has provided a bridge between global negotiations and local realities.

In Sri Lanka, for instance, WEDO has actively supported the mainstreaming of gender into national climate strategies. It has facilitated the initiation of community-based climate adaptation projects, particularly in rural areas, where women are trained in climate-smart agriculture, sustainable water management, and other resilience-building practices. Additionally, WEDO conducts capacity-building workshops aimed at educating women on climate science, environmental governance, and rights-based approaches to adaptation. These programs empower women not only to protect their communities but also to engage confidently in local governance and policy development processes.

The Global Gender and Climate Alliance (GGCA) is an alliance of global entities committed to guarantee that climate policies, programs, and funding mechanisms are gender-sensitive and gender-equitable. It espouses a wide-ranging approach to entrenching gender considerations in climate governance through capacity building, technical support, and specialized training. The GGCA also promotes women-led climate initiatives, recognizing women as essential actors in environmental sustainability. It has conducted workshops to strengthen women's advocacy and resilience skills, facilitating their engagement in national climate policy processes. Furthermore, the Alliance supports community-level projects that enhance women's leadership in environmental conservation, advancing both gender equality and sustainable development (UNFCCC 2019).

UN Women's Climate Change Work supports countries in implementing gender-responsive climate policies. It enhances women's leadership in climate governance and ensures that climate finance mechanisms are accessible to women. UN Women helped to establish women's cooperatives focused on sustainable farming techniques, improving food security and resilience (UN Women 2020). UN Women has partnered with local organizations to provide women access to clean energy solutions, reducing their reliance on traditional biomass (UN Women 2020). UN As stated in Women 2019, women play a vital role in the solar energy sector (UN Women 2018). Initiatives like Barefoot College's Solar Electrification Program train women, particularly grandmothers, to become solar engineers. This empowers them with skills and provides sustainable energy solutions to their communities, addressing gender inequality and climate change (Chowdhury and Suleri 2021). Women's Climate Centers International (WCCI) operates across Africa to empower women through climate-smart agricultural practices and entrepreneurial initiatives. By delivering training and essential resources, WCCI enables women to adapt to climate change and enhance their economic resilience. WCCI, domestically in different countries introduced diverse strategies into climate policies and climate action plans enabling women to take part in those procedures. In Sri Lanka, UN Women advances gender-responsive climate action by promoting women's leadership and access to climate finance. Through partnerships with local entities, it supports women-led renewable energy ventures, contributing to sustainable development and economic empowerment. Additionally, in collaboration with government and civil society, UN Women integrates gender considerations into disaster risk reduction, bolstering the adaptive capacity of women and their communities.

Regardless of the advancement, challenges persist in effectively safeguarding women's rights under the UNFCCC, particularly due to gaps between policy and implementation. Cultural norms and systemic inequalities continue to hinder gender-responsive climate action. Moreover, inadequate cross-sectoral coordination limits effective responses to the complex, multidimensional impacts of climate change on women.

Paris Climate Agreement (Paris Accord)

The Paris Agreement (also referred to as the **Paris Climate Agreement** or **Paris Accord**) was formally adopted in 2015 under the auspices of the United Nations Framework Convention on Climate Change (UNFCCC). This landmark international instrument aims to strengthen the universal response to climate change by maintaining the rise in global average temperature well below 2 °C above pre-industrial levels, and by pursuing efforts to limit the temperature increase to 1.5 °C. It represents a legally binding framework for climate governance, emphasizing inclusive, equitable, and sustainable climate action.

Importantly, the agreement acknowledges that climate change impacts are not gender-neutral. Gender equality and the empowerment of women are critical to the effectiveness and fairness of climate responses. Its preamble explicitly acknowledges the importance of gender equality and women's empowerment in climate action. Furthermore, Article 7.5 specifically stipulates that adaptation efforts must adopt a gender-responsive, participatory, and transparent approach, guided by best available science and traditional knowledge systems.

Building on the gender commitments embedded within the Paris Agreement, the UNFCCC developed the Gender Action Plan (GAP) to facilitate the implementation of gender-responsive climate policies and actions. The GAP outlines key priority areas, including capacity-building, gender balance in climate governance, coherence across UN entities, gender-responsive implementation, and monitoring and reporting mechanisms. The GAP thus operationalizes the gender-related provisions of the Agreement and serves as a roadmap for integrating gender considerations across all levels of climate policy.

A central mechanism of the Paris Agreement is the requirement for Parties to submit Nationally Determined Contributions (NDCs)—self-determined national climate plans outlining mitigation and adaptation goals. In recent years, there has been notable progress in the mainstreaming of gender considerations into NDCs, with countries such as Sri Lanka explicitly acknowledging the differentiated impacts of climate change on women and affirming the necessity of their active participation in climate adaptation and mitigation. This trend reflects a growing awareness that gender inclusion enhances the effectiveness and sustainability of climate actions.

Empirical data further underscores this progress. As of December 2022, approximately 97.5% of National Adaptation Plans (NAPs), 77.1% of Adaptation Communications, and 75% of submitted NDCs contained explicit references to gender. Notably, the proportion of NDCs referencing gender rose from 35% in 2020 to 90% in 2021 and 2022, illustrating the increasing institutional recognition of gender dimensions in climate policy (UNFCCC, 2022).

The agreement also underscores the necessity for financial support, particularly for developing countries, to implement climate measures effectively. In this regard, gender-responsive climate finance is a critical component. Financial mechanisms such as the Green Climate Fund (GCF) have adopted gender policies to ensure that funded projects integrate gender perspectives and provide equitable benefits for women and men. These provisions aim to rectify systemic imbalances in access to financial and technical resources and promote inclusive sustainable development.

Despite these advances, significant challenges persist in fully realizing the gender-related objectives of the Paris Agreement. A principal concern is the implementation gap—the discrepancy between policy commitments and their actual realization on the ground. Translating high-level gender commitments into tangible benefits for women necessitates robust monitoring, evaluation, and accountability mechanisms. Additionally, deep-rooted socio-cultural norms and institutional barriers continue to restrict women's full participation in climate governance and limit their access to resources. Moreover, inconsistency across national contexts remains problematic. The uneven commitment of different countries undermines the global coherence and effectiveness of gender-responsive climate action under the Agreement.

A further impediment lies in competing national priorities and resource constraints. Many governments struggle to allocate adequate financial and institutional resources to gender-specific initiatives, thereby limiting the scalability and sustainability of such efforts. Additionally, the intersectionality of climate change—its impacts across sectors such as health, water, and agriculture—requires coordinated cross-sectoral action, which is often insufficient or absent.

Nevertheless, the Paris Agreement represents a pivotal moment in the evolution of international climate law by embedding gender equality as a core principle of effective climate action. The establishment of the Gender Action Plan, the increasing gender-inclusiveness of NDCs, and the incorporation of gender into climate finance mechanisms collectively illustrate a commitment to inclusive and equitable climate governance.

In conclusion, while the Paris Agreement has laid a strong normative and institutional foundation for integrating gender in climate policy, its ultimate success depends on sustained political will, adequate and equitable funding, and systemic reforms that address structural inequalities. Continued commitment to monitoring, reporting, and capacity-building, along with global solidarity, will be crucial to ensuring that the agreement delivers on its promise to protect the rights and agency of women in the face of escalating climate threats.

Sustainable Development Goals: Intersection of SDG 5 and SDG 13 in Advancing Gender Equality and Climate Action

The Sustainable Development Goals (SDGs) established by the United Nations set a global framework for achieving inclusive and sustainable development. Among them, Goal 5 aims to achieve gender equality and empower all women and girls, while Goal 13 seeks to take urgent action to combat climate change and its impacts. These two goals are intrinsically interconnected and mutually reinforcing.

Goal 5 mandates the elimination of all forms of discrimination and violence against women, promotes equal participation of women in leadership and decision-making processes, and recognizes the value of unpaid care and domestic work. Goal 13, by contrast, emphasizes the necessity of strengthening resilience to climate-related hazards, mainstreaming climate change measures into national policies, strategies, and planning frameworks, and enhancing public awareness and education regarding climate change mitigation and adaptation.

The intersection of these goals reveals a critical nexus: Gender equality is fundamental to the success of effective climate action. As articulated in the UN Women Report (2021) on Gender Equality and Climate Change, women and girls face disproportionate impacts from climate change due to entrenched gender inequalities, including restricted access to resources, education, and decision-making power. The UNDP Report (2020) on Gender, Climate, and Sustainable Development underscores that empowering women through education, economic inclusion, and leadership roles contributes significantly to community resilience and facilitates innovative responses to climate challenges.

Incorporating gender perspectives into climate policies not only ensures equitable outcomes but also improves the efficacy and sustainability of such policies. Women's unique insights and lived experiences serve as invaluable assets in crafting and implementing environmental strategies. Consequently, the achievement of Goal 5 is a prerequisite to the full realization of Goal 13, as gender-responsive climate action fosters global sustainability and resilience.

Among the other international instruments, the Sendai Framework for Disaster Risk Reduction (2015–2030) puts more focus and emphasis on the gender-sensitive disaster risk planning, acknowledging women's specific vulnerabilities and capacities in disaster preparedness and response.

Convention on the Elimination of all Forms of Discrimination Against Women and its Role in Protecting Women's Rights and Addressing Climate Change

The Convention on the Elimination of All Forms of Discrimination Against Women (CEDAW), adopted in 1979 by the United Nations General Assembly, is often described as an international bill of rights for women. It serves as a vibrant tool promoting gender equality, eliminating gender discrimination, and empowering women across the world. Over the years, CEDAW has not only addressed traditional gender discrimination but has also evolved to tackle contemporary global challenges, including climate change. Recognizing the disproportionate impact of climate change on women, CEDAW has taken steps to ensure that women's rights are protected in the face of environmental/ecological degradation and climate-related disasters.

As far as the protection of the rights of women is concerned, CEDAW has 30 articles, each focused on eliminating discrimination and promoting equality in areas such as education, employment, healthcare, marriage, family relations, and political participation. Signatory states are required to incorporate the principles of equality and non-discrimination into their national legislation and report periodically on the progress made. Sri Lanka ratified the convention in 1981 and introduced "the Women Charter," the key policy document on women's rights protection and their empowerment in 1993. One of CEDAW's strongest features is its monitoring mechanism, which requires countries to submit regular reports to the CEDAW Committee. CEDAW has also adopted General Recommendations to interpret specific provisions of the treaty and address emerging issues. For example, General Recommendation No. 19 addresses gender-based violence, and General Recommendation No. 33 emphasizes access to justice for women.

In the discussion on the intersection of women's rights and climate change, it is not wrong to say that climate change is not gender-neutral. Its impacts ranging from food and water uncertainty and insecurity to increased natural disasters disproportionately affect women, especially rural women in developing countries. Women frequently should tolerate the impact of ecological degradation due to pre-existing gender inequalities, limited access to resources, and traditional caregiving roles. In many communities, women are responsible for gathering food, water, and fuel—tasks that become more difficult as ecosystems are degraded. Additionally, women are more vulnerable to displacement, poverty, and gender-based violence during and after environmental crises.

CEDAW recognizes the intersection of environmental issues and women's rights. While the original text of the convention does not explicitly mention climate change CEDAW has responded dynamically to the crisis through General Recommendations. CEDAW's work on climate change is visible in General Recommendation No. 37 (2018), which is the most noteworthy progress, discussing climate change and disaster risk reduction. This recommendation explicitly acknowledges that environmental degradation and climate change are human rights issues that impact women and girls differently due to systemic discrimination and structural inequality through the following key points—Gender-Responsive Policies, Women's Participation and Leadership, Protection of Vulnerable Groups, Access to Resources and Information, Data Collection and Analysis, Accountability and Monitoring, Intersectionality and International Cooperation.

Despite substantial growth, some challenges such as a lack of political will, gender discriminative social-cultural norms, data limitation, and lack of proper, systematic, and strong monitoring mechanisms hinder the effectiveness of CEDAW in addressing environmental and climate justice for women.

National Laws in Sri Lanka on Climate Change and Protection of the Rights of Women

As said earlier, Sri Lanka, as a developing island nation, is susceptible to the adverse impacts of climate change, including rising sea levels, extreme weather events, and disruptions to agricultural productivity. In recognition of these challenges, Sri Lanka introduced a range of laws, policies, and strategic frameworks aimed at addressing climate change and promoting environmental resilience. Expressively, these legal instruments progressively reflect an understanding of the need to integrate gender considerations and ensure the protection and promotion of women's rights within climate action.

The following is an overview and critical analysis of key national legal and policy instruments relevant to climate change in Sri Lanka, particularly regarding their effectiveness in addressing gender dimensions and advancing women's rights.

National Legal Standards

Different statutes and policies regarding climate change have been adopted in Sri Lanka. Namely, they are as follows: The Climate Change Secretariat Act No. 16 of 2016, National Environmental Act No. 47 of 1980, as amended in 2000, Forest Conservation Ordinance No. 16 of 1907, as amended in 1990, Coast Conservation Act No. 57 of 1981, as amended in 2008, National Energy Policy and Strategies 2019, National Adaptation Plan for Climate Change Impacts (2016–2025), National Climate Change Policy (2012), National Disaster Management Policy (2010), National Action Plan for the Haritha Lanka Programme (2009), National Policy on Disaster Risk Reduction and Management (2005), National Policy on Women (1993) and National Policy on Women (2023 Revision).

The National Policy on Women, enacted in 1993, is a central legal policy anticipated to promote gender equality and empower women in Sri Lanka. This policy's relevance encompasses the intersection of climate change and gender equality, recognizing that women often face disproportionate impacts from environmental degradation and climate-related disasters. This policy aimed to eliminate gender-based discrimination, enhance women's legal rights, improve their access to education, healthcare, and employment opportunities, and advocate women's participation in decision-making processes at all social levels (Ministry of Women's Affairs, 1993).

One of the policy's significant strengths is its comprehensive approach to addressing gender inequality within the context of climate change. By focusing on legal reforms and advocating for increased female participation in political and administrative roles, the policy seeks to create a supportive legal environment that ensures women have a voice in shaping policies and programs affecting their lives. This inclusive strategy is crucial for achieving long-term gender equality, as women's perspectives are essential in crafting effective climate adaptation and mitigation strategies (United Nations Development Programme 2018).

However, the policy faces notable challenges. Implementation has been inconsistent due to socio-cultural barriers and a lack of political will. While the policy promotes legal reforms, actual amendments and law enforcement have been slow, limiting its effectiveness in creating tangible changes. Additionally, women's participation in decision-making remains low, highlighting a gap between policy intentions and practice. This disconnect can undermine the policy's goals, particularly in the context of climate change, where rapid and inclusive action is necessary.

While Sri Lanka has implemented various (above) policies and legislative frameworks aimed at addressing climate change and environmental issues (UNICEF 2017), several gaps remain, particularly in terms of protecting women's rights in the context of climate change. The above analysis clearly explained why the above national policies, plans, and acts do not adequately address how climate change impacts women, emphasizing the need for a more inclusive framework that considers gender-specific vulnerabilities and rights. In other words, while Sri Lanka's policies reflect a growing recognition of climate change impacts, they largely overlook the intersection between climate change and women's rights. These frameworks would benefit from a gender-sensitive approach that includes women in decision-making, addresses their distinct vulnerabilities, and protects their rights to health, security, and livelihood. The integration of women's perspectives into climate policies is essential for building resilience and ensuring that climate adaptation and mitigation strategies do not exacerbate gender inequalities.

Although this policy primarily focuses on promoting women's rights and participation in economic, social, and political spheres, it lacks specific provisions addressing the unique impact of climate change on women. Climate change disproportionately affects women, especially in rural and coastal areas, where they are heavily dependent on natural resources for their livelihoods. The policy does not acknowledge these climate-related vulnerabilities, which limits its effectiveness in ensuring women's resilience to climate impacts, such as food insecurity and displacement. A gender-sensitive climate resilience framework is crucial, especially as women in rural areas face barriers to accessing resources, adaptation strategies, and decision-making platforms that could mitigate these vulnerabilities.

Women in Climate Change Adaptation and Mitigation Strategies

Women play an extensive but often underrecognized role in both climate change mitigation and adaptation efforts, at both international and domestic levels. In many regions (including the South Asian region which Sri Lanka belongs to), women are disproportionately and adversely affected by the impacts of climate change due to their responsibilities in caregiving, agriculture, water collection, and natural resource management. Notwithstanding their heightened susceptibility, women are increasingly recognized as vital representatives of change in climate action. They actively contribute to the promotion of renewable energy technologies, the implementation of sustainable agricultural practices, biodiversity conservation, and the development of community-based resilience strategies. Their unique knowledge and leadership are essential for effective and inclusive climate solutions. As stated in UN Women their contribution ranged from grassroots activities to international policymaking, highlighting a variety of characters that women play in combating climate change (UN Women 2020; IPCC, 2014).

International instruments explicitly recognize the critical importance of women's full, equal, and meaningful participation in climate action. These instruments emphasize the promotion of gender equality and the empowerment of women as essential components of effective climate policies and actions. They also highlight the necessity of adopting gender-responsive approaches that reflect the differentiated impacts of climate change on women and girls, and call for the mainstreaming of gender considerations into national and international climate strategies, planning, and financing mechanisms. (UNFCCC, 2015; UN, 2015; UN Women 2020).

Sri Lanka can significantly enhance its legal and policy frameworks to better support women in addressing the adverse effects of climate change. This includes the integration of gender perspectives into national climate laws and policies, the guarantee of women's representation in climate-related decision-making bodies, and the establishment of legal safeguards to address gender-specific vulnerabilities in climate adaptation initiatives. Furthermore, embedding gender-sensitive approaches into environmental impact assessments and promoting women's land and resource rights are vital steps. These measures are crucial for building inclusive climate resilience (Ministry of Mahaweli Development and Environment 2016; Fernando 2018).

Sri Lankan women engaged in climate action face a range of structural and socio-cultural challenges that constrain their effective participation. Deeply rooted gender norms and stereotypical societal roles frequently obstruct women's inclusion in leadership positions and decision-making processes within environmental governance. In addition, women are disproportionately disadvantaged by restricted access to critical resources, including land ownership, financial capital, technological tools, and technical training. These limitations significantly hinder their ability to participate meaningfully and contribute to the planning, execution, and oversight of climate-related initiatives. Addressing these barriers is essential to ensuring that climate action frameworks are both inclusive and equitable. As stated, policy gaps and a lack of gender-sensitive approaches in legal frameworks further impede women's involvement in climate action at both global and national levels (IPCC, 2014; UN Women 2020).

Although women continue to face notable obstacles, there exists a wide array of opportunities to strengthen their role in climate change mitigation and adaptation. Targeted capacity-building programs can equip women with the requisite knowledge, technical expertise, and leadership skills to engage meaningfully in climate governance. According to the UN (2015) and Fernando (2018), empowering women through education, training, and access to decision-making forums is essential for effective and inclusive climate policy. The IPCC (2014) further stresses that women's unique experiences in managing household and natural resources position them as pivotal actors in building climate resilience. Legal and institutional reforms that mainstream gender considerations into national environmental frameworks are vital. Facilitating women's representation in climate-related policymaking bodies, as well as supporting grassroots, women-led climate initiatives, can significantly enhance community-level adaptive capacities and lead to more sustainable, locally informed climate responses. Encouraging inclusive participation not only advances gender equality but also strengthens the overall effectiveness and legitimacy of Sri Lanka's climate response (UN, 2015; IPCC, 2014; Fernando 2018).

Refining Legal Frameworks to Safeguard Women's Rights and Support Women in Climate Change

Legal and regulatory frameworks play a pivotal role in facilitating women's effective engagement and addressing structural gender disparities in climate change mitigation and adaptation strategies. On the international level, legal instruments and global policy commitments provide normative guidance for the integration of gender dimensions into climate-related decision-making and implementation processes. These frameworks acknowledge that women often experience unequal and heightened exposure to climate risks, owing to socioeconomic vulnerabilities and unequal access to critical resources. Consequently, they underscore the necessity for gender-responsive and participatory approaches to climate governance. At the national level, while Sri Lanka has developed policy initiatives (Policy and Strategic Planning Division of Export Developmnet Board, Sri Lanka, 2017) that endorse both gender equality and environmental sustainability, existing legislative structures fall short in conferring sufficient legal authority and institutional backing to women's involvement in climate action. As highlighted in national policy reports, the lack of explicit gender provisions within climate-specific legislation, inadequate monitoring mechanisms, and limited inter-agency coordination continue to hinder progress. To ensure sustained and equitable outcomes, comprehensive legal reforms are needed to mainstream gender perspectives across all stages of climate policy planning, implementation, and evaluation, and to guarantee that women's contributions are formally recognized, protected, and adequately resourced. (UN Women 2020; Ministry of Mahaweli Development and Environment 2016).

Despite widespread international commitments and the existence of legal frameworks intended to advance gender equality in environmental governance (UNDP, 2016), significant shortcomings remain in addressing gender imbalances within climate change responses. One of the most persistent issues is the insufficient incorporation of gender-sensitive analysis in climate policy formulation and implementation. Many frameworks lack concrete enforcement mechanisms, resulting in policy directives that fail to translate into actionable outcomes. Furthermore, women's limited representation in institutional decision-making structures continues to undermine the inclusiveness and effectiveness of climate strategies. In the Sri Lankan context, although gender equality is acknowledged within national policy documents, the absence of systematic implementation, budgetary allocation, and accountability mechanisms has created a substantial gap between policy and practice. These limitations impede women's ability to contribute meaningfully to mitigation and adaptation activities, particularly in vulnerable sectors such as agriculture and disaster preparedness. Research and international assessments emphasize the need for legal reform that not only strengthens institutional mandates but also guarantees equal access to resources, technology, and information for women. Establishing binding obligations and integrating measurable gender targets into climate legislation are essential to realizing equitable and transformative climate action (IPCC, 2014; Fernando 2018).

Sri Lanka holds considerable potential to enhance its legislative and policy frameworks to more effectively support women in confronting the challenges posed by climate variability and environmental stressors. Advancing this objective requires the mainstreaming of gender-sensitive perspectives within national climate laws and development strategies. It is equally imperative to institutionalize women's representation in policy advisory committees and decision-making forums that shape climate resilience agendas. Furthermore, robust legal instruments must be introduced to remedy gender-specific vulnerabilities in climate adaptation programs, particularly in rural and disaster-prone areas. According to national policy guidance, it is also essential to embed gender equity considerations into environmental impact assessment procedures and ensure legal protection for women's rights to land tenure and resource access. These steps would not only promote inclusive governance but also reinforce women's agency and leadership in climate adaptation and mitigation efforts (Ministry of Mahaweli Development and Environment 2016; Fernando 2018).

It is essential to reform and advance legal and institutional structures at both the global and domestic levels to more effectively empower women in responding to the impacts of climate change through both adaptation and mitigation strategies in the case of developing the existing system. This process must include the strengthening of normative legal guarantees that uphold women's rights in environmental contexts, the development of inclusive and gender-aware policy frameworks, and the creation of mechanisms that expand women's equitable access to financial resources, technological tools, and leadership roles in climate governance. As highlighted in international and national reports, targeted capacity-enhancement initiatives for lawmakers, judicial actors, and community leaders can play a pivotal role in supporting the application of gender-sensitive climate legislation. Moreover, effective implementation necessitates coordinated action and dialogue among state institutions, non-governmental stakeholders, and international bodies to formulate coherent legal systems that protect and promote women's interests in environmental decision-making. These collaborative and inclusive efforts are critical to ensuring that legal responses to climate change are not only just and sustainable but also fully reflective of the differentiated experiences and capabilities of women (UN, 2015; IPCC, 2014; Fernando 2018).

Conclusion

The unique challenges discussed earlier posed by climate change exacerbate gender disparities in Sri Lanka, particularly affecting women in rural and marginalized communities. Addressing the interplay between gender and climate change requires an understanding of the specific climate-related impacts on women's lives and rights, and how national and international frameworks currently protect, or fail to protect, these rights.

Women in Sri Lanka, particularly in rural and agricultural sectors, face distinct challenges due to climate change. Rising temperatures, irregular rainfall, and increased natural disasters disrupt their roles in food production, water collection, and natural resource management. Such disruptions intensify the socio-economic pressures on women, often forcing them to abandon sustainable livelihoods and seek low-paying work, thereby increasing economic vulnerabilities. Moreover, climate change worsens health issues by impacting water availability and sanitation, crucial for managing reproductive health and menstrual hygiene. These impacts highlight the need for targeted support to mitigate the gender-specific challenges climate change poses.

Climate change directly threatens women's rights to health, work, and a safe environment. As discussed, the right to health is compromised by increasing water scarcity, which hampers hygiene and sanitation facilities, particularly affecting women during pregnancy and menstruation. Additionally, as environmental stress disrupts traditional work opportunities, the right to work is impacted, limiting women's access to stable incomes and job security. Climate-related displacement also poses threats to personal safety and physical integrity, as women and girls in vulnerable communities face increased risks of violence and exploitation. Addressing these rights violations is critical for gender-equitable climate resilience.

International frameworks such as the Paris Agreement (2015) and the Convention on the Elimination of All Forms of Discrimination Against Women (CEDAW) recognize the need to integrate gender considerations into climate action. While these frameworks underscore the importance of empowering women in climate resilience and adaptation, their effectiveness is limited in Sri Lanka due to the challenges in implementing these guidelines locally. The United Nations Framework Convention on Climate Change (UNFCCC) also encourages gender-responsive climate policies, but gaps remain in enforcing these frameworks at the national level, leading to limited impact on protecting Sri Lankan women's rights.

Sri Lanka's climate policies and legal frameworks, including the National Climate Change Policy (2012) and National Adaptation Plan for Climate Change Impacts (2016–2025), are limited in their gender inclusivity. While these policies promote adaptation and resilience, they do not explicitly address the unique impacts of climate change on women or offer protections for women's rights within climate response strategies. This inadequacy leaves women's socio-economic vulnerabilities unaddressed, and there is a pressing need to revise these frameworks to consider the gendered dimensions of climate change impacts and ensure protections for women's health, safety, and livelihoods.

Women play a pivotal role in climate change mitigation and adaptation due to their knowledge of local ecosystems, community resilience, and resource management. In Sri Lanka, women in rural areas engage in water conservation, sustainable agriculture, and disaster preparedness, contributing to community resilience. However, legal recognition of their contributions remains limited, and they are often excluded from climate decision-making processes. Recognizing and integrating women's roles in legal and policy frameworks can enhance national resilience and ensure that women receive the support and recognition needed to continue these contributions.

Strengthening Sri Lanka's legal and policy frameworks to protect women's rights in the context of climate change is essential. Effective frameworks should include gender-responsive adaptation and mitigation strategies, legal protections for women's rights to resources and livelihood opportunities, and programs that improve women's access to climate-related information and decision-making platforms. Incorporating these elements would empower women, reduce climate vulnerabilities, and enhance resilience. Improved legal frameworks would support women's rights and contribute to sustainable and equitable climate action for Sri Lanka as a whole.

The multifaceted impact of climate change on Sri Lankan women underscores the importance of gender-responsive climate policies. Climate change affects women disproportionately, infringing on their health, safety, economic security, and human rights. International frameworks provide some guidelines, but without robust implementation of national policies, they remain ineffective. By revising Sri Lanka's laws and policies to include gender-specific climate adaptations and protections, the country can ensure that women's rights are safeguarded, empowering them to play active roles in climate resilience and sustainable development.

References

4 Annual Report, Disaster Management Centre of Sri Lanka. Available at https://www.dmc.gov.lk/images/pdfs/AnnualReports/2017_compressed.pdf

Chapter 2 'National Output, Expenditure and Income' Sri Lanka Central Bank Report (2016). Available at https://www.cbsl.gov.lk/sites/default/files/cbslweb_documents/publications/annual_report/2016/en/6_Chapter_02.pdf

Chowdhury N, Suleri A (2021) Empowering Women through solar electrification: the case of bare-foot College in India. Renewable Energy Journal 153:1267–1275. https://doi.org/10.1016/j.renene.2020.12.045

Department of Meteorology, Sri Lanka, Annual Performance Report (2020) Expenditure. Available at https://www.parliament.lk/uploads/documents/paperspresented/1674194396047273.pdf

Fernando S (2018) Women's role in climate change adaptation: Sri Lankan experience, internal displacement monitoring Centre (IDMC), (2019). Displacement Due to Climate Change, Sri Lanka

IUCN Sri Lanka. (2018) Sri Lanka's coastal and marine biodiversity: impact of climate change on local communities and economies

Performance of the Export Sector of Sri Lanka January–December 2016, policy and strategic planning division of Export Development Board Sri Lanka, published in 2017

Ministry of Health, Sri Lanka (2016) Health sector response to climate change in Sri Lanka

Ministry of Health, Sri Lanka (2017) Health impacts of floods and disasters: annual report

Ministry of Mahaweli Development and Environment Report (2016) (published in 2016) on National climate change policy of Sri Lanka Adopted in 2012

Muthukuda Arachchige Dona Shiroma Jeeva Shirajanie Niriella (2025) The intersection of climate change, gender equality and legal framework: analyzing the impact of environmental policies on women's rights and opportunities in Sri Lanka's fragile democracy. In: Wejnert B (ed) The global rise of autocracy–its threat to a sustainable future–democratization and autocratization studies. Routladge, New York

National Disaster Relief Services Centre (2020) Disaster impact report: cyclone Burevi and vulnerable populations in Sri Lanka

Ranasinghe R, Kularatne JDJ, Gunasekara K (2022) Impact of coastal erosion on coastal communities in Sri Lanka: challenges and adaptation measures. J Coast Res 38(1):49–58

UN Women. (2018). Turning promises into action: gender equality in the 2030 agenda for sustainable development, 2018 Report published by United Nations Entity for Gender Equality and the Empowerment of Women (UN Women)

UN Women. (2020). The gender impact on climate change 2020 Report published by United Nations Entity for Gender Equality and the Empowerment of Women (UN Women)

United Nations Development Programme (2018) Gender and climate change strategy for Sri Lanka: understanding Women's vulnerabilities in rural and agricultural sectors. UNDP

United Nations Development Programme (UNDP) (2016) Gender and climate change: overview of linkages between gender and climate change -UNDP Asia-Pacific human development report 2011

United Nations International Children's Emergency Fund (UNICEF) (2017) Climate change and children Sri Lanka, United Nations Plaza, New York, NY 10017, USA

The Role of Tribal Women and Environmental Sustainability in Koraput, Odisha, India

Kapila Khemundu and Niranjan Majhi

Abstract In India, traditionally, women, as agents of nurturing family and community, provide sustenance to the environment and forest resources. As a habitat of forests and with the practice of collecting minor forest products, the connection between tribal women and the forest plays a significant role in preserving and sustaining the environment. They preserve plants for traditional medicinal, cultural, ceremonial, and ritual use by growing them in their home gardens or courtyards. In fact, the practice of certain indigenous activities, such as some traditional agricultural systems, ceremonies, and religious rituals, has fostered a culture of conservation of natural resources and balancing the ecosystem. Additionally, in the twenty-first century, with the environment as a global concern, both government and non-government organizations have initiated many plans and schemes to empower women through the preservation of forests and the environment. The government programs for women farmers, such as Mahila Kisan Sashaktikaran Pariyojana (MKSP), the Mission for Integrated Development of Horticulture, the National Food Security Mission, and the National Mission on Sustainable Agriculture, play a vital role in supporting women's contributions to sustainable environmental development. NGO initiatives, including Women's Environment and Development Organizations, Feminist Action on the Nexus for Economic and Climate Justice, and gender-based climate solutions, also significantly empower women in this context. Tribal women, as changemakers, can be further empowered through the PESA Act, organic farming, and allied activities. This work will highlight the strength of women's contributions and how they can be channeled for the conservation of the environment and its sustainable use for future generations. According to a report (Hindustan Times 2022), nearly 75% of full-time workers on Indian farms are women, including many tribal women. By practicing various

K. Khemundu (✉) · N. Majhi
Department of Sociology, Central University of Odisha, Sunabeda, Odisha, India

B. Wejnert (ed.), *Democracy in Retreat and the Climate Crisis*, Springer Studies on Populism, Identity Politics and Social Justice,
https://doi.org/10.1007/978-3-032-02881-5_10

inherited indigenous activities related to agriculture and environment, tribal women greatly contribute to resource conservation and a sustainable environment. This study primarily aims to explore how tribal women engage with the natural environment and how their indigenous knowledge, social and cultural practices, and religious beliefs help preserve, conserve, and maintain a healthy environment. Additionally, this study seeks to understand various government and non-government policies regarding the environment and their implementation through women, which can be beneficial in addressing environmental issues and promoting sustainability for future generations. In this study, secondary data such as news reports, literature, books, online sources, and government and non-government records are referenced for a comprehensive presentation.

Keywords Tribal women · Environmental sustainability · Government policies · Empowerment · Natural resources · Forest dweller · Conservation of forest

Introduction

The environment encompasses both living (biotic) elements, such as animals and plants, and non-living elements (abiotic) like water, land, sunlight, rocks, and air. In short, it encompasses our surroundings where we live. Primarily, there are two types of environments: natural and artificial or man-made. The tribes in India, primarily forest dwellers, inhabit natural environments while also creating artificial ones to meet their needs. Initially, they adapt their lives to the natural environment. Douglas and Holland (1947) define the environment as the amalgamation of all external forces, changes, and conditions that impact the life, nature, behavior, growth, development, and maturation of living organisms. The natural environment includes the relationship between nature and human beings. Human civilization has maintained an inseparable relationship with nature since ancient times. Human culture varies according to habitat and the surrounding environment. Odisha, a state in eastern India, is divided into 30 administrative geographical units called districts. It is home to 62 different tribal communities (Ministry of Tribal Affairs, Odisha), and the Koraput district is one of the most tribal-dominated geographical areas, featuring diverse cultures and human habitats closely linked to nature (Map 1).

The pre-divided Koraput district, presently comprising Koraput, Rayagada, Nabrangpur, and Malkangiri districts, occupies a unique position on the ethnographic map of Odisha and India due to its large varieties of tribal communities. It is also home to some of the most primitive tribes, including the Bonda, Dangria Kondh, and Koyas, who are well known for their distinct socio-cultural practices. Moreover, they primarily adhere to animism and naturalism as their major religious beliefs. Their traditional knowledge regarding agriculture, natural resource-based livelihood systems, plant-based health systems, and healing practices encourages the tribal communities to regard certain plants as sacred groves, thus aiding in the conservation of the environment (Map 2).

Map 1 Source: https://www.mapsofindia.com/maps/orissa/

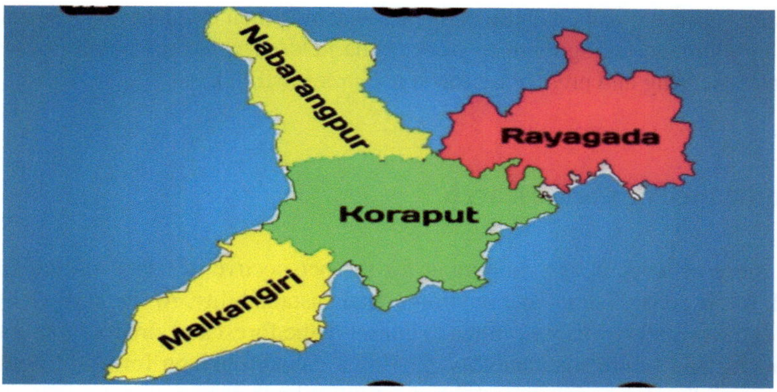

Map 2 The Pre-divided Koraput District. Source: https://slbcorissa.com/geographical-map/

Rationale for the Study

The terrible impact of natural disasters and climate change varies regarding gender relations in any society. According to the International Economic Association (2023), women are disproportionately vulnerable to climatic disasters and their effects, despite their significant involvement in the agricultural sector. Most natural disasters, including cyclones, floods, droughts, and epidemics, have a more substantial impact on women than on their male counterparts. Tribal communities generally inhabit hilly terrain, surrounded by forests or natural environments. Often, they survive and escape such challenging situations by employing traditional survival techniques, especially when it becomes late and difficult for rescue teams to reach them. These natural calamities are occurring due to global environmental issues such as warming and climate change. Tribal women, as major contributors to forest and environmental conservation, are a highly neglected group. In recognition of this, they need to be empowered and inspired in their efforts to preserve, protect, and utilize forest resources. With a close connection to the forest, tribal women not only engage in agricultural farming but also collect firewood and protect trees and plants; they gather various minor forest products, such as gum, tendu leaves, fruits, bamboo, mushrooms, and edible herbs (Das 2022). Mahour (2016) states that throughout human history, men have viewed natural resources primarily from a commercialization perspective to meet their economic needs. In contrast, women focus on conservation and preservation; they collect dry and dead branches for firewood rather than cutting down live trees. However, due to environmental issues and climate change, women are the most affected, exemplified by Nepali women who, even in old age, must climb mountains to collect wood and fodder. Similar practices are observed among the tribal women of Koraput. Their lifestyle fosters a culture of conservation, preservation, and sustainability of the natural environment. The role of Koraput's tribal women in environmental conservation has not been adequately highlighted in research. Like tribal women in other societies, the tribal women of Koraput bear more responsibilities, such as teaching their children about sustainable consumption and taking them to work in the fields, where they impart knowledge about conserving and preserving forest resources (MFPs).

Significance of the Research

This piece of work is very important for the perspective of tribal women's livelihood, forest conservation, and environmental sustainability. At the micro level, the small efforts made by tribal women to conserve the forest and environment for their livelihood and future sustainability provide a valuable lesson for all of humanity. Commercialization, mining, and other activities have proven disastrous to the stability of life on Earth. Therefore, we need more studies and research to validate the role played by tribal women in conserving forests and the environment. This work will highlight the direction towards environmental sustainability.

Research Objectives

The major objectives of this paper are:

1. To understand the role of tribal women in the conservation of forests and the environment.

2. To understand gender-based cultural practices and their significance to nature.
3. To understand various government and non-government plans implemented through women to promote environmental sustainability.

Research Questions

This is an attempt to seek some of the pertinent questions in the context of tribal women, government, and non-government plans/schemes for the conservation of the environment and its sustainability. These are:

1. How do tribal women play a major role in the conservation of the environment and its sustainability?
2. What plans and programs have been implemented by the government and non-government agencies through tribal women to target environmental sustainability?
3. What are the reasons for environmental degradation, and what is its impact on human society?
4. How can tribal women be instrumental in environmental conservation and its sustainability?

Review of Literature on Thematic Issues

Menon (2016), in his article *Role of Tribal Women in Sustainable Development*, articulates that, in comparison to other social groups in tribal communities, tribal women play a significant role in managing the family economy through their practice of collecting minor forest produce. This responsibility makes them dependent on nature. Sherry Ortner, in the U.S. movement, pointed out that women's interests practically coincide with nature, while men's interests align with culture. Tribal women are deeply involved in the practice of worshipping sacred groves, which further serves as an act of preserving nature. Seventy percent of the tribal economy comes from forest resources, making tribal women the best conservationists. The researcher has followed a secondary research method and suggests supporting the rights of tribal people rather than forcing them to relocate for mining purposes that damage their habitats and attachment to nature.

Singhal et al. (2021), in *Role of religious beliefs of tribal communities from Jharkhand (India) in biodiversity conservation*, state that the tribal population in Jharkhand consists of 32 different groups. However, Santal, Munda, and Ho are the dominant tribal groups in the state. Women in these tribal communities play a crucial role in managing the economy, producing, and consuming resources. However, due to rapid industrialization and modernization, their commitment to religious practices and rituals related to nature has weakened, raising concerns for the preservation of biodiversity. Religion serves as a traditional tool to persuade people, which is easily understandable; hence, their religious values and associated taboos must be protected against erosion. A special tribal policy is required to protect and strengthen tribal livelihood practices. Government agencies, corporate organizations, and NGOs should jointly take responsibility for encouraging tribal women in their religious and cultural practices associated with nature.

Sirur (2024): "Dwindling Forest commons leave Koraput tribal communities with little to celebrate." The famous festival Chaita Parab in Koraput district of Odisha is celebrated by the tribes every year. This festival is linked to the collection and consumption of natural resources. The Mahula flowers are crucial for the tribes, as a local brew is made from them, along with fresh mango leaves and other foliage. They perform animal sacrifice before a deity as a traditional ritualistic practice, invoking God for prosperity and a good harvest for the rest of the year. Various tribal groups in Koraput celebrate at least 12 festivals annually, each associated with a deity. The tribes of Koraput believe their deity governs the forests, streams, grazing pastures, and riverbeds. The Nuakhai Parab is linked to tamarind trees, pumpkin leaves, tubers, and maize, while Pus Parab is associated with cattle, millets, and tubers. Despite this rich cultural heritage, the degradation of various flora and fauna is occurring. For instance, earlier, the tribes of Koraput used to cultivate numerous rice varieties; however, today, due to the public distribution system, which provides only one type of rice, all marginal farmers are dependent on it.

Das' (2022), *Role of women in environmental protection,* states that women's day-to-day activities are closely related to the environment. They collect water and firewood directly from nature, especially in developing countries. Since their activities are intertwined with nature, they possess extensive knowledge about the environment—soil, plants, trees, water, and other natural resources. Thus, their participation is vital. The involvement of women in environmental activities is most evident in developing countries; however, it raises the question of what occurs in developed countries, where a large part of the population is engaged in industrial and service sector activities. The researcher has offered some suggestions aimed primarily at women in developing countries, suggesting training for proper implementation and promoting values for the protection and conservation of nature, but this focus does not extend to women living in mainstream society.

Tribal Women and Indigenous Knowledge

The Koraput region of Odisha state is recognized as an important agricultural heritage system (GIAHS); the Food and Agriculture Organization (FAO) of the United Nations recognized in 2012 the maintenance of unique tribal traditional agricultural practices, which further plays an important role in the conservation and utilization of forests. In the process of traditional agricultural practices, women primarily play a vital role by investing their interest, time, and energy in cultivating diverse plants as their traditional and communal harmonious food (Mishra et al.

2018). The tribal women of the Koraput region take responsibility for the mixed cropping pattern when their landholding is small.

Mix cropping field showing 5 different crops grown. Source-https://www.etribaltribune.com/index.php/volume-5/mv5i2/agrobiodiversity-conservation-and-tribal-women-of-koraput-odisha#datamethods

In tribal families, labor is usually divided based on past practices; men are primarily engaged in ploughing fields and sowing seeds, while women focus on uprooting, transplanting, weeding, rouging, and storing. Thus, tribal women invest more time in traditional agricultural processes.

According to the UN Climate Change report (2023), tribal women contribute to global environmental protection. For example, in the Kandhamal district of Odisha, India, the Kandha tribe engages in producing minor millet, which can be cultivated in less irrigated areas, especially in uplands, without the use of chemical fertilizers and pesticides (Map 3).

The UN climate change group's report states that in tribal (Kandha) families, the women are the main breadwinners; however, they are rarely recognized within the social hierarchy of their communities. Women never abandon the practice of traditional agricultural methods, thereby fostering an environment in their villages that further promotes organic manure prepared by themselves. They utilize upland and barren land, increase income, and protect the surrounding forests from deforestation. Thus, these activities are carried out by women and their communities, addressing basic fodder needs, which has become a significant benchmark for sustaining natural resources and ensuring a healthy and resilient climate.

According to Easter Boserup, the African farming systems illustrate that the responsibility for food production-related tasks falls to women. Schultz's agricultural study found that 90% of women manage the natural resources upon which

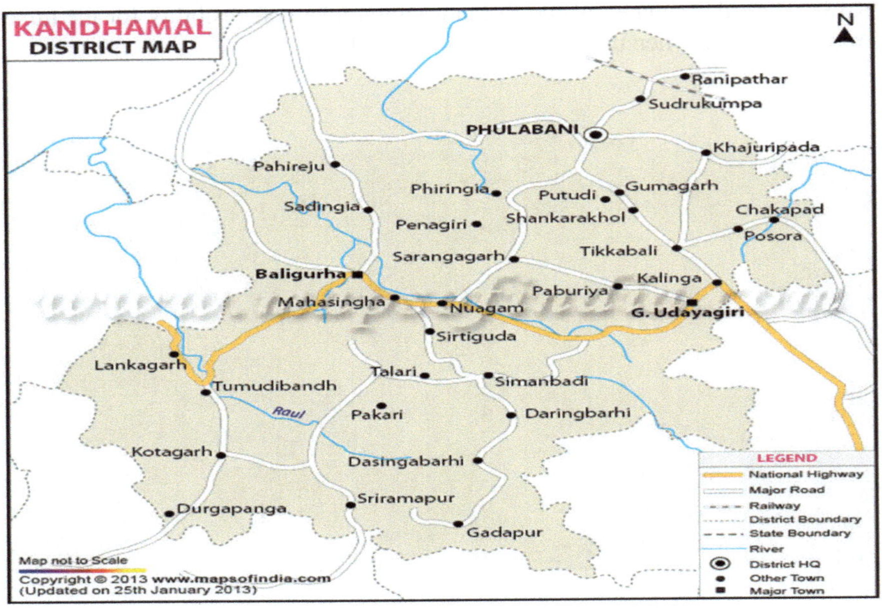

Map 3 Kandhamal District. Source: Maps of India. https://www.mapsofindia.com/maps/orissa/districts/kandhamal.html#google_vignette

their survival depends. Women oversee 30% of households in most developing countries; 80% of food production is conducted by women in Saharan Africa, 60% in Asia, and 50% in Latin America (Mahour 2016). Thus, a deep connection exists between women and their surroundings.

Findings and Discussions

Women are primarily involved in the protection and conservation of natural resources and the environment by engaging in natural resource-based activities such as agriculture, livestock management, and forestry (Brixi et al. 2022). According to the World Bank Group 2022, tribal women are at the forefront of making significant efforts to adapt to and mitigate the consequences of climate change, utilizing their indigenous knowledge and practices. For example, people living in hilly areas have a distinct culture compared to those in plain areas. With the growing population and rapid industrialization, environmental degradation and the depletion of forest resources are increasing. At this critical juncture, tribal women residing in hilly regions play a vital role in protecting and sustaining the environment through their various indigenous rituals, community norms, and agricultural practices. Minor forest produce (MFP) significantly boosts the tribal economy, contributing around 70% of their total income. Additionally, tree conservation is vital for their medicinal needs, as certain plants' roots, stems, and leaves—such as Vanda tessala, Alternanthera sessiles, and the roots of Cassia adnate, Sida cordata, and Bauhina purpurea—are used to heal broken bones within an expected recovery time of 10 to 15 days (Menon 2016).

According to the National Committee on Forest Rights Act, 2011 report, an estimated 100 million people rely on collecting and marketing minor forest produce (MFP) for their livelihoods.

As reported by the World Bank in 2011, MFP supports approximately 275 million people in rural India, significantly comprising tribal individuals, with women playing a major role. The importance of MFPs in tribal livelihoods and in the protection of natural resources can be illustrated in the following Table 1.

The above table demonstrates not only the importance of MFPs but also the significance of nature and its resources, illustrating how people depend on natural resources and the environment for their livelihoods, and how this indigenous culture and lifestyle consequently foster conservation and sustainability of nature and its resources. The table also highlights the importance of the relationship between human beings and nature. The Ministry of Tribal Affairs (2020) also states that the MFPs are an important source not only for those not engaged in agricultural activities but also for providing income during slack seasons for those involved in agricultural activities (Tables 2 and 3).

Table 1 Plants grown and used for several traditional ceremonies and rituals

MFPs collected	Economy
Lac (resin), *mahuwa*, flower and tamarind	During the month of January to march, more than 75% of tribal households in Odisha, Madhya Pradesh, and Andhra Pradesh collect mahuwa flower and earn Rs. 5000 a year
Tendu leaves, Sal seeds and *chironji*	During the month of April to June, more than 30 million forest dwellers depend on seeds, leaves, and resins from Sal trees; people get 90 days of employment that includes around 7.5 million people, and a further three million people are involved in bidi processing
Chironji, mango, *mahuwa* fruits, silk cocoons and bamboo	During the month of July to September, around ten million people depend on bamboo for livelihood, and 1,26,000 households are involved in Tussar silk cultivation
Lac, *kullu* gum, resins used in incense sticks	During the month of October to November, around 3 lakh people get employment from the collection of gums
Finger millet	The tribal women are engaged in growing millet in several states (Andhra Pradesh, Madhya Pradesh, Odisha, Chhattisgarh) to use in various rituals and traditional ceremonies

Source—Ministry of Tribal Affairs, Govt. of India https:/trifed.tribal.gov.in/non/timber/msp-mfp

Table 2 Tribes, their population & literacy rate of Koraput, Rayagada, Malkangiri, and Nabrangpur districts as per the 2011 Census

Sl. No	Name of the district	Tribes	Population (ST) as per 2011 census	Literacy rate (%) As per 2011 census
1	Koraput	Kondh, Paroja, Gond, Gadaba, Omanatya, Bhumia, Bhottadas, Soura, Durua	697,583	50.56
2	Rayagada	Soura, Kondh, Kandha Gauda, Bagata, Jatapu, Paroja, Bhumia, Dangria Kandha, Lanjia Soura, Mirdha, Pentia, Sounti	541,905	36.69
3	Malkangiri	Koya, Paroja, Bhumia, Dharua, bonda, Didayi, Gadavas, Bhunjia, Holva, Mirdha, Parenga, Pentia, Soura, Soun	354,614	35.23
4	Nabrangpur	Gond, Bhottada, Omanatya, Dharua, Paraja, Kandha, Bhatra, halva, banjara, Bhunjia	681,173	38.54

Source: https://stsc.odisha.gov.in/sites/default/files/2021-03/Population_%26_Literacy-2011.pdf, https://repository.tribal.gov.in/bitstream/123456789/73776/1/SCST_2018_book_0016.pdf

Table 3 Natural resources used by the tribes in Koraput, Rayagada, Malkangiri, and Nabrangpur districts

Sl. No	Name of the district	MFPs used by tribes
1	Koraput	Hill grass, tamarind, *Sal*, *mahua*, *mahua* flower, tej patta, siyali leaves, shikakai, honey, myrobalan, sericulture (Muga/Tassar), shellac (lac), gum (karaya), *Barda saag*, mango, *mahua* seed, small *khajoor*, *charphal,* jackfruit, *jamun*, *kendu* leaves, *kendu* fruits, *nangalkandha*, *saal* seeds, mushrooms (*bino chatti*, *swan chatti*, *Bati chatti*, *Bali chatti*), *karkat* phal, *peeta kandha*, *karke kandha*, *jhuna*, babe (a kind of grass used to produce ropes), *amla*, *beharra*, *harra*
2	Rayagada	*Kendu* fruits, *kendu* leaves, *mahua*, *sala*, *kusuma leaves*, *jambo leaves*, *mango*, *dumru*, *bela* leaves, *tulashi*, *dhalaphula*, *kuruchi*, gum, *mandara* flower, *malli* flower, *aparajita* flower, *Katha Champa* flower, *malati* flower, *dudura* flower, *panasha* fruit, *sdabahari* flower, lemon, *solopa* wine, bamboo, *menjuati* leaves, *neem* leaves, *chandan* bark, *rashi* fruit, lotus, *duba* tree, banyan tree, *barakoli* fruit, *siali* leaves, custard apple, mushroom, guava, tamarind, red gram, castor seeds, broom sticks, honey, *mahua* flower

(continued)

Table 3 (continued)

Sl. No	Name of the district	MFPs used by tribes
3	Malkangiri	Tulashi leaves, *kendu* fruits, *mahua* fruits, *mahua* flower, bamboo sticks, *sala, pia sala, kusuma* fruits, *jambo* fruits, *bela* leaves, *bela* fruits, *aparajita* flower, *malati* flower, *dudura* flower, *neem* leaves, *solopa* wine, date wine, *khunto chhhatu* (mushroom), *menjuati* leaves, *chandan, ghasi* plant, *padma* fruit, mango fruits, *barokoli* fruits, *rashi* seeds, *duba* plants, banyan trees.
4	Nabrangpur	*Girili, giridi* flower, *kalama* herbs, *baunsa chattu, mahua* flower, *mahua* seeds, jackfruit, neem leaves, *barada saga, kollari saga, sajana saga, bhuin khajuri* (small dates), *solopa* wine, *saru, kansiri saga*, bamboo sticks, *pita kandha* (bitter potato), *kendu* fruits, *kendu* leaves, *kanta leutia* leaves, *madaranga* keaves, *bela* leaves, *masia Kanda, lengudi saga, hidimicha* leaves, *dumar, dumri* fruits, *barakoli, kuta chhatu* (*kuta* mushroom), *panisigada* (*Onagraceae*), *hunka chhatu* (*hunka* mushroom), *balli chhatu* (sand mushroom), *sravana chhatu* (mushroom), *nada chhatu* (mushroom), *tolaka* fruits, tamarind, *jamu* fruits, *jayanti* leaves, *kathalu, simuli Kanda, pasaruni* leaves, *gandhuli* leaves, *rimili* fruits, *nimbura mai* ruits, *sarupatri mai* fruits, *limbur* fruits, *kusuma* fruits, olua tuber, etc.

Source*:* Data taken from various research articles on the above district

Tribal Women Engaged in Organic Farming

According to a 2021 report by the Indian Express, under the agency of Mission Jeevika of Koraput ITDA, tribal women from SHG members are rapidly learning about and focusing their attention on the organic farming system. They utilize fertilizers made from locally available cow dung and leaves instead of pesticides and chemical fertilizers. This practice protects biodiversity without harming other ecosystems in the natural environment.

According to the Seva Kendra Silchar (SKS), an NGO in Assam, in the past, the tribes of hilly regions in some parts of India practiced slash-and-burn agriculture to grow traditional crops. However, this traditional method of cultivation partially contributes to the degradation of the green environment. Although the slash-and-burn practice enhances soil fertility, its effect is temporary, and it contributes to deforestation while depleting natural resources around their villages. With rapid population growth, they require land and a healthy environment to survive. Aware of this, the Karbi tribe, one of the major ethnic communities from a small village called Boalcherra in the Cachar District of Assam (24 families), has abandoned jhum cultivation in favor of horticultural production methods. Households that once cultivated crops such as rice, maize, pumpkin, brinjal, chillies, beans, and gourds have started growing areca nuts, coconuts, betel leaves, oranges, litchi, and bananas to restore their ecological balance and improve their income, which suffered during the practice of jhum cultivation. The NGO Seva Kendra Silchar also encourages and trains Karbi women in certain villages to cultivate traditional seeds and crops using modern agricultural practices.

The Indian Council of Agricultural Research published a case study of many tribal women engaged in organic farming. Mrs. Anitha M., a tribal woman from Bettampaddy village, practiced a traditional farming system, cultivating various crops such as coconut, pepper, paddy, areca nut, and vegetables on her 4.5 acres; however, it yielded very poor economic returns. Subsequently, as an active member of the NGO Shri Kshetra Dharmasthala Rural Development (SKDRDP), she abandoned traditional farming methods and adopted organic farming on her land, which has become a sustainable system free from fertilizers and chemical additives that polluted the soil and water and harmed the ecosystem. The organic farming system helps create a healthy environment and provides good returns to Smt. Anitha. After gaining knowledge of organic farming, she now offers advisory services to 400 rural farmers, most of whom are tribal women.

Marak and Singha (2023) states that today tribal women are leading an organic revolution in India's villages and protecting a healthy environment and ecosystem while the agricultural world struggles to find sustainable farming methods that prevent environmental pollution. Keeping this in mind, in 2021, the Ministry of Woman and Child Development, Government of India, launched "Pradhan Mantri Jana Jatiya Vikas Mission (PMJVM)" for 2021–2022, which includes two important components: first, "Mechanism for marketing of minor forest produce (MFP) through Minimum Support Price (MSP) and Development of Value Chain for MFP," and second, "Institutional Support for Development and Marketing of Tribal Products." This scheme aims not only to enhance the tribal economy but also to conserve and sustain forest resources for the next generation and achieve a healthy, sustainable environment. The introduction of MSP by the government on certain minor forest products has become a latent function in boosting the tribal economy and their interest in engaging with the forest and safeguarding the plants and trees.

Tribal Women and their Religious Activities

Religious practice varies among different communities, societies, and locations. Religious rituals have evolved and been practiced by mankind since ancient times. Significantly, the religious practices of certain tribes have a close bond with the forest. Niyogi (2018) says that "without women, no ritual ever takes place." The 'Sarna' is one of the tribal religions practiced among the forest and hilly dwellers of Jharkhand State, India, which maintains a strong connection with nature. Followers of this ritual worship natural entities such as forests, mountains, lands, and rivers, and protect local biodiversity through their religious and cultural practices. This is primarily expressed through festivals, taboos, rituals, and various ongoing traditions. Sarna followers believe they will face curses and misfortune if they destroy certain sacred trees. This indigenous practice promotes conservation and a sustainable environment and resources (Khan 2024).

Tribal woman prays at a shrine at Hathipakhna village in Koraput. Source: https://pressinstitute.in/grassroots/dwindling-forest-commons-leave-koraput-tribal-communities-with-little-to-celebrate/

In Koraput district, the most famous festival, called 'Chaita Parab,' is celebrated annually in April. During this festival, tribal women perform various religious rituals deeply connected with nature, reflecting the collection and consumption of shared natural resources. These rituals include the preservation of sacred groves and natural resources (Sirur 2024).

In India, many tribes traditionally use various plants in rituals and ceremonies. Plant resources play an important role in their social and cultural lives. For example, in Uttarakhand, India, a tribe called 'Tharu' from Udham Singh Nagar grows several useful plants in their home gardens and fields. They also venture into deep forests in search of leaves from the *sal* tree, *neem* tree, *belpatri*, *sarson*, and others. They cultivate several flower plants in their home courtyards, such as Ocimum basilicum L, Ocimum tenuiflorum L, *chandini*, *gainda*, *mehendi*, *mirchigurhal*, *kela*, and more, which are used for worshipping their gods and goddesses and decorating during marriage ceremonies and various traditional rituals (Lata et al. 2022). Thus, various traditional beliefs and practices encourage the Tharu tribe to grow these plants and continue following their traditions. The ongoing practice of cultivating these traditional plants plays a crucial role in the conservation of diverse flora and a healthy environment.

Major Policies and Programs on Environment and their Implications on Women

During the 1970s and 1980s, environmental concern began globally with international obligations, including a robust political and legislative framework that transformed the environmental outlook and triggered a shift in India's domestic environmental policy (Ballal et al. 2021). In the 1980s, the government of India and development agencies became increasingly aware of the challenges women faced regarding environmental and gender issues. The National Environment

Policy (2006) represents a comprehensive developing strategy on the environment that addresses challenges related to livelihood security, health care, empowerment of marginalized groups, agriculture, natural resources, and economic, political, and social aspects (National Environment Policy 2006, Government of India).

The National Forest Policy (1988) and Joint Forest Management (JFM) programs of the 1990s emphasized that women should have 33% membership in the *Vana Samarakshana Samitis* as decision-making bodies in both rural and urban areas. This included the interaction and involvement of local communities in addressing deforestation. Mohan and Sharma (2022) state that 'The Gujarat Water Policy (2002)' focuses on water management and supply programs where women are involved in the usage of water as well as its management concerning consumption, health, sanitation issues, and the agricultural sector. Renewable energy policies also play an important role for women by providing biogas plants and solar cookers, which produce clean and green cooking energy. Self-help groups (SHGs) have spread to every corner of the country (India), helping women engage in planting and growing crops and trees. Today, India ranks 46th out of 72 nations in the environment and gender index (Sethi 2013).

Tribal women in the Koraput district are empowered through sustainable cereal production by implementing the Self-Help Groups (SHGs) program at the grassroots level. The SHG assists tribal women in investing in sustainable agriculture and increasing employment opportunities (Mohanty et al. 2024).

Conclusion

Women play a significant role in both the family and society. The knowledge they gain from their families and communities is eventually shared in the broader context, particularly in tribal communities. Tribal women are more inclined to visit forests and gather minor forest produce for their families; as they collect firewood, they discover many useful plants and protect them. Their various traditional rituals and ceremonies are connected to plants, flowers, trees, and leaves, which they preserve and cultivate for their own use. However, due to modernization and the rapid growth of mining industries, their ritualistic behaviors gradually weaken, and their agricultural lands are encroached upon by several government development projects; for example, 'Bharat Mala Road' has taken away huge hectares of land. Therefore, there is a significant need to acknowledge the ethnic cultural practices and their lands, which play a major role in conserving plants and their ecosystems. In many tribal areas, women should be encouraged and trained to adopt organic agricultural systems. This will allow them to practice both traditional and organic agriculture, which will promote a healthier environment in the future.

References

Ballal et al (2021) The evolution of India's environmental policy. Indian Forester 147(8):743–753. https://doi.org/10.36808/if/2021/v147i8/165166

Brixi HJ, Sara JJ, Peschka PM (2022) People and planet together: why women and girls are at the heart of climate action. Published on Development and a Changing Climate. Retrieved from https://blogs.worldbank.org/en/climatechange/people-and-planet-together-why-women-and-girls-are-heart-climate-action

Das U (2022) Role of women in environmental protection. Int J Pol Sci Gov 4(2):125–128. https://doi.org/10.33545/26646021.2022.v4.i2b.183

Hindustantimes (2022) Empowering India's women farmers: bridge the gap. Retrieved from https://www.hindustantimes.com/ht-insight/economy/empowering-india-s-women-farmers-bridge-the-gap-101647511783243.html

Indian Express news report (2021) Odisha: Tribal women sow the organic way, earn high, https://www.newindianexpress.com/states/odisha/2021/Nov/01/tribal-women-sow-the-organic-way-earn-high-2378277.html

Khan AM (2024) Nature worshippers: how Sarna rituals propagate forest conservation to new generations & other communities. Down To Earth. Retrieved from https://www.downtoearth.org.in/forests/nature-worshippers-how-sarna-rituals-propagate-forest-conservation-to-new-generations-other-communities-93656

Lata S, Sharma S, Maurya C (2022) Traditional uses of plants in various rituals and ceremonies among Tharu tribe of Udham Singh Nagar, Kumaun Himalaya Uttarakhand, India. Plant Arch 22(1):334–342. https://doi.org/10.51470/PLANTARCHIVES.2022.v22.no1.053

Mahour K (2016) Role of women in environment conservation. J Adv Lab Res Biol 7(1):17–26

Marak C, Singha P (2023) Tribal women ushers in organic farming in India's last village. Village Square. Retrieved from https://www.villagesquare.in/adivasi-woman-ushers-in-organic-farming-in-indias-last-village/

Menon KP, Seema (2016) Role of tribal women in sustainable development. Ind J Appl Res 6(5):272–274

Ministry of Tribal Affairs (2020) Importance of minor forest produces (MFPs). Retrieved from https://trifed.tribal.gov.in/non/timber/msp-mfp

Mishra et al. (2018) Agrobiodiversity conservation and tribal women of Koraput, Odisha. The Tribal Tribune. https://www.etribaltribune.com/index.php/volume-5/mv5i2/agrobiodiversity-conservation-and-tribal-women-of-koraput-odisha#datamethods

Mohan R, Sharma P (2022) The role of women in the conservation of environment with special reference to Assam. J Positive School Psychol 6(3):5849–5855

Mohanty S et al (2024) Women empowerment through sustainable cereal production in tribal dominated regions of Koraput district of Odisha. Journal of Xidian University 18(5):1081–1090. https://doi.org/10.5281/Zenodo.11241842

National Environment Policy (2006) Ministry of environment & forest. Retrieved from https://ibkp.dbtindia.gov.in/DBT_Content_Test/CMS/Guidelines/20190411103521431_National%20Environment%20Policy,%202006.pdf

Niyogi GD (2018) Culture has helped millets survive. Down to Earth Retrieved from https://www.downtoearth.org.in/food/culture-has-helped-millets-survive61730#:~:text=In%20Madhya%20Pradesh%2C%20madiah%20(ragi,evil%20spirits%20in%20tribal%20communities

Sethi N (2013) India 46th in environment and gender index. Retrieved from https://www.livemint.com/Politics/15u3Ro6JA4f718MKI4ZOlN/India-46th-in-environment-and-gender-index.html

Singhal V et al (2021) Role of religious beliefs of tribal communities from Jharkhand (India) in biodiversity conservation. J Environ Plan Manag 64:1–23. https://doi.org/10.1080/09640568.2020.1861587

Sirur S (2024) Dwindling forest commons leave Koraput tribal communities with little to celebrate. Press Institute of India. https://pressinstitute.in/grassroots/dwindling-forest-commons-leave-koraput-tribal-communities-with-little-to-celebrate/

United Nations Climate Change (2023) Tribal women farmers as "change agents" for environmental protection–India. Retrieved from https://unfccc.int/climate-action/un-global-climate-action-awards/winning-projects/activity-database/tribal-women-farmers-as-change-agents-for-environmental-protection-india

World Bank Group (2022) Indigenous women leaders persevere amid a changing climate. Retrieved from https://www.worldbank.org/en/news/feature/2022/08/08/indigenous-women-leaders-persevere-amid-a-changing-climate#

Dr. Kapila Khemundu is Associate Professor and the Head of the Department of Sociology, Chief Warden, and Honorary Director of the Centre for Tribal Welfare and Community Development, Central University of Odisha, India. Prior to his current position, he worked with the National Foundation for Communal Harmony (an Autonomous Organization under the Ministry of Home Affairs) GoI, New Delhi. He also worked with ISS, New Delhi, ISID, New Delhi, and Gateway Research Worldwide. He is the author of a book and 35 peer-reviewed journal articles on tribal welfare and sociology issues.

Niranjan Majhi is a doctoral candidate in Sociology and a Research scholar in the Department of Sociology, Central University of Odisha.

Climate Change and Populism: Examining Climate Governance in Contemporary India

Jyotasana

Abstract Climate change is a critical and pressing challenge for humanity in the twenty-first century. Climate change and populism are closely intertwined, often promoting climate action by endorsing climate governance. As an ideology, populism is expressed through discourse by both political actors and the media within the political system. This chapter examines the complex relationship between climate change and populism in India's contemporary political landscape. The rise of populist politics worldwide, including in India, has presented climate governance with new challenges and opportunities. The chapter also examines the impact of populist beliefs and rhetoric on climate policy and environmental governance and also understand the climate initiatives, political statements, and media discussions in India. The empirical foundation of this chapter is based on a qualitative research method that involves content analysis of political speeches and environmental policy papers.

Keywords Climate change · Populism · Governance · Policy · Political rhetoric

Introduction

Populism is a dominant feature of modern democratic politics and is increasing in prevalence globally. The term 'populism' has been used to describe contemporary politics worldwide, from the political process to authoritarian leaders. Populism has a contradictory relationship to democracy. On the one hand, promising empowerment and promoting public participation appeals to democratic principles, but conversely, populism undermines democratic institutions, civil liberties, and minority rights, encourages authoritarian rule, and refuses to recognize the legitimacy of political contestation (Basu 2018, p. 86).

Jyotasana (✉)
School of Liberal Arts and Humanities, Woxsen University, Hyderabad, Telangana, India
e-mail: jyotasana.verma@woxsen.edu.in

B. Wejnert (ed.), *Democracy in Retreat and the Climate Crisis*, Springer Studies on Populism, Identity Politics and Social Justice,
https://doi.org/10.1007/978-3-032-02881-5_11

In recent years, the emergence of nationalist populism has posed an important challenge to representative democracy and has had a profound influence on the liberal international order. This phenomenon has become one of the most noteworthy occurrences in international politics (Lofflmann 2022). Benjamin Arditi describes the discomfort that populism creates for democracy as akin to the drunken guest at a dinner party who disrupts table manners and rules of sociability (Arditi cited in Basu 2018: 87–88). Hence, the paper examines how populism, as manifested in the governance approaches of India's prime minister, influences climate policy and governance.

The election of Donald Trump as the 45th President of the United States and the success of the Brexit referendum campaign in the European Union (EU) referendum campaign in the United Kingdom in 2016 are two notable instances of populist upheaval in international politics in recent years (Lofflmann 2022, p. 403). The global rise of populism and the prevalence of populist leaders in various governments around the world, including Jair Bolsonaro (Brazil), Rodrigo Duterte (Philippines), Recep Erdogan (Turkey), Boris Johnson (UK), Narendra Modi (India), Donald Trump (USA), and others, has sparked increased interest in the international and transnational aspects of this phenomenon. This includes studying the impact of populist leaders on climate change policy and governance and the interaction between populism and climate change within the liberal international order (Lofflmann 2022).

Research Objective

The following research objectives form the basis for the study:

- To analyze the impact of current populist politics in India on climate governance.
- To explore the political rhetoric and media representation that influence climate governance in India.

Research Methodology

The study is based on the qualitative research approach. Qualitative methods are essential in comprehending the contemporary era and populism. The study is based on the content analysis method. Content analysis, as described by Klaus Krippendorff, is a research method used to draw reliable and accurate conclusions about the settings in which texts (or other relevant material) are used (Krippendorff 2004, p. 18). Content analysis is advantageous for several applications. For instance, deliberate variations in the substance of communication can be revealed. Intentions can be discerned in conjunction with many attributes of the communicator, propaganda, and cultural norms of organizations, institutions, and societies (Rai 2022).

Both primary and secondary data were collected for the study. Primary data is collected from government reports and other documents. For secondary data, relevant literature, news articles, magazines, and non-governmental reports were reviewed, and also collected from various international organizations such as the United Nations, the UN Environment Program, Oxfam, Action Aid, the World Health Organization's (WHO), and other valuable sources. The prime focus of the research is to critically analyze the various organizational reports and government and non-profit organization reports in India and the globe. Apart from this, other organizational reports worked on climate change and populism, especially in climate governance in contemporary political regimes. One of the main aims of the research seems to be to find factual evidence of the increased degree of political rhetoric and media representation of how populist rhetoric influences public perception of climate change in India.

Contextualizing Populism and Climate Change

Climate change refers to global warming, the continuous rise in the average temperature of the Earth's atmosphere. It also encompasses the wider impact of this temperature increase on the Earth's climate system, including historical changes in its climate over extended periods. The changes have had a significant and far-reaching effect on developing nations, especially India, which faces additional difficulties due to its large dependence on natural resources. In India, where a significant portion of the population relies on agriculture and rural livelihoods, the accessibility of vital resources like water and wood fuel is of utmost importance. Climate change has disturbed conventional weather patterns, resulting in erratic rainfall, extended periods of drought, and severe weather occurrences. These disruptions pose a significant risk to the economic well-being of millions of people and exacerbate the vulnerability of communities that lack the necessary infrastructure to adjust.

Therefore, climate change has become an urgent global problem requiring prompt action and collaboration among nations. Nevertheless, the emergence of populism poses a substantial obstacle to endeavors aimed at addressing climate change since populist movements often prioritize immediate public desires at the expense of long-term sustainability (Singha and Singha 2024).

The previous research has indicated a significant connection between populism and the climate crisis. Specifically, right-wing populists have been found to reject the existence of environmental risks that extend beyond national borders. Furthermore, they strongly oppose any measures taken to address the climate crisis. Although these studies offer valuable insights into support for climate action, they often fail to distinguish between political ideology and populism, with only a limited number of research making a distinction, such as between left- and right-wing populism (Huber et al. 2021).

Huber et al. (2021) provide two theoretical explanations that specifically pertain to populism and are not influenced by the political ideology that may shape the impact of populism on climate skepticism. Initially, there was widespread skepticism among the general public towards political institutions, elites, and established norms that play a crucial role in the democratic decision-making processes regarding managing the climate problem. The second is a different category of prestigious establishments that have been the target of populist rhetoric and are assumed to wield influence on the matters of 'truth' in public discussions on climate change. These establishments include academics, scholars, and specialists (Huber et al. 2021).

Further, to a certain degree, climate change and environmental deterioration are susceptible to being targeted by populist movements that oppose globalization. Climate change's abstract and intricate character enables populists to downplay these problems as elite initiatives (Norris and Inglehart 2018). The major emphasis of populism on political elites as the principal players in policymaking provides a unique perspective that is sometimes overlooked in discussions about climate skepticism and positions on environmental preservation. Expanding the scope of political ideology and delving into previously overlooked aspects helps our comprehension of climate and environmental sentiments. In certain situations when party loyalty, political ideology, and populism may be mixed due to the influence of political parties, such as in the United States after the emergence of Donald Trump as a candidate, the impact of party loyalty may be exaggerated (Oliver et al. 2016).

There has been a noticeable rise in political engagement and action around climate change in recent years, with individuals, organizations, and governments adopting more extreme and unconventional approaches. Right-wing populists globally have significantly contested the narrative of climate change as a worldwide issue that is based on intricate interconnections, cumulative greenhouse gas emissions, and a peril to the global population as a whole. Prominent figures such as former US President Donald Trump, Philippine President Rodrigo Duterte, Brazilian President Jair Bolsonaro, and Indian Prime minster Narendra Modi were prominent in leading right-wing populist movements that opposed measures to address climate change (Marquardt and Lederer 2022). Held and Fane-Hervey contend that modern liberal democracies frequently have difficulties efficiently addressing global collective action issues, such as climate change. Controlling the climate problem gives rise to numerous democratic issues and disputes, especially during populism (Held and Hervey 2011, p. 90).

The 'populist moment' in climate change politics can be characterized in three distinct ways: Initially, we demonstrate that the emergence of populist ideologies on both the left and right ends of the political spectrum has resulted in a renewed emphasis on the political aspects of climate change in various nations globally. Furthermore, this renewed focus on the political dimensions of climate change is likely to persist in the long term (Marquardt and Lederer 2022, p. 747). Furthermore, politicians from both right and left ideologies have recognized climate change as a significant arena for political conflict. They employ populist arguments to support their ideas. Although they all strive to protect 'the people,' the definition of 'the people' varies significantly depending on the context (Marquardt and Lederer 2022). Third, regarding populism and democracy, our contributions reach different results. Some have a somewhat favorable view of climate justice movements, while others have a negative vision of the future of liberal democracy (Marquardt and Lederer 2022).

Populism in Contemporary India

Modern politics is filled with populist movements that revolve around elections, political parties, and leaders. These tendencies profoundly impact our understanding of contemporary democracies and their institutions. They primarily indicate a pervasive lack of confidence in traditional democratic establishments, a growing alienation between the public and elected officials, and an escalating aversion to typical political practices (Chakrabarti and Bandyopadhyay 2019).

The emergence of populism in India has substantially influenced the state of democracy in the nation. Populist leaders have endeavored to capitalize on widespread complaints and use them to attain power, sometimes disregarding democratic values and processes. Consequently, there has been a decline in the effectiveness of the nation's democratic establishments and a gradual deterioration of popular confidence in the political system. Populist leaders have faced allegations of employing polarizing language and indulging in hate speech to advance their political agendas (Madar and Basavaraja 2023).

In India, populism has emerged within a distinct socioeconomic framework characterized by swift economic growth, which sets it apart from the Western environment. Narendra Modi, the Prime Minister of India, gained prominence during a time when India was transitioning from a period of rapid economic expansion, sometimes referred to as India's "Gilded Age," characterized by the rise of wealthy individuals. When Modi emerged as a prominent figure in national politics before the 2014 elections, he effectively utilized modern digital media technologies and communication methods to establish a politics centered around his leadership (Tillin 2024). Over the past decade, there has been a noticeable rise in the prominence of Hindu nationalist ideology at the national level, accompanied by an increasingly authoritarian stance towards politics.

The concept of right-wing populism, which associates the desires of the ethnic/racial/religious majority with the public will, is a relatively new phenomenon at the upper levels of Indian politics (Varshney et al. 2021). Populist leaders have employed their language's persuasive force to captivate the public's emotions and thoughts and influence political conversation. They have advocated for many policies and projects, including economic changes, augmented social expenditures, and safeguarding minority rights (Mudde & Kaltwasser 2017).

These policies and activities have significantly impacted the democratic process in India. Populist politicians in India have successfully influenced how public discussions are carried out. They have enhanced the level of personalization and successfully conveyed the worries of the individuals in a more impactful and captivating manner. This has enabled them to captivate the affections and intellects of the populace and influence public sentiment. As a result, they now have more power to impact the political process and determine the government's policy agenda (Subramanian 2007).

Populism's ascent has exerted a substantial influence on India's democratic system. To begin with, populism has led to a greater prevalence of majoritarianism in Indian politics. This has resulted in increased polarization throughout the population, with individuals being divided between those who endorse the populist narrative and those who are opposed to it. This has exacerbated the governing challenge and burdened the country's institutions. Furthermore, populism has impacted the operation of the Indian political system. Populist rhetoric is frequently employed to bypass legal procedures and enact legislation that does not prioritize the public's welfare. This has undermined legal principles' authority and facilitated the government's ability to take independent actions. Furthermore, populism has diminished opportunities for expressing disagreement and engaging in discussions (Kenny 2017).

Therefore, the rise of populism in India is significantly impacting democracy. Populist politicians are using rhetoric to appeal to the masses and employ socio-economic divisions to further their agendas. This has resulted in increased political polarization, a rise in hate speech, and the incitement of violence. Populist policies have been found to restrict democracy by prioritizing specific groups' interests over society's collective interests.

Climate Governance in India

Climate change is the result of either natural internal processes or external forces. It is defined as a shift in climate that can be directly or indirectly attributed to human activity, specifically the alteration of the global atmosphere's composition. This change is in addition to the natural variability of climate observed over similar periods. It impacts the social and environmental factors that determine health, such as clean air, safe drinking water, adequate food, and secure housing.

According to the National Action Plan for Climate Change & Human Health (2018), the prevalence of climate-sensitive diseases is rising due to climate change and extreme weather events, both via direct and indirect effects. The United Nations basis Convention on Climate Change (UNFCCC) and its Kyoto Protocol, established in 1997, serve as the legal basis for the worldwide process of addressing climate change. The Conference of the Parties (COP) to the Convention convenes yearly to engage in negotiations and deliberations about the global climate change agenda and the commitments made by different nations. SDG 13, the sustainable development goal 13, highlights the importance of promptly addressing climate change and its consequences.

The first National Action Plan on Climate Change (NAPCC) of India was unveiled by former Prime Minister Manmohan Singh on June 30, 2008. The document delineates current and forthcoming policies and initiatives to mitigate and adapt to climate change. The strategy outlines eight fundamental 'national missions'. Further, following the conclusion of the 21st Conference of Parties (COP 21) under the United Nations Framework Convention on Climate Change (UNFCCC) in Paris, Prime Minister Narendra Modi expanded India's efforts to address climate change by launching four new missions, including one focused on 'Health,' in 2014. The planned 'Mission on Health' will tackle the health-related elements of climate change using a comprehensive and multifaceted strategy.

Climate change, a dynamic global and regional change, is causing environmental degradation, pollution, deforestation, and resource depletion. Due to economic expansion and lifestyle changes, India faces increasing pressure to establish a strong climate and environment policy. This growth is causing increased greenhouse gas emissions, impacting agriculture, rainfall, forest coverage, coastal erosion, and health and livelihoods. India must establish a strong climate and environmental policy to address these challenges (Varshney 2023).

Hence, it is crucial to have a comprehensive national plan that focuses on India's growth profile and prioritizes the implementation of environmental legislation. The development of the 'National Action Plan on Climate Change' in 2008 (NAPCC) has resulted from all these factors. Implementing this strategy required a challenging adjustment in our development approach towards achieving both 'sustainable growth' and addressing climate change. With this consideration, eight missions were developed. Their identities are:

- National Solar Mission
- National Mission for Enhanced Energy Efficiency
- National Mission for Sustainable Habitat
- National Water Mission
- National Mission for Sustainable Himalayan Ecosystem
- National Mission for Green India
- National Mission for Sustainable Agriculture
- National Mission for Strategic knowledge for climate change (NAPCC)

India faces the challenge of balancing economic growth with reducing greenhouse gas emissions. Effectively tackling climate change and ensuring sustainable development requires a comprehensive national policy. India has various options to develop such projects and is positioned for progress. The goal is to establish a self-sustaining society that protects the interests of current and future generations (Varshney 2023). The National Action Plan hinges on the development and use of new technologies. Its implementation would be through appropriate intuitional mechanisms suited for effectively delivering each mission's objectives, including public–private partnerships and civil society action.

India's approach to climate change is seen in several social and economic development initiatives. India is successfully executing its national plans on climate change. The government aims to decrease the emissions intensity of gross domestic product (GDP) by 20–25% by 2020, compared to 2005. The Twelfth Five-Year Plan has selected many key focus areas for this goal. India's Twelfth Five-Year Plan, from 2012 to 2017, aimed to achieve quicker, more inclusive, and sustainable economic growth. The Indian government has implemented many policies and programs to research and implement options for low-carbon development.[1]

The Prime Minister's Council on Climate Change (PMCCC), first established in June 2007, was recreated by the Government in November. The Council's primary responsibilities include developing a unified approach to addressing climate change challenges at the national level, supervising the creation of action plans for assessing, adapting to, and mitigating climate change, and regularly evaluating significant policy choices.[2] Prime Minister Modi has the chance to position India as a leader in adopting a new low-carbon development strategy. This would showcase India's ability to be a global environmental leader while achieving economic growth (Antholis 2024).

Nexus between Climate Governance and Populism

India, with the highest population in the world, is also very susceptible to climatic dangers. This is not just because of the intensified heat and floods caused by global warming but also because a significant portion of the country's population, which amounts to 1.4 billion individuals, is already susceptible to such impacts. Most individuals worldwide are impoverished and lack a social safety net (Sengupta 2024). In this regard, Populists utilize several platforms to disseminate their beliefs, including conventional political campaigning, but they are increasingly focusing on social media endeavors. Populists frequently employ assertive rhetorical claims to bolster their political goals. Populist leaders and movements have always adjusted their communication strategies to suit the available media platforms (Nadler 2019).

The Indian government, led by Narendra Modi, has gained international recognition and a reputation as a prominent advocate for climate action. This is evident from Modi's speech at COP26 in Glasgow, where he pledged India's commitment to achieving net-zero emissions by 2070. During his G20 presidency last year, Modi announced a target to triple India's renewable energy capacity, which was further emphasized during the initial 'global stocktake' at COP28. Nevertheless, despite the growth in renewable energy capacity, the Indian government's swift development of coal and Prime Minister Modi's associations with fossil fuel companies have been criticized as troublesome by many and regarded as practical by others (Chandrasekhar 2024).

India was ranked as the world's third-highest emitter of carbon dioxide in 2020, according to the UN's Emissions Gap Report (EGR). Although its per capita emissions were lower than the global average, India's target for achieving net zero emissions by 2070 lags significantly behind the deadlines set by other nations and does

[1] India, Ministry of Environment, Forest and Climate Change, India's Progress in Combating Climate Change, December 2014, P. Sec. I; ii

[2] Lok Sabha Unstarred Question No. 3809 dated 16.12.2014

not align with the Paris Accord. The Paris Agreement advocates achieving net zero emissions by 2050 to limit global temperature increases to 1.5C. India has committed to producing 50% of its power from non-fossil fuel sources by 2030 and aims to preserve forest cover to act as a carbon sink. Prime Minister Narendra Modi has also set a goal of achieving 500 GW of energy capacity from non-fossil fuel sources by 2030 (Menon 2022).

During the 26th session of the Conference of the Parties (COP26) to the United Nations Framework Convention on Climate Change (UNFCCC) in Glasgow, United Kingdom, in November 2021, Prime Minister Modi announced his commitment to strengthen India's Climate Action Plan (CAP). He presented five key components, referred to as the '*Panchamrit,*' which outline India's approach to addressing climate change. In addition to setting a five-pronged target for India and committing to achieving zero emissions by 2070, PM Modi emphasized the importance of adopting a sustainable lifestyle. He also called for making 'Lifestyle for Environment' (LiFE) a global mission through more ambitious actions by the global clean energy community.[3]

Prime Minister Narendra Modi introduced a proactive foreign policy, known as the 'Modi doctrine,' with key policies such as 'India first' and 'neighbourhood first' emphasizing the country's culture and heritage. This approach includes soft power tactics like the International Day of Yoga and the 'Made in India' initiative. Modi also aims to involve the Hindu diaspora through cultural and emotional ties to India. The adoption of the International Day of Yoga by the United Nations was presented as a step towards achieving world peace and countering climate change (Kinnvall 2019).

Yoga is an invaluable gift of our ancient tradition. It is not about exercise but to discover the sense of oneness with yourself, the world and the nature. By changing our lifestyle and creating consciousness, it can help us deal with climate change. (Modi 2014).

This emphasis on culture and heritage has resulted in a more precise definition of Hindu nationalism as Indian nationalism. The narrative of India as the 'world's largest democracy' and a 'rising' power must be viewed in this context. This narrative is influenced by the notions of India's exceptional status among 'great powers' (Mehta 2009, pp. 224–225). It is a celebration of a magnificent culture with profound historical origins. Mentions include those of 'India's responsibility and inherent pacifism' (Chacko 2014, p. 329). During the Glasgow summit, Prime Minister Narendra Modi announced India's commitment to achieving net zero emissions for the first time. Net zero, or achieving carbon neutrality, refers to the state when greenhouse gas levels in the atmosphere do not increase (BBC, November 2021).

India's participation in COP-28, under the leadership of Prime Minister Shri Narendra Modi, demonstrates a thorough and aggressive strategy in addressing climate change. Prime Minister Modi underscored the pressing need for readily available climate funding for developing countries while stressing the need to implement

---

[3] See details: https://pib.gov.in/PressReleaseIframePage.aspx?PRID=1961797SS

Nationally Determined Contributions (NDCs) and expressing concerns on behalf of the Global South. India's dedication to multiple techniques for sustainable development is demonstrated via several projects, such as Lead IT 2.0 for industrial transformation, the Green Credits Programme, which incentivizes pro-environment acts, and the focus on the fragility of the Himalayan area (Forest and Climate Change 2023).

It is crucial to analyze the climate concerns presented in Indian media within a populist framework since these form public opinion and impact policy discussions. Due to its potential to produce national differences in the public's awareness of this global environmental concern, disseminating climate change information from scientists and policymakers to the public through mass media has attracted considerable attention (Billett 2009). The formulation of climate policies involves a thorough analysis of the present condition, the strategic positioning of climate change as a matter of policy, and a clear vision for a favorable future. These policies are intricately intertwined with intricate power dynamics and societal perspectives (Fischer and Forester 1993; Lakoff 2010).

Conclusion

The populist focus on swift economic growth and meeting the expectations of a wide voter base has occasionally led to concessions on environmental constraints. In certain cases, policy choices prioritizing industrial expansion and infrastructural development may clash with the rigorous environmental safeguards required to address climate change properly. Furthermore, populist language has the ability to influence public discussions on climate matters in a manner that either minimizes the need for immediate environmental action or presents it in a way that is more compatible with nationalistic objectives rather than worldwide environmental obligations. Eventually, India's Prime Minister Narendra Modi has substantially impacted climate policy through his political communication. Conflicting populist goals still influence the nation's attention towards climate change. The challenge is to bridge this divide and guarantee that climate change receives the immediate and continuous focus it needs in policy and governance structures.

References

Antholis JW (2024) India's climate change policy in a Modi government. Retrieved from https://www.brookings.edu/articles/indias-climate-change-policy-in-a-modi-government/

Basu, A (2018) Narendra Modi and India's Populist Democracy. *I*ndian Politics & Policy 1(1): 83–106. https://doi.org/10.18278/inpp.1.1.5

BBC (2021) COP26: India PM Narendra Modi pledges net zero by 2070. Retrieved from https://www.bbc.com/news/world-asia-india-59125143

Billett S (2009) Dividing climate change: global warming in the Indian mass media. Climatic Change, 99 (1):16. https://doi.org/10.1007/s10584-009-9605-3

Chacko P (2014) A new ‘special relationship?’ Power transitions, ontological security, and India-US relations. Int Stud Perspect 15(13):329–346
Chakrabarti K, Bandyopadhyay KK (2019) Populism in contemporary Indian politics. Working papers series: populism in Asia 5. Retrieved from file:///C:/users/Dell/downloads/5.Populism_India%20(2).Pdf
Chandrasekhar A (2024) India election 2024: what the manifestos say on energy and climate change. Retrieved from https://www.carbonbrief.org/india-election-2024-what-the-manifestos-say-on-energy-and-climate-change/
Fischer F, Forester J (eds) (1993) The argumentative turn in policy analysis and planning. Duke University Press. https://doi.org/10.2307/j.ctv1220k4f
Forest and Climate Change (2023) India at COP-28: highlights of 28th conference of parties. Ministry of Environment, India. Retrieved from https://static.pib.gov.in/WriteReadData/specificdocs/documents/2023/dec/doc20231212285701.pdf
Held D, Hervey A (2011) Democracy, climate change and global governance: democratic agency and the policy menu ahead. The Governance of Climate Change. 89–110
Huber RA, Greussing E, Eberl JM (2021) From populism to climate scepticism: the role of institutional trust and attitudes towards science. Environmental Politics 31(7):1115–1138. https://doi.org/10.1080/09644016.2021.1978200
Kenny PD (2017) India’s turn to populism. In: Kenny PD (ed) Populism and patronage: why populists win elections in India, Asia, and beyond. Oxford University Press, Oxford, pp 100–122. https://doi.org/10.1093/oso/9780198807872.003.0006
Kinnvall C (2019) Populism, ontological insecurity and Hindutva: Modi and the masculinization of Indian politics. Camb Rev Int Aff 32(3):283–302. https://doi.org/10.1080/09557571.2019.1588851
Krippendorff K (2004) Content Analysis: An Introduction to its Methodology. New Delhi: Sage Publications.
Lakoff G (2010) Why it matters how we frame the environment. Environ Commun 4(1):70–81. https://doi.org/10.1080/17524030903529749
Löfflmann G (2022) Introduction to special issue: The study of populism in international relations. The British Journal of Politics and International Relations 24(3):403–415. https://doi.org/10.1177/13691481221103116.
Madar D, Basavaraja G (2023) The impacts of rising populism on democracy in India. Int J Sci Soc Sci Res 1(3):291–298
Marquardt J, Lederer M (2022) Politicizing climate change in times of populism: an introduction. Environ Poli 31(5):735–754. https://doi.org/10.1080/09644016.2022.2083478
Mehta PB (2009) Still under Nehru’s shadow? The absence of foreign policy frameworks in India. India Rev 8(3):209–233
Menon S (2022) Climate change: what emission cuts has India promised? Retrieved from https://www.bbc.com/news/world-asia-india-58922398
Ministry of Health & Family Welfare. (2018). National Action Plan for Climate Change & Human Health. Government of India. https://ncdc.mohfw.gov.in/wp-content/uploads/2024/04/27505481411548674558.pdf
Modi N (2014) Text of the PM’s speech to the United Nations. Retrieved from https://www.narendramodi.in/text-of-the-pms-statement-at-the-united-nations-general-assembly-2-2868
Mudde C, Kaltwasser CR (2017) Populism: A very short introduction. Oxford University Press. https://doi.org/10.1093/actrade/9780190234874.001.0001
Nadler A (2019) Populist communication and media environments. Soc Comp 13(8):1–10
Norris P, Inglehart R (2018) Cultural backlash: trump, Brexit, and the rise of authoritarian-populism. Cambridge University Press, New York
Oliver JE, Rahn WM, Bartels LM (2016) Rise of the Trumpenvolk: populism in the 2016 election. Ann Am Acad Pol Soc Sci 667(1):189–206. https://doi.org/10.1177/0002716216662639
Rai D (2022) Contested Representation: Dalits, Popular Hindi Cinema, and Public Sphere. New York: The Rowman & Littlefield Publishing Group.

Sengupta S (2024) Modi will feel the heat in a third term. And Not Just Politically. Retrieved from https://www.nytimes.com/2024/06/04/climate/modi-climate-india-election.html

Singha S, Singha R (2024) Populism and climate change. In: Chacko Chennattuserry J, Deshpande M, Hong P (eds) Encyclopaedia of new populism and responses in the 21st century. Springer, Singapore. https://doi.org/10.1007/978-981-16-9859-0_445-1

Subramanian N (2007) Populism in India. The SAIS Review of International Affairs 27(1):81–91

Tillin L (2024) The political economy of populism in India. Retrieved from https://www.kcl.ac.uk/the-political-economy-of-populism-in-india

Varshney S (2023) Climate change Laws of India. IARJSET 10(9):352–355. https://doi.org/10.17148/IARJSET.2023.10942

Varshney A, Ayyangar S, Swaminathan S (2021) Populism and Hindu nationalism in India. St Comp Int Dev 56:197–222. Retrieved from. https://doi.org/10.1007/s12116-021-09335-8

Dr. Jyotasana is an Assistant Professor of History at the School of Liberal Arts and Humanities, Woxsen University, Hyderabad, Telangana, India. She pursued a Doctorate of Philosophy from the Central University of Gujarat, Vadodara, Gujarat, India. Her areas of expertise are political history, populism, modern Indian thinkers, caste, class, gender, political communication, the socio-economic history of modern India, and business history. Her book reviews and research articles have been published in South Asia Research (Sage), National Identities (Taylor & Francis), History: The Journal of the Historical Association (Willy), Contemporary Voice of Dalit (Sage), Our Heritage, Knowledge Consortium of Gujarat, Sab log, and Radha Kamal Mukherjee: Chintan Parampara.

Role of Women in Local Governance. A study on Panchayati Raj, Chiraura, Patna, Bihar, India

Divya Sharma

Abstract The goals of sustainable development emphasize gender equality, which entails equal rights, opportunities, and treatment for all genders in social, economic, and political aspects. This paper explores women's equality from political and social perspectives in Chiraura, a village in Patna, India. Despite the constitutional democratic rights granted to women, the ground reality remains different. Women's role in governance, especially at the micro level, was studied to determine the necessary improvements. The objective of the research paper was to analyse how decision-making in planning and implementation is influenced or co-opted. Social conditions and stereotypical mindsets were examined, along with women's involvement in the workforce and their financial reliance on men. This exploratory study utilized a mixed-methods approach, combining qualitative and quantitative techniques. It gathered insights through an open-ended questionnaire and observational data, supported by anecdotes and documented observations. The data was visually represented using pie charts and bar diagrams, illustrating the proportion of women in the village's Panchayati Raj system. The findings reveal that women elected as Panchayat heads are often leaders in name only, while their husbands or male relatives wield the actual power, making decisions on their behalf. This practice undermines their authority and limits their real influence in governance. Women were often uninformed about political and governmental affairs. Their involvement seemed tokenistic, with men retaining actual decision-making authority. The cultural stereotype persisted—how could a mother or wife step out when she had never ventured outside before? Women's economic situation was characterized by financial dependence on men. Societal conditioning and cultural expectations shaped women's perceptions from childhood, confining their roles to that of a wife and

This study was presented as a paper at the ninth Annual Conference of the World- CWS, Central University of Odisha, Koraput, Odisha, India. seventh to ninth December 2023.

D. Sharma (✉)
Faculty of Economics and Business Studies & Convener, Department of Commerce and Social Science, Manava Bharati International School, Patna, India

B. Wejnert (ed.), *Democracy in Retreat and the Climate Crisis*, Springer Studies on Populism, Identity Politics and Social Justice,
https://doi.org/10.1007/978-3-032-02881-5_12

mother. Understanding gender discrimination, stereotypes, patriarchal attitudes, and subjugation, along with recognizing their impact on women's lives, is crucial. It involves questioning assigned roles and fostering awareness of the importance of women's equality in society.

Keywords Women · Decision-making · Gender discrimination · Governance · Tokenistic · Co-opted Gender · Stereotype · Subjugation

Introduction

> Women have a vital role in environmental management and development. Their full participation is therefore essential to achieve sustainable development. (Principle 20, Rio Declaration)

The concept of sustainable development first received international recognition in 1972 at the UN Conference on the Human Environment held in Stockholm. It later became popular in the year 1987, the United Nations Brundtland commission report also referred to as Our Common Future, defining sustainable development as 'development meeting the needs of the present without compromising the ability of future generations to meet their own needs', alerting about the negative environmental concerns. In September 2015 the United Nations adopted the Sustainable Development Goals (SDGs) as a universal need for actions to end poverty, protect the planet, and to ensure that all people enjoy peace and prosperity by 2030. Four areas of sustainable development including human, social, economic, and environmental have to be in proper equilibrium in order to protect our ecosystem and preserve natural resources. The United Nations conference on environment and development (UNCED) in 1992 exerts women contribution and full participation in sustainable development.

The survival of mankind depends on resources, and injudicious and over-utilization of resources causes severe environmental threats. Development depends mainly on how efficiently resources are utilized, not only by one section of society but by all. Ensuring the participation of both men and women in the allocation of resources is not only the path to the betterment of the economic and social aspects of women but, when shifted towards sustainability, benefits society as a whole, including SDGs. The fifth goal of SDG talks about gender equality and empowering all women and girls. SDG indicator 5.5 ensures women's full and effective participation and equal opportunities for leadership at all levels of decision making in political, economic, and public life. SDG indicator 5.5.1 Proportion of seats held by women in national parliaments and local government monitors the percentage of women in elected roles within local governments. It measures progress toward gender equality in leadership and governance. Women and development and women development are commonly given importance by government and other bodies reflecting the share of development yet to be received by women. Economic participation not only contributes towards a country's GDP growth but also reduces poverty among women and raising household income, providing independence. By providing women with opportunities and support, we foster economic, social, and environmental strength through their unique contributions and perspectives.

This paper focuses on the understanding of women's involvement in the grass-roots level of governance, at the panchayat gram sabha level. A village-level assembly of voters that serves as the legislative body for a *gram panchayati* as the *Gram Sabha*. It composes of all the registered voters within a village or group of villages under a gram panchayat and serves as the primary decision-making body for local governance. It is a permanent body accountable and responsible for overall functioning of the panchayat. Bihar is located in the north eastern part of India, divided into 38 administrative districts. Patna is the capital of Bihar, being one of these administrative districts. The study of this paper is from Chiraura village which is located in Naubatpur subdivision of Patna district in Bihar, India. The study uses a mixed-methods approach, combining both qualitative and quantitative data collection methods. This includes open-ended questions to gather detailed insights and observational data to provide additional context and understanding. The paper highlights several dynamics, such as limited decision-making opportunities, pervasive gender inequality, cultural stereotypes, ingrained beliefs, and the presence of proxy candidates influenced by patriarchal norms. These factors collectively reflect a lack of awareness and reinforce traditional power imbalances (Map 1).

Women have been deprived for ages in various forms, facing inequalities, lack of equal opportunities, decision-making power, and still facing challenges globally. Mostly women are deprived in rural areas; no importance is given to improvement. The political equality of women recognized in the Indian Constitution Act is a landmark in the political history of India. The 'National Policy for the empowerment of women, 2001' ensured to take all measures to guarantee women equal access and full participation in decision-making at every level. The policy aims to bring advancement, development, and empowerment of women. Women's active participation in governance enhances women's decision-making power in society, providing equal opportunities and combating gender discrimination and injustice against women.

Rationale for the Study

The political equality of women recognized in the Indian Constitution Act is a landmark in the political history of India. The 'National Policy for the empowerment of women, 2001' ensured to take all measure to guarantee women equal access and full participation in decision making bodies at every level. The policy aims to bring advancement, development and empowerment of women. Women's active participation in governance enhances the decision-making power of women in society providing equal opportunities to combat gender discrimination and injustice against women.

Map 1 Map of Bihar, India. Source: www.mapsofindia.com

India forms a decentralized government, divided into three levels: National, state and local. Distribution of legislative, executive and administrative powers at these three levels emphasizes men's and women's direct and inclusive participation. Research on this level has been minimal, reflecting the knowledge deficit. To understand the representation of women and all levels of governance. Major focus is given to India's highest legislative body, the parliament, but the story of the grassroots level remains untold. Measuring women's representation in governments through their vote percentage in legislative election results is straightforward, but assessing their actual involvement in the decision-making process is much more complex. This paper highlights the actual participation of women in local government.

Significance of the Research

The study of this research paper is important from the viewpoint of rural women who are deprived and face inequalities in society. The role of women in governance especially at micro level must be studied in order to understand whether policy intervention is needed to improve it. 50% of the population is women so their effective involvement has inclusive impact on government. Society as a whole will be benefitted with increased involvement of women in governance in terms of inclusive and equitable society. Women do not only need equality but deserve equality. More study and research is required to estimate the actual participation of women in decision making process in local governance. This paper will throw lights on awareness required in taking women involvement in local governance.

Research Objectives

This research paper aims to present a concise overview of research on women decision-making patterns in the *Panchayat Gram Sabha* focusing on environmental issues and their priorities in micro planning for the *Panchayat*, based on a case study of Chiraura village situated in Naubatpur block of Patna district in Bihar.

The key objectives of the research paper include:

1. Understanding participation and the role of women in grassroots governance.
2. Capturing the dynamics of co-option of decision-making related to planning and implementation.
3. Women's participation in the workforce. Their financial dependency on men.
4. Societal condition, stereotypes, and the mindset of people.

Literature Review

Women, also known as caregivers, play a key role in nurturing their families, children, and communities, acting as the backbone of every society. It is the mother who welcomes both genders as children, wishes for both genders to attend school without discrimination, and plays a primary role as an educator in boosting child development. As women, known for maintaining balance in life and taking better care of their surroundings, they play a significant role in shaping society but are still left behind and remain in a vulnerable position. Women are considered the backbones of our communities; yet, their voices are silenced in this patriarchal society. Even when women are recognized by governmental bodies, they are still deprived of basic initiatives, including inequality, lack of opportunities, exclusion from decision-making power, political rights, and many more. The same pattern and tradition can easily be seen among local village governments, namely *Panchayat Raj* and *Gram Sabha*, where the participation of women is highly encouraged, but their interference in administration is highly discouraged. As for sustainable development, it is a broader aspect that focuses not only on the efficiency of resource utilization but also on ensuring participation. Both men and women are involved in the allocation of resources.

The Status of Women in Local Governments, by Shaul (1982), argues in her paper that achieving equal political power for women will enable them to contribute more effectively to society, develop their unique abilities, and address societal needs. This will lead to a better, more inclusive, and diverse society where the interests of everyone, not just the dominant group, are represented.

Women in Local Government: A Study of Maharashtra, by D'Lima (1983), clearly shows that many women were elected to political positions due to the influence of their husbands or other powerful politicians. Furthermore, 58% of these women were unable to participate actively in important decision-making within the panchayat. This underscores the urgent need to tackle gender inequality in the political framework of PRIs. Only 42% of the women actively engaged in their political roles and fulfilled their responsibilities. The study also highlights a connection between women's social development and their involvement in local self-government.

Women Better Placed for Local Government, by Smith (1995) examines women's participation in local governance. The author notes that despite some women holding decision-making roles, their representation remains limited. Smith emphasizes women's close connection to social issues in their communities, highlighting their unique sense of community and exceptional ability to identify and address local governance challenges.

Understanding the Policy Impact of Electing Women, by Swers (2001), found in her study that women elected to state legislatures prioritize policies related to women's issues, such as welfare, health and education. The author argues that women bring a distinct perspective to their role as representatives, with female legislators showing a stronger sense of responsibility to advocate for women's interests compared to their male counterparts.

Women's Participation in Politics through Reservation by Benke Survana (2011) emphasizes the effectiveness of reservation as a tool for women's political participation and highlights its potential to improve intercommunity understanding.

A Case Study on Women leadership in Panchayati Raj Institutions (PRI) at the Gram Panchayat level by Narender Paul emphasizes that Smt. Mamta Devi is currently serving her second term as the President of her *Panchayat*. This *Panchayat* is located in the Dharamshala block of District Kangra in Himachal Pradesh. This case study explores how elected women leaders in the PRI succeed in their role despite the highly patriarchal and traditional social norms prevalent in the region.

Women's empowerment traces its roots to the nineteenth-century feminist movements, with the initial wave emphasizing political rights, notably the right to vote. Subsequent waves in the 1960s and 1970s prioritized social and economic rights like equal pay and educational access. The 1990s saw the third wave, spotlighting the diverse experiences of women and the intersection of gender, race and class. Optimizing the utilization of women's skills and abilities can yield widespread benefits, such as boosting economic growth globally; reducing poverty levels worldwide; enhancing the cost-effectiveness of healthcare and social programs; enabling government policies to better address the diverse needs of citizens; and mitigating environmental harm resulting from unsustainable practices. Women are presumed to be better planners, as they are more inclined to collaborate, seek input from diverse perspectives, fostering a more inclusive planning process.

The Study

The foundation of this research is part of Manava Bharati International School's annual Community Engagement program in Patna, which centres on land use and sustainability according to the CBSE framework. Chiraura is a village located on the outskirts of the state capital, Patna. Data are collected through a questionnaire developed by teachers and students, aligning with experiential learning principles in the National Education Policy (2020) and CBSE guidelines, involving both teachers and students.

The study is based on the case of the *Gram Panchayat* of village Chiraura. The *Gram Panchayat* is the local self-government authority situated within the Naubatpur Panchayat Samiti, which falls under the jurisdiction of the Patna Zila Parishad. The Chiraura *Gram Panchayat* exercises administrative control over three villages. The *Panchayat* of Naubatpur is broadly subdivided into 14 wards and comprises a total of 32 elected representatives. The village has a total population of 3414 residents, including 1775 males and 1639 females, covering an area of 358 hectares, according to the 2011 Population Census. The average sex ratio of Chiraura village is 923, which is higher than the Bihar state average of 918. The total literacy rate is 75.17%, with male literacy standing at 83.18% and female literacy at 66.59%, as per the state census of 2011 (Figs. 1 and 2).

The village is renowned for its diverse flora and fauna. In Chiraura village, a relatively small proportion of the population, specifically 11.77%, is engaged in agriculture as either cultivators or agricultural laborers. Out of the total land holdings, only 15% is owned by women, showcasing skewed gender biases.

Women make up half of the population in India, yet their representation in political, economic and social fields is not proportional to their numbers. This phenomenon has been emphasized in the provisions of the Indian Constitution, which aim to incorporate women into the mainstream through reservations for women on political grounds. Indian Constitution Clause (3) of Article 243D ensures women's participation in Panchayati Raj institutions by allocating no less than one third of the total number of seats filled by direct elections and offices of chairperson of Panchayat

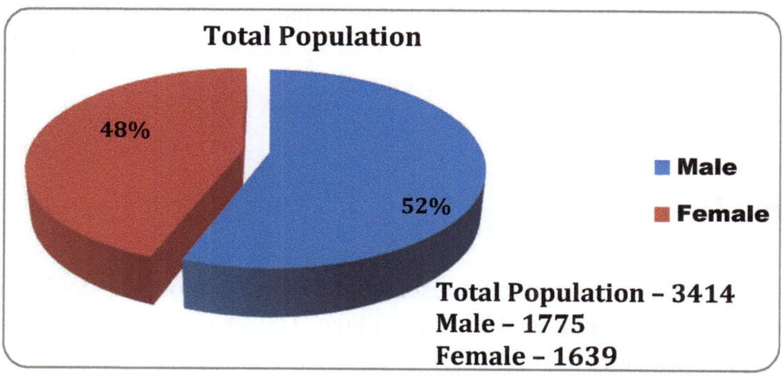

Fig. 1 Total population of Chiraura, Patna, Bihar, India. Source: Population Census 2011 India

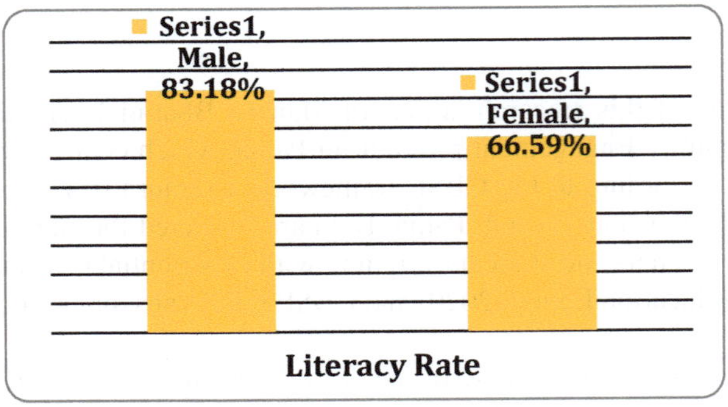

Fig. 2 Literacy rate of Chiraura, Patna, Bihar, India. Source: Population Census 2011 India

to women. Twenty states in India, including Bihar, have enacted provisions for 50% reservation for women in Panchayat Raj institutions (PRI) in their respective Panchayat Raj acts. The total number of PRI representatives in Bihar is 136,573, of which the total number of elected women representatives (EWRs) is 71,046.

Indian politics is well known primarily for its male members and for having restricted opportunities for women to participate, especially at the local level. Women's participation in *Panchayati Raj* institutions, breaking down barriers and conventions of family and society, improves decision-making power and involvement in society.

Method

This exploratory study employs a mixed-methods approach, incorporating both qualitative and quantitative methods. Utilizing an open-ended questionnaire, along with observational data, the research captures insights through anecdotes and documented observations. The investigation relies on a combination of primary sources, including the open-ended questionnaire responses and observations, as well as secondary sources such as the District Census Handbook, NSSO (National Sample Survey Office), and various reports. The primary focus is on gathering information related to women's participation in Panchayat.

Research Design

This study adopts a cross-sectional research design to explore the decision-making paradigm among women in the Chiraura village. The focus is on understanding how this paradigm influences the relationship between micro planning and sustainability. By taking a snapshot, the research aims to grasp the current perceptions and characteristics of women ward members and their broader societal impact. Observation was made of the critical natural resources a pond an embankment used by the pedestrians and a program related to drainage system.

Sampling Strategy

The study employs stratified sampling to ensure representative participation of women in the village. Using secondary data from NSSO, the village's population categorized in male and female. Random selection was facilitated to mitigate biases and enhance credibility. A set of participants from different ward member including both men and women, residing in Chiraura. Demographic information was collected, including age, gender, and caste affiliation. The responses are based on the Focused Group Discussion (FGD) of the ward members and also family members. Data collection involved a structured questionnaire with 30 questions, pre-tested for clarity and adjusted based on feedback. Informed consent was obtained, ensuring confidentiality and anonymity. Self-administered questionnaires encouraged candid responses.

Variables and Measures

The primary variables investigated included population, average sex ratio, child sex ratio, literacy rate, land use, occupation and ownership of land among women and men, which were quantified.

Data Analysis

Quantitative data was obtained in reference to the universe of study, emphasizing the division of the democratic setup on the basis of gender for the village Chiraura. This diagrammatic representation of data is shown with the help of a pie chart and a bar diagram. It elaborates on the ratio of women in the village's *panchayat raj* system. Skewness in land ownership majorly favours the male section of the society (Fig. 5). While going through the primary data collection, ethical considerations were paramount, and informed consent was obtained, and participants were assured of the voluntary nature of their participation.

Limitations

It's essential to acknowledge the study's limitations, which could be related to the sample size, the representativeness of the selected groups, or potential biases. To enhance the study's validity and reliability, specific measures such as content validity and test-retest reliability could have been implemented.

Findings and Discussions

Women elected as *Panchayat* heads often serve as symbolic leaders. At the same time, their husbands or male relatives take control of their roles and responsibilities, effectively undermining their authority and reducing them to figureheads. While reservation policies have increased the number of women in leadership roles, these policies sometimes result in token representation rather than real empowerment. Patriarchal norms discourage women from actively taking up leadership roles. Male dominance often overrides their authority. Women are unaware of their rights and duties; they feel incapable or reluctant to fulfil their roles.

Women Participation at Ward Meetings

- General meetings of the ward *samiti* are organized on a regular basis for their day-to-day activities; this was found to be absent from the universe of the study. Out of a total of 14 ward members, six are women representatives (Fig. 3).

- Women don't perceive attending meetings as significant, and patriarchy plays a key role here, suggesting that our women will not step out of the house. Mukhiya Pati (husband of mukhiya) takes pride when people from upper castes visit him to obtain the initials of his wife, lady mukhiya, the elected head of the *panchayat*.
- Women lack awareness about occurrences in governance. They are least interested in the planning and implementing plans; for them, it's men who actively participate and make decisions.
- Local people are uninterested in attending *Gram Sabha baithak*, which is used to discuss local governance and development plans of the village; everyone wants direct benefits without attending meetings.

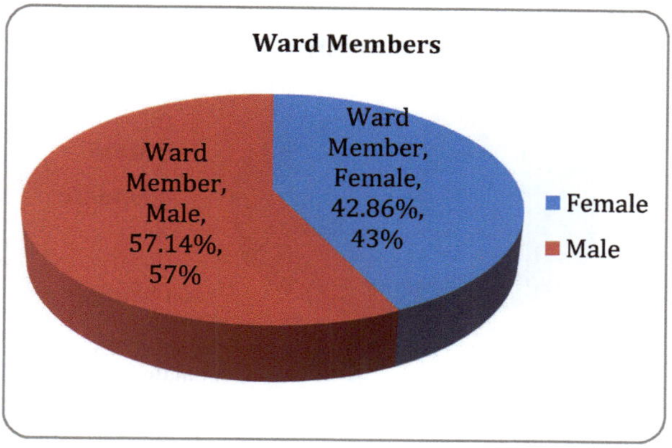

Fig. 3 Percentage of ward members on the basis gender. Source: Population Census 2011 India

Proxy Representative

- Women ward members, especially those from high caste communities, often have their spouse or son co-opt their participation and decision making.
- • Men were the representatives in the past; since the seat has been reserved for women, their wives or mothers contested the elections and became proxy candidates and representatives.
- • Female participation appears largely symbolic, as decision-making power remains predominantly in male hands.

Biased Stereotype Mindset

The cultural stereotype works—how can a mother/wife step out of her house—she has never been outside.

- Women are confined to the customs and traditions of their families as they are considered to be a reflection of the family's reputation, giving freedom to women degrades the family's reputation. They step out of doors for a medical emergency only.
- Women's contribution to the overall workforce is significantly lower than that of men. Traditional gender roles persist, with many viewing women primarily as responsible for household chores.
- There's a prevailing mindset that women prioritize domestic duties as they are born to do housework.

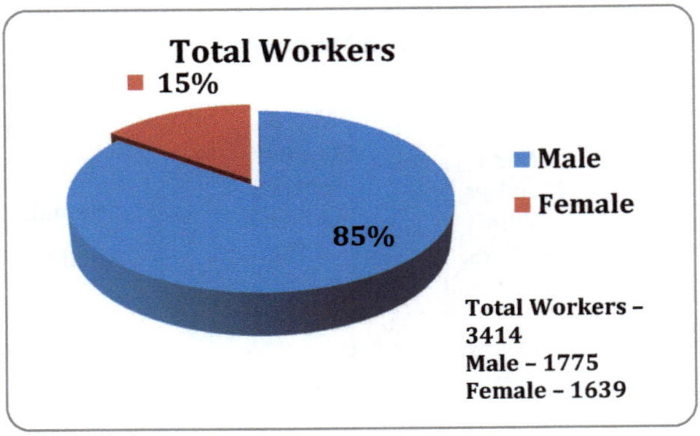

Fig. 4 Percentage of total workers based on gender. Source: Population Census 2011 India

Economic Vulnerability

Financial dependence on men characterizes women's economic situation.

- Some men, perceiving themselves as financially secure, question the need for women to work outside the home and contribute financially.
- Women have no independent income source; only 15% of the workforce comprises females (see Fig. 4). They are paid less than men for any work in the village.

- Female participation in society is significantly low compared to that of males.
- The depiction of land distribution among different gender segments within society in the adjoining pie chart analyses the key findings related to the study. Most land ownership is held by men in the village (see Fig. 5).

Outlook at the Village People

Everybody within the Panchayat agrees with the prevailing norms for women and is happy about it—what the women members can do is the real refrain.

- Abiding by the social norms and customs, people don't see it as gender discrimination; in fact, women, without questioning, accept the given role.
- Societal conditioning and cultural expectations shape women's perceptions of their limited role as wives and mothers from childhood.
- Some women perceive seat reservation for women in the *Panchayat* as a burden rather than beneficial, due to cultural influences and a traditional mindset.

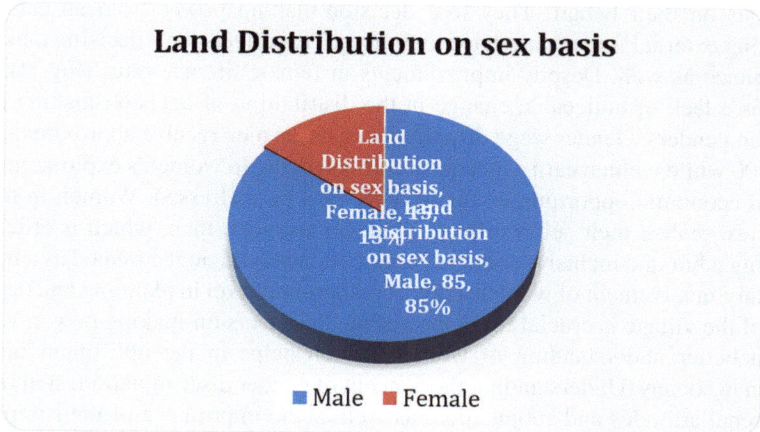

Fig. 5 Percentage of land distribution based on gender. Source: Population Census 2011 India

Sustainability

Utilization of renewable energy sources in the village.

- Using solar energy to generate electricity for street lighting.
- Cow dung used for cooking, agriculture, and as manure.
- A neglected well causing groundwater pollution; the hand pump is dry with no water.

Conclusion and Discussion

The study reveals that while women are involved in *Panchayat*-level activities, their actual impact on decision-making is minimal and often lacks substantial contribution. With 48% of the population being female (Fig. 1), the role of women remains confined. The dominant influence of ingrained stereotypes, patriarchal attitudes, subjugation, and adherence to cultural and societal norms plays a vital role in decision making. The disproportionate distribution of domestic work makes women's participation in political affairs difficult, as they spend the majority of their time fulfilling family needs, which becomes their first priority. Subjugation plays a significant role here, manifesting through various forms such as economic exploitation, social hierarchies, and political power dynamics, as well as restricted access to education and limited opportunities outside traditional roles. Due to extreme pressure and criticism from relatives, neighbours, and families since childhood, restrictions on girls are enforced. Affected by this stigma, several girls abandon their freedom and education, binding themselves in a vicious circle of servitude. Before marriage, their parents, and after marriage, their husbands and in-laws, make all the decisions on their behalf. They lose decision-making power from an early age, accepting external forces that shape their lives. This affects their decision making in governance as well. Despite improvements in female literacy rates (Fig. 2), there remains a lack of noticeable change in the distribution of decision-making power between genders. Gender wage disparity persists, as men receive approximately Rs. 350–400 while women earn a meager Rs. 168, leading to women's exploitation. The limited economic opportunities for women must be addressed. Women in the village must realize their self-worth; they are no less than men, which is crucial for fostering a fair and inclusive society. Participation should not be treated as a burden; voluntary involvement of women at the actual ground level in planning and development of the village is crucial for empowering their decision-making power. Women have a better understanding of women, which helps in the upliftment of other women in society. Understanding the concept of gender discrimination, stereotypes, patriarchal attitudes and subjugation, as well as the importance of their impact on their lives, is essential for not accepting the given role without questioning. Encouraging awareness is key regarding the importance and equality of women in society. Resources are not managed sustainably. The visited pond is not appropriately maintained; the dressing of the surrounding area and maintenance of the common embankment are needed. The well frequently used by women is in a deteriorated and polluted state. There is minimal participation of women in resource management. Nobody nurtures the environment better than women by engaging in pond management, water quality monitoring, sustainable fishing practices, and aquaculture, ensuring the well-being of the aquatic ecosystem. Women are involved in the practice of sustainable farming, promoting agroecology and organic cultivation to preserve the local ecosystem. Encouraging composting organic waste and increasing the use of renewable energy, such as solar or wind energy solutions for power, as well as promoting biogas plants for cooking, is essential. Enhancing access to quality education for girls and promoting vocational training for locals is crucial. Raising awareness about menstrual hygiene and sanitation is essential, as women have a better understanding of these issues. Implementing rain harvesting systems ensures access to clean and safe drinking water. Encouraging women's participatory decision making and promoting community-based conservation initiatives are vital. Women's involvement in the process of economic development is not only ethical but also beneficial for the economy, making them both agents and beneficiaries of development programs. Promoting the empowerment of women is key to creating a sustainable future.

References

Berevoescu I, Ballington J (2021) UN women. Working paper women's representation in local government: a global analysis. New York, December

Chandramouli C (2011) Population Census. The Census Organization of India conducted a national census survey. Chiraura Population–Patna, Bihar. https://www.census2011.co.in/data/village/245538-chiraura-bihar.html

Faculty and students of Manava Bharati International School, Patna (2023) Weakening of the symbiotic relationship between land and people: Case from a Patna village. ILDC, FLAME University, Pune, 1st to 3rd November

Huq SMS (2016) Women's participation in local government: a tool for sustainable development of Bangladesh. Euro J Bus Soc Sci 4:169–183

Inter-Agency and Expert Group (n.d.). Sustainable development goal indicators (E/CN.3/2016/2/Rev.1) Annex IV Final list of proposed Sustainable Development Goal indicators https://sustainabledevelopment.un.org/content/documents/11803Official-List-of-Proposed-SDG-Indicators.pdf, https://wcd.nic.in/sites/default/files/National%20Policy%20for%20Empowerment%20of%20Women%202001.pdf

Ministry of Panchayati Raj (2020) Government of India. Representation in Panchayats

Paul N (n.d.) A Case Study on Women leadership in Panchayati Raj Institutions (PRI) at the Gram Panchayat level. Office of National Director CORD and CORD Training Centre, Sidhbari (2) https://nirdpr.org.in/nird_docs/casestudies/cord/cord1.pdf

Posted On: 23 SEP 2020 2:42PM by PIB Delhi https://pib.gov.in/PressReleaseIframePage.aspx?PRID=1658145

Press Information Bureau (PIB) (1987) Ministry of Panchayati Raj. 23rd September 2020, Delhi Sustainable development as defined in the World Commission on Environment and Development's report "Our Common Future"

Reshi IA, Sudha DT. (2022) Women's empowerment: a literature review. 2(6) https://www.oecd.org/social/40881538.pdf

Vishwakarma BK (n.d.) Impact of women's participation in local self-government: a case study of Baghmara block, Dhanbad. University Department of Political Science, Binod Bihari Mahto Koyalanchal University, Dhanbad. https://shodhgangotri.inflibnet.ac.in/bitstream/20.500.14146/15517/1/baikunth%20kumar%20vishwakarma_pol_sc.pdf

Warth L, Koparanova M (2012) Empowering women for sustainable development. UNECE discussion paper series, no. 2012.1 January

Divya Sharma holds an MA in Economics from Indira Gandhi National Open University. She is a teaching faculty member of economics and business studies at Commerce at Manava Bharati International School, Patna, and the Convener of the Social Science and Commerce Department. Prior to this position, she taught economics at the Vedanta Academy, Kolkata (2022) and St. Augustine's Day School, Kolkata (2016). She also serves as a mentor in a community engagement program.

Climate Change Politics Through a Humanities and Social Science Lens

Along with the *Unsung Heroes* Who Opened the Way to Rome: Sustainable Development Goals *Ante Litteram*

Maria Anita Stefanelli

Abstract This chapter focuses on the 1951 film directed by Polish filmmaker Krzysztof Łukaszewicz, *Czerwone Maki na Monte Cassino*, which evokes the military dramatic adventure of the Polish Second Corps in their march from Russia—where they had been deported—to the Middle East, the Mediterranean Sea, and the Italian city of Taranto wherefrom the Army moved along the Apennine ridge to reach Monte Cassino Abbey on the site that would become, after the corps' heroic fighting, resilience, and bloodshed, a tomb tinted with blood for many of them.

The curtain opens on the outbreak of World War II and the formation, in 1941, of the Army under General Władysław Anders' command on Russian soil. With time and a lengthy bargaining with the Soviets, Anders insisted on having, besides soldiers and officers, political prisoners, amnestied fellows, people that had been reduced to slavery in Siberia, and civil citizens of all ages, different religions and different nationalities.

During the exhausting march, a number of what would be defined by UN, in our times, "Sustainable Development Goals," were targeted and reached by the members of the Corps. The Monte Cassino mission was successful in opening the way to the city of Rome, but the outcome was not—unfortunately—*freedom* as expected. The participating nations to the 1945 Yalta Conference, in fact, agreed to Poland falling within the Russian sphere of influence. The role of the heroes, *unsung* for more than one reason, was neglected and their hopes nullified.

Keywords World War II · Russia · Second Polish Corps · Sustainable development goals · MonteCassino · Unsung heroes

M. A. Stefanelli (✉)
Department of Anglo-American Language and Literatures, Roma Tre University, Rome, Italy
e-mail: mariaanita.stefanelli@uniroma3.it

B. Wejnert (ed.), *Democracy in Retreat and the Climate Crisis*, Springer Studies on Populism, Identity Politics and Social Justice,
https://doi.org/10.1007/978-3-032-02881-5_13

Red Poppies

"When it moves, *red*—the color of passion, power, pleasure, and beauty—makes the air sparkle. The sex life of plants is part of the picture. Look for red corollas if you want to see happy hummingbirds disporting themselves" (Anton 2023, p. 73).[1] Far from marking an element of the astonishing diversity in sexual characteristics of plants, *red* in the film *Czerwone Maki na Monte Cassino* (1951), directed by Polish filmmaker Krzysztof Łukaszewicz, is the color of the poppy, a poignant icon of loss and memory, sacrifice and remembrance; the sign is associated with the bloodshed, between August 1944 and May 1945, of those belonging to the Polish Second Corps who fought the final battle on the ruins of the sixteenth-century monastery of Monte Cassino to open the way to Rome for her liberation. They had marched from Russia to various ports of the Middle East, transported across the Mediterranean Sea to Taranto, and moved along the Apennine ridge to reach the Monte Cassino Abbey that had served as the German's enclave in the last hours after Hitler's suicide and Germany's unconditional surrender.[2] Once bombed, the site tinged with the fallen soldiers' blood is associated to the flower that grew abundantly on the battlefields of World War II, a fact that tied the film to the historical contextualization of courage and impending deaths necessary to earn freedom from tyranny. The poppy is a symbol, for American poet Louise Glück, of beauty and fragility like the fleeting moment of life; at the same time, the lively imagery of the poppy opening to the sun conveying the memory of the past as a transient human condition evokes a spiritual enlightenment like a journey through suffering to a moment of clarity and articulation. So be it also for the Polish film which focuses on a war fought with anguish and torment with the objective—finally achieved in spite of the many losses—of opening the way to the liberation of all those who had been, and still were, in the hands of the enemy.

Having contextualized the red poppies as today's call to the dark days subsequent to the threatening time of Adolf Hitler and Iosif Stalin, two men who drove the western and eastern world to strenuously defending their freedom, in 1939, following Winston Churchill's initiative of directing a passionate address to the British people to declare war on Nazi Germany through the alliance with USA and USSR, a narrative follows that goes back to the beginning of Hitler's brutal design to dominate the world by betraying his former ally, the Soviet Union, and take Moscow. So it happened in the end that an Army on the verge of defeat—the Second Polish Army whose destiny was that of being made of, eventually, *unsung heroes*—was led to the victory, in a sense, of democracy over autarchy.

---

[1] The quotation is taken from the chapter "La Vie En Rouge. The Cardinal Flower and the Ruby-Throated Hummingbird." Anton is a Belgian writer and artist. She attended botany courses at the Jardin des Plants in Paris, where the National Museum of Natural History is located, and botanical illustration lessons with Elisabeth Vitou at the Sociétć e: Nationale d'Horticulture de France.

[2] The reference is to the site of the sixth-century abbey—destroyed and rebuilt more than once—founded by Benedetto da Norcia, to become the cradle of Western monasticism.

The project had a high price, given the high cost of global destruction, traumatic warfare, trampling of human rights, and worldwide bereavement. The war of autarchy—epitomized by the German leader, and, for a short time, his ally Stalin (who would later join forces with his former opponent)—against democracy cost universal terror along with the fatal destiny of soldiers and their families. The development of its narrative is rightly recorded in *Czerwone Maki* as a fatal immolation to the survival of humaneness amidst carnage through the rhetorical image of the red poppies.[3]

The Explosion of World War II and the Formation of the Second Polish Army

"So, we are at war." Thus, the Polish Radio, on Friday September 1st, 1939, announced the German invasion of the country. Two days later United Kingdom Prime Minister Neville Chamberlain declared war on Nazi Germany and, on the same day, France followed suit.[4] Both the UK and France had declared themselves ready to defend Poland in any event that would threaten the country's independence. A few days before the announcement, however, the Third Reich, with the first Slovak Republic, had signed a Nonaggression Pact (Molotov–Ribbentrop)[5] with the Soviet Union. A "secret Protocol" included in the pact defined Europe's borders of Soviet and German spheres of influence across Poland, Lithuania, Latvia, Estonia, and Finland. Sixteen days after the German invasion of Poland from the west, the Soviet Union's invasion followed from the east, the result of which was the division of Poland into the "spheres of influence" of two hegemonic powers, Hitler's and Stalin's. While Hitler's target was a Polish corridor to Danzig for the expansion of Germany towards east, the Soviet leader made 13.5 million Polish citizens Soviet subjects and some 320,000 of them prisoners of war. The regime soon ordered repressions of military officers and police, deportation of men, women and children, and murders of political prisoners, among other violations of personal freedom.

Winston Churchill became Prime Minister on May 10, 1940, on the same day as Germany invaded France starting the "battle of France" with the occupation of

[3] The word "rhetorical" that, in time, has gone through a development in a negative sense that betrays its original meaning coming from *rhetorics* as the systematic study of the art and practice of discourse, is used here in its positive meaning as explained, as many as 140 years ago, by Charles Baudelaire who predicates: "[...] il est évident que les rhétoriques et les prosodies ne sont pas des tyrannies inventées arbitrairement, mais une collection de règles réclamées par l'organisation même de l'être spirituel; et jamais les prosodies et les rhétoriques n'ont empêché l'originalité de se produire distinctement. Le contraire, à savoir qu'elles ont aidé l'éclosion de l'originalité, serait infiniment plus vrai." (Baudelaire 1859/1885, 13; first appeared several years earlier in Revue française).

[4] The headline on the first page of *Paris-Soir* is: "La Guerre Est Declarée."

[5] V.I. Molotov and Joachim von Ribbentrop—the Soviet and German foreign Ministers—signed the Pact on August 23, 1939.

Belgium, Luxembourg, and the Netherlands, too. On the eighteenth of June, he delivered a speech in the House of Commons that would broadcast via radio to the British, from which the gravity of the situation deriving from the realistic danger of an incumbent Nazi invasion was presented to the Nation. The tone of the speech was clearly intended to acquire public support for a war against Germany. It was the end of a "Phoney War" and the beginning of a World War[6]—the second one since 1918. From then onwards, Churchill's richly idiosyncratic and performative speeches aired across the nation to reach European nations, and spread out throughout the world with defiance and occasional touches of English humor. His words, real weapons against the Wehrmacht, were his perseverance and indignation that would help to make democracy win over despotism.

Fatally, in the summer solstice of 1941, Germany's earlier pact with the Soviet Union came to a conclusion, with the consequence of Hitler invading the Soviet Union. Called 'Operation Barbarossa' after Fredrick I, Hitler's ultimate goal was earning more *Lebensraum*, just one word for Germany to grow stronger extending its power, through an arrogant military offense against the Soviets, towards the east, and impose, sided by most of his Axis Allies (Romania, Finland, Italy, Hungary, and Slovakia), Nazi autocratic rule in a *purified*—in the German leader's language—land through the deportation of native and ethnic elements to the Northern-Eastern regions for a repopulation of those lands with Germans. As a consequence, alliances and balance of political powers changed considerably and, after a June 22 broadcast speech rich in rhetorical expression to avoid creating alarm in the population, Stalin and Churchill signed, on July 12, the Anglo-Soviet agreement that would include the Soviet Union in the Allied Coalition.

As a consequence of the earlier (Russian) and later (German) incursions on Poland, the Polish government and army were established, initially, in France, a country who later suffered Hitler's arrogation. The Polish government had to relocate in Britain as Polish government in exile. With Hitler's invasion of the Soviet Union, his alliance with Stalin obviously came to an end with a new destiny for all those—both soldiers and civilians, including women, the elderly, the ill, disabled people, infants, and children—who had been imprisoned or deported by the Soviets to areas with road constructions, the *kolkhoz* (theoretically *collective* agricultural experiments), forced labor camps in Siberia, or the desert of Kazakhstan. Their work would be in mines (like Kolyma, for lead and gold) where they would suffer, in an environment of up to 68 centigrade below zero, a thermal shock, the loss of body parts, starvation, or poisoning from lead oxide, and would often die from hardships or be shot by firearms for ineptitude. It was an extermination of human beings.

The idea of an Allied offensive against the Third Reich from the southern direction—i.e., through the Balkans—was mature at the time of the Second Polish Corps being gathered under the command of General Władysław Anders, who was freed from his imprisonment when a conflict Soviet Union-Third Reich (that Anders had

[6]The *Phoney War* was a prelude to the conflict that would involve Western and Eastern countries, both Europe and the World. It was a period of plans for the universal devastation and the great death tolls that followed the fighting.

strongly wished and anticipated) broke out.[7] A Polish Army under Anders' command was organized on August 22, 1941, headquarters in Buzuluk, with General Anders officially nominated Chief Commander of the Second Polish Army (Anders 1949, 2014, p. 69). At the beginning only a "General commanding a non-existing Army," as recorded in his Memoir, the General, a Colonel in a 1939 cavalry unit, assumed, when the Russian-Polish agreement was finally signed, the General Command of the Second Polish Army to be created on Russian soil (Anders 1949, 2014, p 67).

With much difficulty and the overcoming of a complex set of obstacles for questions, also, of reciprocal trust between the Polish and the Soviets, General Anders proceeded in the project of providing assistance, formation, and education for his Army, and insisted on taking members of the civil society, including women, older people, and children along with soldiers. This last proposal took longer to be accepted by the Soviets, but in the end the General prevailed.

It was the greatest ever formed community of people who had been severely and ferociously pulled out of their country, deported, imprisoned, reduced to slavery, slayed or assassinated to be led out of a brutal and inhumane state of affairs towards a culturally lively context of physical and mental sustainability. It was necessary for most of them to be psychologically cured from the experience of suffering trials, being supplied with the necessities of life and provided with whatever means were available for their sanity, given a reason to engage in the preservation of existence. "The Polish Armed Forces in the West"—as one Army became known—comprised ground, sea, and air forces; "The Polish Armed Forces in the East" became the Polish Second Corps.

The Polish Second Corps, besides enlisted soldiers and officers, would comprise family members of military staff, citizens evacuated at the time of Poland's invasion, unwanted people by Soviets for religion or political reasons, members of cultural institutions to give advice, teachers to provide courses for children and adolescents, and artists who would diffuse music, painting, dramatic performances and cinematic products. The cultural context was a primary element with intellectuals and writers interacting with the military members of all ranks. The company was

---

[7] As protagonist of the narrative concerning the Second Polish Corps' activities being dealt with in this paper, General Władysław Anders receives attention for his strategic planning and organization of the Army through his Memoir (2014) wherefrom the history of the Second Polish Corps is taken. On that basis the General's intentions and goals are made explicit and his own reports considered authentic. The text used for the purposes of this paper is Anders' Memorie 1939–1946.

Anders' report, by himself, of his arrest and detention follows: imprisoned by the Soviets not only for organizing a clandestine movement against the Soviets in Lviv; on account of his having fought against the Bolsheviks in the 1918–1920 war, for the Polish alliance with England and his belief in her victory; for his fighting the Soviet troops in September 1039 and having caused the several losses of Russian trucks and men, and finally, for being a spy because of his involvement in the building of a Polish government and his not accepting the offer to be enlisted in the Red Army (Anders 1949, 2014, 35–36). Those and the many other events in General Anders' experience as commander of the Second Polish Army are recorded in his Memoir, wherefrom page number/s are supplied whenever a straight translation from the Italian edition of *Bez ostatniego rozdziału*, is used of his words.

inspired by events that made up the history and geopolitical significance of the international and transnational experience that would characterize not only military operations, but also actions, events, publications, exhibitions, and civil commitments originating from their dedication in doing well—in an exceptional and extraordinary set of happenings—for the good of all.

Freeing Human Beings from an "Inhumane Land"

From Krystyna Javorka's essay on less known aspects of the Second Polish Corps' activities, we learn that the officer responsible for the Army's Intelligence eventually wrote to Józef Czapski that the hardest span of time for General Anders (then dead) had been the one "of our permanence in Russia. The following phases for the General, in my opinion, were child's play compared with the period in which he was to create an Army in that 'inhumane land'" (Fondazione Romana 2014, p.39). Anders does not refrain, in his Memoir, from stigmatizing the Soviets' failure to honor agreements, their outright lying and untrustworthiness, their reducing to slavery thousands of Polish officers, civil citizens (including fragile members of the population), and directly or indirectly murdering them by coercing them to forced labor in inhospitable regions.

The climate factors, particularly the harsh environmental conditions of the Soviet Union, were unbearable to the General who reports of his intervention at the December 3, 1941 conference with Stalin, "[…] with the current state of food supplies, provisions and medical supplies, as well as with the difficulties of the climate, it is not possible to think of organizing any unit suitable for combat. It is a wretched, miserable life, where each individual's energy and efforts are spent only to survive and stay alive in terrible conditions [...] This is why it is necessary to transfer the Army to another climate, where food and supplies are granted to us to allow us to move forward," and concludes on the most suitable hospitable land being Iran (Anders 1949, 2014, p. 101). General Władysław Sikorski, at the time Prime Minister of Poland in exile and promoter of the conference with Stalin, more firmly than his colleague in his confrontation with the leader of the Soviet Union, went down hard insisting: "[…] here there are insufficient conditions to organize our army and I don't want our men to die needlessly. I don't present an ultimatum, but when there is a harsh winter, with wind and frost that kill my men, then I cannot turn a blind eye or remain silent" (Anders 1949, 2014, p. 102).

The Soviet Union utilized forced labor extensively, particularly in the Gulag system, where prisoners, including those considered enemies during WWII (such as captured soldiers, political prisoners, and other detainees), were sent to labor camps often located in remote, chilly regions of the USSR, such as Siberia, the Arctic, and other areas with extreme climates. If by some twist of fate some of those detained were amnestied (as all Poles should have been once the alliance between Poland and the USSR to date from July 30, 1941 was in force—but in fact were not [Anders 1949, 2014, p. 115]), they would come back with reports of waking up in the morning with frostbitten noses and ears, being forced to freeze in light tents and left with no footwear, or, once back from the Kolyma fields, having lost a limb, nose, or ear, turned blind or mutilated beyond belief. During the journey of those amnestied and ready to be trained for the army, casualties were frequent, given the snowstorms and chill wind of an untimely winter causing the thermometer to drop to 52 °C below zero and the overall situation still in the grip of food shortage, hardship, and epidemics (Anders 1949, 2014, p. 116). Given the richness of natural resources like timber, gold, and minerals that the state wanted to exploit in those regions, it is obvious that subjecting the invaded population to such a freezing climate was a strategic choice. It is not difficult to realize that no soul could have escaped from a remote and harsh environment that appeared to be a haunted natural barrier of terror to any escape attempt.

Social, political, and historical systems of oppression, as we have learnt, have impacted upon the environmental degradation that we are facing today. Extracting practices, in particular, have contributed to the climate crisis that was almost unknown 80 years ago; what was not ignored, however, was that forced labor was inefficient and the gulag system—with harsh working conditions, insufficient food, and lack of comfortable housing—inhumane. Mining, for instance, caused extensive environmental degradation due to unregulated extraction methods, lack of environmental oversight, and a focus on maximizing output at the expense of the natural world. The damage from lead, uranium, nickel, and other mineral extraction activities resulted in polluted air, contaminated water sources, deforestation, soil degradation, and long-term health issues for local populations. Although Anders was not aware of what the future would hold for us nor did he have any climate change concerns, he did realize that the power dynamics was ethically inequitable and a culture of "whiteness"[8] in the sense of servile forced labor in freezing areas *unsustainable* in the long run. Although *sustainability* was not part of military strategy at the time, promoting *sustainable* attitudes, abilities, behavior, and values was, from examining his memoir, a well-articulated project that envisaged psychological resilience, high morale, emotional control, and education to a civilian life, after the battlefield, tending to survival in peacebuilding.

[8] "Whiteness studies," as the study of structures that produce white privilege dates back to the late twentieth century and has roots in the work of writers like W.E.B. Du Bois and James Baldwin. Here, its meaning is limited exclusively to the economic exploitation of sources and the hard labor of people often confined to glacial arctic zones. The case of the Soviets (Chapter 2 in *An Environmental History of Russia*) supplies an account of environmental and human costs of slave labor camps of the gulag system.

The Troops, a New Climate, and Biodiversity

Climate factors in Russia and the Soviet republics are related, in our times, with the contemporary consequences of climate change. Environmental changes due to global warming, for instance, would cause, with time, permafrost to thaw with consequent damage to the infrastructure that often dates back to the Soviet epoch and to the fact that increased human presence disrupts migration patterns and the wildlife in general. Permafrost thawing may also provoke environmental degradation through the new activities put into motion by the thawing with significant ecological impacts that directly and indirectly affect biodiversity in these sensitive regions.

Though nobody could be aware of the repercussions of the extreme climate factors on the environment in the following years that led, as noted above, to challenging contemporary consequences, the threatening impact of the climate on the availability of natural resources such as food and water during extreme weather events, on the social opportunities for those cut off from neighboring communities, and on applying adaptation strategies to the adverse effects of the weather was so evident that could not be ignored. The accurate description of the suffering and traumatic consequences for those who had experienced the harshness of the climate proved how strong was the General's resolution to leave Russia as quickly as possible with the aim of training the troops in Iran. Among the multiple causes of the General's pressure on the Soviet authorities to be allowed his Army to leave Russia for Iran on behalf of the soldiers' training, his preoccupation for his soldiers' welfare was that of the potential freezing climate, with temperatures well below zero, expected in the winter ahead of them had they remained longer in Russia. Anders' arguing with Stalin was so strong that the Soviet leader was finally defeated in his trying to obstructing Anders' plans.[9] Only on August 22 of 1942 the General managed to have a meeting with Churchill at Cairo about the move of the Army to Iran. So, the Army's journey could finally begin.

It was in the new environmental area that a new guest joined the troupe and became before long an element of the unique bond between humans and animals during the hardships of war (British Library, online).[10] From an Iranian boy who found a Syrian brown cub bear came the suggestion to the soldiers to adopt the creature for its own needs and, maybe, in lieu of a domestic pet. With the General's consent, the Army welcomed the new "member" of the company. The animal, baptized Wojtek, soon made friends with the soldiers and became a symbol of integration. It would be impossible in this context to venture into the significance, for the troop, of a bear lost in some Iranian forest. Today's claims are for an important role

[9] Anders records his agreement with Stalin in its complexity (Anders 1949, 2014, pp.119–132).

[10] Although the European Studies blog has it that Wojtek's recruitment was a "legend" it does support the fact that the British authorities approved the bear's travel warrant to embark on the ship from Egypt to Italy (British Library, 2015). More facts associated with Wojtek are accounted for in two contributions (by D. Kulawiak and A. Riccomi, respectively) to the proceedings of a Conference held in Rome (2013, Fondazione Romana Umiastowska, 2014, pp. 3, 327–330, 331–335).

of biotic factors in climate regulation: the degradation or destruction of ecosystems have concurred, as a matter of fact, to a climate crisis. Though the memory of the environment where the bear cub was found has not been preserved, the historical episode records, in a most successful encounter of humans with a wild animal, a parameter of biodiversity. On the subsequent approach to Monte Cassino in Italy, in order to bypass the difficulty of the bear's transportation to the final destination, Wojtek was recruited as a member of the community and enlisted officially as a private of the Polish Army. An icon of environmental balance of social and cultural needs, the bear became a symbol of camaraderie and resilience. The concept anticipates, in a way, the goal of a sustainable use of the terrestrial ecosystems through the identification of local species to protect, the opening of wildlife corridors for the conservation of biodiversity, the relationship between the human and non-human life on the earth, and the construction of a life on earth harmonized with nature—all processes to which climate change today poses serious challenges. After the war, Wojtek was moved to Scotland's Edinburgh Zoo, and ceased to live in 1963.

At the first lengthy stop in British - controlled Iran, Polish civilians who had been placed in refugee camps had access to food and medical care, while trainee soldiers lived in a military encampment. General Anders's actions were driven by the immediate need to save lives rather than being devoted to long term sustenance. It was during that harsh journey, started with the stage in Iran, that important aspects of his leadership since the evacuation operations could be called *sustainable*. Not precisely as we use the word today after the United Nations call to transform our world with an "action programme for people, the planet and prosperity" through the *Agenda 2030 for Sustainable Development*; instead, supposably adopting, for the manifold community, different practices from what had happened to them till then, and trying to meet their (civilians' and soldiers') needs of the present in all concerned dimensions—environmental, social, cultural, educational and, to a certain extent, economic.[11] With UNESCO's worldwide call for an urgent action regarding the famous seventeen sustainable development goals, aspects such as resource (however limited) management, food, medical supplies, and transportation had to be dealt with efficiency, avoiding waste and preserving the different members' (especially children's and elderly's) health.[12] The extreme conditions suffered in the URSS were a thing of the past; sufficient nourishment and a sense of inclusion were promoted; education and culture were pursued with careful planning and adequate staff; gender equality as concerned women and old-age members of society was a fact; human rights received respect; a future with peace was envisioned. In sum, the General's strategic leadership, the soldiers' motivation and adaptability after their

[11] United Nations General Assembly of 21 October 2015.
https://www.un.org/en/development/desa/population/migration/generalassembly/docs/globalcompact/A_RES_70_1_E.pdf Retrieved July 15, 2024.

[12] UNESCO (United Nations Educational, Scientific and Cultural Organization), Education Sector (January 2017). Sustainable Development Goals: Learning Objectives.
https://www.unesco.org/en/articles/education-sustainable-development-goals-learning-objectives Retrieved July 15, 2024.

experience on Soviet enclaves, the community's welfare as their leader's main goal, everything contributed to heal the wounds and restore dignity to the soldiers' and civilians' life. Their past gave way to their future when, in Anders' words, those "emaciated, real skeletons … partly covered in ulcers …. those miserable human remains," at the mere thought of "leaving that hell," would reveal to him their "bright eyes …. animated by 'faith and will'" (Anders 1949, 2014, pp. 77–78).

An Ample Community Along with the Unsung Heroes

Anders' task of forming a social group to lead through countries that were foreign for nearly everyone entailed issues related to situations with some degree of unexpectedness that might surprise, scare, and/or excite most members as regards age, sex, gender, or religion. Nationalities were varied, since the roughly 80,000 evacuees did not only rank among the deportees from Poland (without distinction of race or nationality), but were also—thanks to Anders' stubbornness in opposing the Russians—Jewish, Ukrainians, and Byelorussians (Anders 1949, 2014, pp. 109, 141). Grouping and training an army with the presence of family members and being able to provide a livable life for all was an ambitious project for the General for the potential difficulties and obstacles lurking ahead. Requirements for the community were not disjointed from a good amount of patience, wish to learn, skills in problem solving, knowledge of dangers, ethical values, and, as far as Anders was concerned, a positive attitude to empower every subject for the attainment of specific goals in replacement of a life of hard work far from home in the hands of the Soviets. Each member, whether civil of military, must have been spurred to consider all that.

Questions like, "would they want to come to terms with what was truly relevant to their lives?"; "were they ready to learn how to survive the risks, hazards, threats, uncertainty and destabilizing factors that time and the environment might have in store for them?"; "were they aware that all that would mean transformations in their ways of life?" were fundamental for the immediate future of Anders and Co.

General Anders' ability to inspire his troops and lead them through unknown circumstances, his personal commitment to their well-being, and his socio-cultural project that would include a group not belonging to armed forces were instrumental in the progress of the soldiers' confidence in themselves and keys to ensure the corps' sense of belonging, openness, connection, empathy, and personal values. The provision of medical care was an added benefit that inspired confidence and opened the way to the Corps' adaptability to the harsh conditions the soldiers would be encountering in the middle east environment and the rocky hills of the Apennine regions of the Italic peninsula ahead of them. Several of the goals that are today listed as sustainable by the UN Educational, Scientific, and Cultural Organization were indeed in Anders' mind since his efforts to save as many souls as he could from the atrocities perpetrated by the invaders of their country.

Without any further details of a background intended to support the gathering of a real social community within the limits of a free space for assistance, protection, education, instruction, formation, and entertainment of different age groups, attention must be drawn to a community in physical, psychological, and cultural progress to be engaged in a military and civil journey through countries of Europe and Middle East with the purpose of freeing them from nazi-fascism and allowing them to meet their democratic aspirations.

The community of soldiers and civilians was to leave Russia, reach Iran, cross Iraq, Syria, Palestine, Egypt, or Libya, then cross the Mediterranean Sea to land in fascist Italy at Taranto, and, through the Apennine ridge with Monte Cassino more than 500 m above sea level, advance towards their final target: the descent to Rome. Meanwhile all civilians, struck by the traumatic situation they had lived through in the previous months, would dream of a home when and where it would become possible for them to settle, during the war or at its conclusion. The Corps and its annexed population demonstrated adaptability on their way from the harsh conditions of the Middle East to the mountainous terrain of Italy, and the different conditions arising from the risks to which they became exposed—even though after the conditions they had suffered in Siberia, Kazakhstan, or remote regions in the Ural Mountains, more rigid conditions would have been improbable. The transition took place from extreme hardship to a more stable situation where everybody, including women, children and elderly people stopped suffering malnutrition, disease, or exposure to the elements, although the many who survived were weakened and traumatized by the past experiences. In the UK, Canada, the United States, or Australia, where the majority of the Polish civilians later would be able to settle, the experience of the journey, or a part of it if they managed to identify a possible home before its conclusion, became something never to be forgotten, actually, a real seal of the Polish diaspora.

At a very early time General Anders, who had been strongly discouraged by Soviet authorities from taking such weakened people—tens of thousands—with his Army, ignored the directives or bans, and left as soon as the move to British protected Iran became practicable for a journey that he knew would be harsh for scarcity of food and inadequacy of sanitary conditions. Providing satisfactory medical care was essential for sustaining the troops. In the long run the corps ensured organized medical units for wounded soldiers to receive timely and effective treatment, which helped maintain overall troop strength. Besides, the Polish government-in-exile would secure political and material support for the II Corps by way of the Allies' diplomatic efforts. As to those last, the efforts by the Polish government-in-exile to secure political and material support from the Allies were crucial. They ensured that the Corps was not isolated and received the necessary backing to continue its operations. Education was crucial for the youngsters, but was also, for the senior members of the group, a help to lifelong learning and strategies to take informed decisions and act responsibly for environmental integrity, economic viability, and a just society, for present and future generations. As to logistics and supplies, due to Allied support and efficient organization the Polish Second Corps was well-supplied with food, ammunitions, medical supplies, and effective treatment when needed.

Whiteness and Redness

The rigid climate made a substantial difference to the Polish prisoners, many of them engaged in extracting limited natural resources in an oppressive environment. The white snow and ice evoke a sense of desolation, isolation, and hardships that those forced into labor under extreme climatic conditions had to face. Appreciative of the degradation of people encapsulated within extreme temperatures and unable to leave, Anders, in his face to face with Stalin, rejected the latter's pretensions, in his usual paternalistic way of acting in diplomatic negotiations to show good will and make promises he wouldn't keep, and insisted for the evacuation of all, soldiers and civilians, who, from having been slaves, became free citizens.

The whiteness of the frozen gulag, from which all those who managed to escape enrolled in Anders' Second Corps, is opposed to the redness of the poppy, a flower that grows freely in open fields and is symbolically opposed to the icy areas representing restrictions, captivity, and erasure of identity. The red poppies are a sign of remembrance of the sacrifices made by Anders' Second Corps that often meant death—yet, it was a personal death offered in exchange for somebody else's life. The white non-color, instead, translatable as it is into the *white page*, means absence, silence, forgetfulness, and also death, a death not inscribed in the annals, however, but often piously (perhaps) hidden below, in a mass grave (as was the case with the massacre of Katyn).[13] Also, red as the blood pouring from the heroes' veins is in contrast with white as a starting over, a pure new beginning for all those lucky enough to find a new home where they would live in peace.

One would expect that the redness of the sacrifice, being both a sign of death but also, as mentioned at the beginning of this paper, a sign of life, would persist in the people's memory. In the case of the Polish Second Army, people have spoken of *unsung heroes* for the bravery and dedication made by all those involved in its operations throughout the campaign, and, above all, in Monte Cassino, when faced directly with death. A filmed narration of the bombardment is published online with the following words:

> Historic monastery was one of the key anchoring elements of the Gustav Line, a system of fortifications running across Italy's narrowest point. Ribboned with mountains, ridges, rivers, and swamps, these geographical features made it ideal terrain for the German defenders.

[13] The Katyn massacre, ordered by Stalin, happened in April and May 1940. In April 1990, the Soviet Union/Russia admitted they were responsible. In 1990, Mikhail Gorbachev officially apologized.

> In the words of historian Matthew Parker, the Gustav Line "was the most powerful defensive system encountered by the British and Americans during the war."[14]

Unfortunately, however, they did not receive the recognition they deserved, and that was in spite of the crucial role they played in the breaking of the Gustav Line and the launching of one of the final military assaults against German forces along a 32-km-long front. On 18 May 2024, on the occasion of the 80th anniversary, a Polish flag and the British flag were raised over the ruins.

Many of Anders' decisions focused on ensuring the survival and well-being of his troops and the civilians he had at his own command, in terms of the campaign against the Germans and the liberation of Italy, the well-being of soldiers and the families for whom education, cultural apparatus, and entertainment for the very young and those who did not fight were made available. The resistance to the Soviet domination and fight for freedom worked in favor of the reinforcement of a sense of identity and a long-term sustainability against the soldiers' previous experience of captivity, lost independence, and demotivation.

General Anders' Polish Second Corps should be remembered as an episode of a sustainable army. Surviving against odds, maintaining identity, adapting to all sorts of difficulties, and trying to secure a future for the soldiers *at home* after the war reflect the search of sustainability no more in military operations but in a world of "peace, justice and strong institutions"—as one of the UN sustainable development goals recites—but *in their own country*. Regrettably, there was no *free* country for those people to go back to. Their free, democratic country, according to the winners of the war (UK, USA, and URSS) meeting in Yalta was Poland become a de facto one-party country and satellite state of the Soviet Union—autocracy won over democracy. Poland transitioned from a potential democracy to subjection under the Soviet Union's autocratic influence.

Ensuring that the Army could contribute meaningfully to the Allies animated the soldiers, whose story is a model of sustainability in the context of military history, in the sense of being able to endure, adapt, and remain effective over time, despite significant challenges. Though Anders' nearly visionary, though appropriate, project of how to go about the military campaign he would lead had *no reference* models to *refer* to, nonetheless he envisioned, as has become clear, many sustainable development goals like the importance of health with safe nutrition and medical care, equality of sexes, inclusion within an environment granting respect for all, integration of civilians, biodiversity, struggle against unfit climate, peace, and justice in the future. A consolidation of the last goal (in the list of the 17 SDGs of the UN), however, would be necessary today in the light of the tremendous increase in conflicts that offer no sign of a resolution. Specifically, it would be invaluable to ensure, in the course of peacebuilding, the protection of human rights standards, the opposition to violence during campaigns, the urge to address deep causes like inequality or

[14] TVP World broadcasts "Remembering the Battle of Monte Cassino," Polish Television, 17/05/2024, 2024. https://tvpworld.com/77585836/remembering-the-battle-of-monte-cassino Retrieved September 3, 2024.

political instability, the need of limiting the impact of the conflict on the environment and ecosystems, the plans for post-conflict reconstruction, and, finally, the provision of mental health care for soldiers and civilians to guarantee a long-term psychological resilience.

Unfortunately, the outcome of the war was far from positive for the Polish Army; despite the sacrifices and the bloodshed, those heroes went *unsung* for the very reason that had caused their deportation to the Soviet Union: the invasion of a free country.

The 1945 Yalta Conference—with British, Americans and Soviets participating—agreed to Poland falling within the Russian communist sphere of influence. For Poland it meant losing, along with its independence, the democratic jurisdiction for which General Anders and his troops had fought. For the winners, the sacrifice of Poland was exchanged for the gratitude to the URSS to have played a major role in the defeat of Germany. With the Cold War geopolitics that records the dominance of the Soviet Union on Eastern Europe, the role the Polish Corps played in the famous battle was, all in all, neglected.

General Anders' link with the Polish government in exile in London and his obdurate planned exit from Russia for his troops' training, gave the Polish Second Corps an anti-communist label that produced the Soviets' total disregard for the *heroes*. Indirectly, the general political atmosphere in post-war Italy, where the strength of the Communist Party (PCI), growing rapidly after the fall of the fascist government, could not accept the Polish resentment against Russia, may have caused a lack of enthusiasm for the standing out of the Polish Second Corps at Monte Cassino and a belated recognition of Polish heroism in Italy.[15]

A further sustainable development goal and the fact that a lasting peace is inseparable from sustainable development would enrich the concept of sustainability development when *peace* is at stake and should be integrated into the global agenda. A goal that General Anders might have conceived—when desperately (and uselessly) trying to provide safety and security for the Poles abroad who, since the 1939 invasion of their nation, had hoped to return to an independent Poland (Anders 1949, 2014, 337–390) receiving only sanctimonious reactions—in his own consciousness.[16]

[15] For an account of the relations between the Second Polish Corps and the Italian Communist party, see Strzałka (2014), pp. 55–77.

[16] I take the opportunity here to thank Cavaliere del Repubblica (OMRI) Nicola Merucci, an Italian peace officer residing in Ireland, for driving me, on the 20th of September 2025, to Cassino War Cemetery, to pay respects to General Anders' burial ground.

References

Anders W (1949) An army in exile: the story of the second Polish corps. MacMillan, London

Anders W (2014) Memorie 1939-1946. La storia del II Corpo Polacco. Imola, Bacchilega

Anton J (2023) Sexus botanicus. The love lives of plants (trans: Butler E). MIT Press, Cambridge, MA

Baudelaire C (1859/1885) Le salon de 1859, Revue française, n 159. Paris 20 Juin, p 1043; published later in L'Art Romantique. In: Œuvres complètes de Charles Baudelaire, tome III. Calmann Lévy, Paris

British Library (2015) Wojtek, the soldier bear from the Polish Army. European Studies Blog. 25 Nov 2024. https://blogs.bl.uk/european/2015/11/wojtek-the-soldier-bear-from-the-polish-army.html. Retrieved 9 Aug 2024

Fondazione Romana Marchesa J.S. Umiastowska (2014) Ricordare il 2°Corpo d'Armata polacco in Italia (1943-1946). Inter arma non silent Musae. Italian translation. Djin edizioni, San Giuliano Terme (PI)
Glück L (1992) The red poppy. In: The Wild Iris. Ecco Press, Hopewell, NJ
Gorodetsky G (1988) The origins of the Cold War: Stalin, Churchill and the formation of the Grand Alliance. Russ Rev 47(2):145–170
Headline (1939) La Guerre Est Declarée, Paris-Soir, Septembre 4
Jaworka K (2014) Alcuni aspetti poco noti delle attività del Secondo Corpo. In: Fondazione Romana Umiastowska, pp 38–53
Josephson P, Dronin N, Mnatsakanian R, Cherp A, Efremenko D, Larin V (2013) Stalinism. Creating the socialist industrial, urban, and agricultural environment. In: An environmental history of Russia. Cambridge University Press, Cambridge, MA, pp 71–135
Kulawiak D (2014) L'orso Wojtek e la storia del 2° Corpo Polacco: come costruire un movimento vivo. In: Fondazione Romana Umiastowska, pp 327–330
Riccomi A (2014) Tramandare la memoria del 2° Corpo Polacco. La collaborazione fra scuola e istituzioni. In: Fondazione Romana Umiastowska, pp 331–335
Strzałka K (2014) Il 2° Corpo polacco in Italia negli anni 1944-1946. I rapporti con la società italiana e il Partito Comunista Italiano. In: Fondazione Romana Umiastowska, pp 55–77
TVP World (2024) Remembering the battle of Monte Cassino. Polish Television, 17 May 2024. https://tvpworld.com/77585836/remembering-the-battle-of-monte-cassino. Retrieved 3 Sept 2024

Maria Anita Stefanelli, Ph.D. (Edinburgh), former Chair Professor of Anglo-American Language and Literatures at Roma Tre University and Visiting Fellow of Trinity College Dublin, is Vice-President of "World Center for Women Studies." She has published extensively, in Italy and abroad, on eighteenth century American literature, modernist poets, feminist North American theater and cinema. Her most recent publication is the co-edition, with a chapter, of Scars of War. Migration, Security and Sustainable Future for Emerald (2025). In 2016 she received an Outstanding Author Contribution Award by Emerald Literary Network for the chapter 'Human into Birds: Italian Interpretation of Gender in Canadian Theatre'.

Is There Pro-Climatic Potential in National Symbols? An Essay on History, Memory, and Nature

Izabela Skórzyńska

Abstract The pro-climatic potential of national symbols stems from the historical use of plant and animal motifs as religious and coat-of-arms signs. They were then transferred by establishing emblems and/or national flags, as was the case of the rise of modern nations in the nineteenth and twentieth centuries.

In the text, I draw on symbolic forms of national memory in the context of the deepening climate crisis. To this end, I have analysed the designation of plant and animal motifs in national symbolism, and its status in the emergence and practice of environmental patriotism. I advance a thesis that national symbolism (which applies to hundreds of countries and national communities around the world) shows the potential for climate concern. However, when subjected to abuse, trivialisation and/or commodification (with a special contribution of populists), it becomes futile.

So, do national symbols that make abundant use of plant and animal symbols still have pro-environmental potential, and if so, what does it consist of? I consider this question by referring to contemporary conceptions of places and landscapes of remembrance, and rescue history in the context of the debate on environmental patriotism.

Keywords Climate crisis · Plant and animal motifs in national symbolism · Memorials · Memorial landscapes · Rescue history · Environmental patriotism

Introduction

> Sugar maples set up on Concord…They are worth all and more (…) if only because they have filled the open eyes of children with their rich color unstintedly so many Octobers…. No annual training or muster of soldiery, no celebration with its scarfs and banners, could import into the town a hundredth part of the annual splendor of our October. We have only

I. Skórzyńska (✉)
Department of History, Adam Mickiewicz University, Poznan, Poland
e-mail: izabela.skorzynska@amu.edu.pl

B. Wejnert (ed.), *Democracy in Retreat and the Climate Crisis*, Springer Studies on Populism, Identity Politics and Social Justice,
https://doi.org/10.1007/978-3-032-02881-5_14

> to set the trees, or let them stand, and Nature will find the colored drapery,—flags of all her nations.
>
> — *Henry Thoreau*[1]

I interpret the eponymous national symbolism in a broad way, as officially recognised state flags and emblems under legal and penal (constitutional or statutory) protection in most countries around the world, and as symbols ascribed to states and nations, not officially observed as their identifying sign but commonly associated with their history, culture and identity.

National symbols play a key role on the map of collective references to the past. Semantically, they are signs with specific references: a river, an eagle or a lotus flower, and multiple meanings: life, power, spiritual harmony, bravery, freedom, and so on. Metaphorically, they are places and landscapes of memory that carry a wealth of past experiences and perceptions of the world, in response to the question of who we are as a nation and citizens.

Among the historically shaped religious signs and coat-of-arms, present at least since the early nineteenth century in national symbolism, animal and plant motifs prevail. This is because, as Ewa Behrens-Nowakowska rightly stated, at an early stage of human communities, plants and animals were omnipresent in the human environment. Thus it is not surprising that they acquired numerous meanings and became objects of veneration for these communities.[2]

Thus, the plant and animal motifs (present in national symbols of hundreds of countries and communities around the world) show the potential for concern for the climate, provided that, in addition to their symbolic meaning, their natural sense is revived. It is crucial from the point of view of environmental identity focused on the subjective relations of the human and non-human worlds that co-create diverse homelands (regions, countries, continents, the world). Therefore, in the text I ask about the status of plant and animal motifs in national symbolism in the context of diverse theoretical approaches (memorials, memorial landscapes, rescue history), and social and political movements and practices for which environmental patriotism is or could be the common denominator.

---

[1] Thoreau, H., The natural history essays. Salt Lake City: Peregrine Smith Press, 1980, pp. 160, 165. Quoted by Cafaro Ph., Patriotism as an Environmental Virtue, "Journal of Agricultural and Environmental Ethics", 2009/ 23 (1–2) p. 205.

[2] Behrens-Nowakowska E., Symbol w lustrze. Obieg symboliczno-handlowy róży we współczesnej kulturze [Symbol in the mirror. The symbolic and commercial circulation of the rose in contemporary culture], Master's thesis written under the supervision of Kasia K., Ph.D., Warszawa 2021, p. 25 (typescript pdf https://upcoming.asp.waw.pl/wp-content/uploads/2021/12/asp-rose-behrens-nowakowska.pdf [access 03 Oct. 2024].

Plant and Animal Motifs in National Symbolism: An Intersubjective Review

Of the 205 UN countries, no less than 124 have plant motifs in their symbolic repertoire, and at least 27 countries use and/or are strongly associated with animal symbolism.[3]

Among woody plants, various types and species of *palms* are most common, representing Saudi Arabia, the Bahamas, Benin, the Dominican Republic, Fiji, Haiti, Jordan, Cuba, Maldives, Malta, Micronesia, Nauru, Paraguay, Peru, Saint Vincent and the Grenadines, Senegal, Seychelles, Trinidad and Tobago, Venezuela, Côte d'Ivoire, Marshall Islands, São Tomé and Príncipe. Equally popular are oak tree motifs, characteristic of England, Azerbaijan, Bulgaria, Estonia, Cuba, Lithuania, Latvia, San Marino, Germany and Italy. The *olive tree* is traditionally included in national topics of Cyprus, Eritrea, Israel, Colombia, Malta, Paraguay, Samoa, Tonga and Uruguay.

Various types and species of *cereals* feature in the symbols employed by Azerbaijan, Northern Macedonia, Myanmar, Turkmenistan, Vietnam, Jordan, Pakistan, Tajikistan, Uzbekistan and Zimbabwe. *Maize* is symbolic of Burkina Faso, Kenya, Mozambique and Zimbabwe, while rice is a symbol of Indonesia, North Korea, Laos and East Timor. The *cotton bush* is symbolic of Indonesia, Pakistan, Tajikistan, Turkmenistan, Uganda, Uzbekistan and again Zimbabwe. The *sugar cane* is symbolic of Antigua and Barbuda, Fiji, Mauritius, Saint Lucia, and the *banana tree* of the Dominican Republic, Fiji and Tuvalu.[4]

Meanwhile, woody plants (abundantly used in national symbols) have fallen victim most to the long process of modernisation, including demographic development, industrialisation, urbanisation and transport (clearing), and as means used in the construction, shipbuilding and mining industries, the thermal power industry, an object of trade in industrial raw materials extracted from trees for consumption purposes (deliberate felling). So, with the adverse climate change, is humanity's love for trees—national symbols—also a natural resource indispensible to breathing? Relatively, yes. Here are some examples.

Referring to the symbolism of the oak tree, Germans have been struggling with the effects of deforestation since the early eighteenth century. The process affected most long-growing deciduous trees, like the Fagaceae, including the beech and oak tree. While at the dawn of the modern era, 80% of the territory of Germany (the German states) was covered with forests, today they account for 33% with prevailing coniferous trees, and oak occupying only the fourth position in the German

[3] More on the subject in National symbol(s), The World Factbook, https://www.cia.gov/the-world-factbook/field/national-symbols/ [access 12 Sept. 2024].

[4] Tomasz [no name], Narodowe kwiaty—roślinne symbole państw [National flowers—plant symbols of countries], ABC Florysty https://abcflorysty.pl/kwiatowe-symbole-krajow/ [access 12 Sept. 2024].

ecosystem.[5] Ultimately, the oak is better off in monetary symbolism (the German euro) than in German forests, even though reforestation started as early as in the early nineteenth century owing to the concept of sustainability formulated only a little earlier by Hans Carl von Carlowitz, author of the first forestry textbook.[6] With that, the German forest ecosystems, restored throughout the nineteenth century, were then exploited on an unprecedented scale by the Third Reich for military purposes. After WWII, Germany restored forests at a rapid pace but focused primarily on coniferous trees of economic importance. Restoration programmes for deciduous forests are therefore relatively new, and we still have to wait for oaks, since this tree reaches adulthood at around 60–70 years of age!

A similar fate has befallen various species of oil palm which has become an industrial crop on a global scale at the expense of its natural ecosystems like rainforests. A vivid example is Indonesia,

> (…) once enjoying 99% forest cover, now (2015) 49.8% as a result of intensive deforestation since the 1970s. FAO data show that the country's forest area has decreased from 118.5 million ha in 1990 to 91.0 million ha in 2015. Palm oil plantations have been established in Indonesia since as early as 1911. Initially, is was a low intensity movement. Between 2005 and 2011, the rate of expansion averaged 514,000 ha per year, and by 2010 had covered 7.7 million ha.[7]

The situation is not very different in other regions of the world where most of water providers and CO_2 absorbers (the rainforests) are being cleared and replaced by palm trees in the name of the national economy and the local population whose standard of living directly depends on the palm tree and the oil it produces.[8]

As a result of climate change and rising temperatures, coupled with economic competition and deliberate arson, olive groves burn year after year in southern Europe. Greece is particularly affected, with extensive fires in the first two decades of the twenty-first century, including the Peloponnese peninsula in 2007, especially its western and southern parts, and the south of the island of Euboea; in 2009 in mainland Greece around Athens; in 2012 the island of Chios; in 2018 the coastal areas of Attica; in 2019 the island of Euboea. Finally, in 2021, when the peak

[5] Hallfahrt Ph., Lasy dziedzictwem kulturowym Niemiec [Forests as a cultural heritage of Germany], 24.09.2018 https://www.deutschland.de/pl/topic/srodowisko/ile-lasow-jest-w-niemczech-liczby-i-fakty [access 20 Sept. 2024].

[6] Hans Carl von Carlowitz and "Sustainability", Environment & Society Portal https://www.environmentandsociety.org/tools/keywords/hans-carl-von-carlowitz-and-sustainability [access 09 Sept. 2024].

[7] Węgiel A., Olej palmowy a lasy tropikalne [Palm Oil and Tropical Forests], Uniwersytet Przyrodniczy Poznań, http://www1.up.poznan.pl/orangutany/?q=lasy_tropikalne [access 20 Oct. 2024], no pagination.

[8] European Parliament resolution of 4 April 2017 on palm oil and deforestation of rainforests (2016/2222(INI)), https://www.europarl.europa.eu/doceo/document/TA-8-2017-0098_PL.html [access 23 Sept.2024]. Cf. Naukowcy: plantacje palmy olejowej-katastrofą dla azjatyckich lasów [Scientists: Palm Oil Plantations—A Disaster for Asian Forests], 12.10.2026, https://naukawpolsce.pl/aktualnosci/news%2C411573%2Cnaukowcy-plantacje-palmy-olejowej-katastrofa-dla-azjatyckich-lasow.html [access 12 Sept. 2024].

temperature in Greece reached 47.1 °C, many areas of the country were affected simultaneously.[9]

Meanwhile, climate change has only contributed to the ease with which fires ignite and spread. In Greece, the main cause of fires is:

> (...) a spark, either from human activities or natural occurrences like lightning (…). Typically, wildfires originate from agricultural practices such as burning crop residue or to stimulate pasture grass growth, or from clearing unwanted vegetation in scrubland and grassland areas.[10]

Thus, it is only the fires, or more precisely their effects on the ecosystem, that are significantly altering the local climate.

Another group of symbolic references in the national topics are floral plants. These include *roses* characteristic of Finland, Iran, Saint Kitts and Nevis, Saint Lucia, USA, UK. Second in order of occurrence are *lilies*, characteristic symbols of France, Canada, Saint Kitts and Nevis and Saint Lucia, and the *lotus* which applies to Bhutan, Egypt, India, Mongolia and Sri Lanka.[11]

Prominent among the flowers in national symbolism are the rose, and two related countries using a very similar, despite the differences, strategy of obtaining roses namely importing them on a large scale, and to pay high environmental cost.

In the U.S., the rose became a national symbol relatively recently, with the decision of President Ronald Reagan in 1986, who announced it in Proclamation 5574. In his argument, the president invoked the otherwise familiar significance of the rose as a bearer of memory, treating it as a symbol of life, love and devotion, beauty and eternity, and associating it with the national characteristics of Americans.[12] In the USA, the rose—a symbol—does not translate into rose production. While in 1985–1999, between 263 and 379 ha were devoted to cultivation of bouquet roses, after 1999 it virtually disappeared.[13] American roses are therefore not American, which, in view of their popularity, contributes to significant environmental pollution. According to Ewa Behrens-Nowakowska, *on Valentine's Day, an additional 9000 thousand tonnes of carbon dioxide are produced in America, all this due to red roses, cut down in Africa and South America, and transported by plane to the States. Faced with climate change, Colombian growers have set up the Floraverde organisation to promote good standards in the floriculture industry: the use of renewable energy sources, social packages for workers, and a ban on pesticides. The question*

[9] Radlak A., Pożary po grecku [Fires in Greek], "Przegląd Pożarniczy" online, https://www.ppoz.pl/czytelnia/rozpoznawanie-zagrozen/Pozary-po-grecku/idn:3104 [access 22 Oct. 2024].

[10] Fabrègue B.F.G., Greece vis-a-vis climate change: tourism and agriculture affected, 20 Nov. 2020, https://www.blue-europe.eu/pl/pl-analysis/pelne-raporty/grecja-wobec-zmian-klimatu-wplyw-na-turystyke-i-rolnictwo/ [access 22 Oct. 2024].

[11] Tomasz (no name), Narodowe kwiaty…

[12] More in: 101 STAT. 2038 Proclamation 5574—Nov. 20, 1986 https://www.govinfo.gov/content/pkg/STATUTE-101/pdf/STATUTE-101-Pg2038.pdf [access 23 Sept. 2024].

[13] Jabłońska, L., Zmiany na rynku róż ciętych w Unii Europejskiej [Changes in the cut rose market in the European Union], "Zeszyty Naukowe SGGW—Ekonomika I Organizacja Gospodarki Żywnościowej", 2013/102, p. 101.

is how helpful these changes can be in the face of such a huge carbon footprint caused by crops located at a large distance from the market.[14]

The British are no different in sourcing their beloved roses, relying heavily on Dutch exchanges and these, in turn, on crops grown in Africa. In 2010, the UK was Europe's fourth largest importer of roses sourced from Dutch re-exports.[15] Meanwhile, in England, enjoying a centuries-old tradition as a national symbol (the Wars of the Roses, 1455–1485), the rose is a garden flower, and gardens are the apple of British families' eye.[16] A distinguishing feature of the place and role of roses in the British ecosystem is therefore a rose garden, coupled with the co-occurrence of roses with vegetable or edible gardens, in reference to British World War II entrepreneurship in the 'Dig for Victory' campaign[17] and contemporary environmentalism. This, however, has two faces. On the one hand, organic gardens are created in the UK for the sake of the climate, while roses are bought in *mass* in the Netherlands, where, however, "(…) rose production has fallen by as much as 46.5%"[18] by 2011. On the other hand, "African countries are growing very dynamically (...), in most cases with the support of Western European capital, knowledge and experience. Due to the high energy and labour intensity of rose cultivation, production in these countries is accompanied by much lower costs".[19] They only increase disproportionately when the climate costs are taken into account, but these, as in the USA, and also in the UK, are not yet calculated in the context of climate costs.

Let me now proceed with animals in national symbolism. Among animal motifs, birds prevail, especially *eagles* (the white-tailed eagle, the bald eagle, the eastern imperial eagle), symbols of Poland, Canada, Germany, and Nigeria. This is followed by mammals, with the *lion* symbolising the UK, Israel and Kenya, and the *bear* (the brown bear, the polar bear, the panda), associated with Russia and China. There are individual cases of both birds and mammals, including the *kangaroo* (Australia), the California *condor* (Mexico), the *jaguar* (Brazil), the *buffalo* (USA), the *red-crowned crane* (Japan), the *Bengal tiger* (India), the grey wolf (Italy), the *rooster* (France), the bull (Spain), the *dolphin* (Greece), the *Canadian otter* (Canada), the *African elephant* (Egypt), the *steinbock* (Iran), the *oryx* (Jordan), the antelope (South Africa), the moose (Sweden), the golden *jackal* (Turkey), the *turul/ falcon* (Hungary), etc.[20]

---

[14] Behrens-Nowakowska E., Symbol…, p. 50.

[15] Jabłońska, Zmiany…, p. 106.

[16] Titchmarsh A., as interviewed by Bukalska P., Brytyjczycy kochają ogrody [Britons love gardens], "Polityka" 37.2011 (2824) 18.09.2011, https://www.polityka.pl/tygodnikpolityka/swiat/1519266,1,brytyjczycy-kochaja-ogrody.read [access 22 Oct. 2024].

[17] Titchmarsh A., Bukalska P., Brytyjczycy kochają ogrody…

[18] Jabłońska, Zmiany…, p. 109.

[19] Jabłońska, p. 109.

[20] Zwierzęce symbole państw świata [Animal symbols of countries in the world], https://wikizoo.pl/ciekawostki/zwierzece-symbole-panstw-swiata/ [access Oct. 10, 2024], also, The World Factbook…

Unlike plants, the meaning of animals in national symbolism, very rarely refers to economics or does not refer to them at all, and they represent universal virtues such as strength, independence, firmness, greatness, prowess. Sometimes also predation, and at other times gentleness and intelligence.[21]

In this context, let me discuss the case of the white-tailed eagle, which, as it occurs in nature, is a *Haliaeetus* (and therefore not necessarily an eagle), which officially appeared in the Crown Crest of the Kingdom of Poland in 1295. The white bird with a golden crown against a red background, the modern form of the Polish coat of arms, was established in The Republic of Poland by law in 1927.

The white-tailed eagle, like the golden eagle (also associated with the Polish coat of arms) are the two largest birds of prey in the Polish ecosystem, subject to strict protection dating back to the late nineteenth century. The third eagle is the osprey, which in the third decade of the twenty-first century has been protected by the EU, but still is the most endangered population in Poland because it feeds on fish farms.[22] To protect Polish eagles, in 1981 the Eagle Protection Committee was established, an informal organisation of ornithologists and volunteers tasked with protecting the eagle's habitat in view of the bird's ever-decreasing population. In 1991, the Committee was registered as an association (non-governmental organisation). Today, it attracts nearly 500 experts who, by protecting eagles in the Polish ecosystem, also contribute to increasing their number. They are facing new challenges, like the sometimes erroneous location of renewable energy sources (windmills) in the vicinity of the eagles' habitats, and the deliberate killing of birds of prey, also eagles, feeding on fish from breeding ponds.[23] Given the spectacular nature of Polish national symbolism, with the eagle dominating a multitude of ceremonies of varying rank, the bird's fate is all the less enviable. At the same time, its protection has been handled with great success by a grassroots organisation, which bodes well for the future.

National Symbolism as a Site of Memory

Plant and/or animal motifs in national symbolism carry a wealth of meanings relating to the national character of its beneficiaries, disposers and their historical (including natural) heritage. Meanwhile, the contemporaries' attitude to this symbolism varies, depending on their worldview and mentality: religious and/or instrumental, stemming from an imperative, choice or fashion, reserved by law, encased

[21] Blog of International Fund for Animal Welfare (IFAW), Animals with symbolic meaning around the world, May 6 2024, https://www.ifaw.org/international/journal/animals-symbolic-meaning-world [access 12 Oct. 2024].

[22] Osojca-Krasiński G., Rybołów—ginący orzeł znad wód i lasów [Osprey—the vanishing eagle of waters and forests], "Dzikie Życie", 2009/178/4. On-line article, https://dzikiezycie.pl/archiwum/2009/kwiecien-2009/rybolow-ginacy-orzel-znad-wod-i-lasow [access 1 Oct. 2024].

[23] About the Eagle Protection Committee, http://www.koo.org.pl/o-koo [access 21 Sept. 2024].

in ritual. Its interpretations are emotionally and politically charged. The presence of national symbols in social life is therefore situational. The patriotic agitation accompanying national and state holidays, social protests, sporting competition, even more so armed conflicts, usually result from a genuine need for identification through national symbols. In times of peace and prosperity, people attach less importance to them, remaining indifferent to their official presence in the public space.

However, especially in the twentieth and twenty-first centuries, in an era of accelerated modernisation and upheaval, including revolutions, wars and genocide, migration and the radical remodelling of social structures, the hyper-consumption of the global north and the poverty of the global south, the very idea of a national community and its identity is called into question when subjected to the pressures of rapid civilisation and cultural change on a global, national or local scale. The community's crisis is changing the status of its symbols. But what does (or could) this change consist of with regard to their plant and animal motifs, and what does it mean for the climate?

National symbols play a key role as places of national memory. They are therefore time-honoured carriers of historical memory.

In the more than a century-old theoretical discussion on historical memory, it was first separated from history and, following this distinction, contrasted as living history (memory) and dead history (historiography). Constitutive of this distinction was the recognition that memory (living history) comes, develops and operates in the element of experience and emotion of a specific community in response to the needs of the present, and the challenges of the future. Inanimate history, on the other hand, develops gratuitously in the element of science and cognition in response to the question of what past we represent.[24]

Nowadays, however, the above distinction is no longer self-evident, and memory is increasingly defined as revolted memory or salvage history.[25] This relatively new understanding of memory refers to intellectual and political movements of minorities of all kinds, in response to marginalisation or even exclusion thereof from the national or local community. Such minorities include contemporary pro-climate or environmental movements. Their representatives increasingly refer to the community of people, land and nature as inalienable elements of national landscapes of memory.

The point is that the status of memorials, including national symbols and their natural references, stems from a sense of loss. A place of remembrance, therefore, is anything that we experience as lacking, in a symbolic and material sense. This also applies to nature and its resources.

---

[24] More in Halbwachs M., La mémoire collective, Presses Universitaires de France, 1950; Nora P. (ed.), Realms of Memory: Rethinking the French Past, Vol. 1—Conflicts and Divisions Columbia University Press, 1996; Nora P., Czas pamięci [time of the memory], "ResPublica Nowa" 2001/7, transl. Dłuski W., pp. 37–43.

[25] More in Domańska E., Historia ratownicza [Rescue history], "Teksty Drugie" 2014/5, pp. 12–26.

In response to the exponential natural barrenness of vast areas of the world and the climate crisis in the form of a steady annual increase in average temperatures, droughts, fires, floods and destructive winds and storms, their progressive restoration is being planned and supported by states and governments.[26] These initiatives happen to be tentative and overdue while natural resources are increasingly protected as their memorials: memorial gardens, trees, and national or landscape parks, etc. These places are currently constitutive of memory landscapes.

National Symbolism as a Landscape of Memory

National landscapes are the effects of a specific approach to memory identified by Simon Shama as landscapes of memory[27] refers to "an environment or a place where traces of the past are preserved. It is an assemblage of bygone eras, with reference to past memories, latest customs, or remains unchanged in its own way despite the passage of time".[28] National landscapes are defined as:

> (...) a territory politically and/or emotionally close to a nation with needs, values, perceptions and traditions associated with that nation or defining characteristic elements of a landscape through national values (...).The idea of this type of landscape often revolves around a country-specific (...): 1) fauna and flora, 2) elements of topography and local history, 3) geography, 4) climate, 5) (...) modes of land use (...), 6) approaches to the environment. Iconic imagery is involved in the creation of such landscapes, illustrating the relationship between a people or a nation, and the territory and nature they inhabit.[29]

Therefore, both the natural traces of the past in the national symbolism and their actual resources are inherent in national landscapes of memory, sometimes present in the native landscape for centuries. However, nowadays they require protection and support so that they do not become a relic of the past. This is how I see the task for environmental patriots. Am I right?

Rescue History as Environmental Patriotism

The notion of environmental patriotism is discussed all the more readily in academic and activist milieus when a crisis of a nation coincided with the climate crisis. In the face of this first crisis, Pierre Nora has developed the concept of memorials together with a team of historians and historical anthropologists. They decided to consolidate the French heritage of the past by writing a series of monographs on the

---

[26] Radlak, Pożary...

[27] More in Schama, S., Landscape and Memory, New York: Vintage Books, 1996.

[28] Landscape of Memory, Vocabulary of the Landscape Studies, http://studiakrajobrazowe.amu.edu.pl/vocabulary_tag/krajobraz-pamieci/ [access 14 Sept. 2024].

[29] Krajobraz narodowy [National Landscape], Vocabulary of the Landscape Studies,

subject.[30] In the similar way, Simon Schama has described and analysed the landscapes of memory, which consists in explaining "how human imagination creates national sanctities out of simple nature. Landscape as an expression of ecological awareness is also an important phenomenon".[31]

The difference is that nature has taken a key place in his area of interest.[32] It is very likely that the sense of loss of cultural and natural resources attributed to a nation, building up over decades, has given the latter a new life in turbulent times of identity crisis and in even more turbulent times, a climate crisis, as Urlich Beck developed his concept of risk society.[33]

While experiencing climate catastrophes on a global scale, humanity does seem to be re-affirming national communities, including the enduring values behind them, invariably carried by national symbols. Nowadays, their beneficiaries are nations and citizens, and their legal disposers are the state and the movers and shakers. However, the attitude of nations and citizens to national symbols depends not only on the experiences, emotions and perceptions (the heritage and resources of a specific community) but also on the doctrine and ideology adopted by those in power. Thus, conservative political groupings and social movements are keen on upholding the national community in the spirit of traditional religiosity, sovereignty and patriotism. For them, national symbols are sacred. In contrast, liberal groupings and movements place these symbols within the bounds of tradition but leave citizens more freedom to decide whether and how they want to refer to them. Meanwhile, attitudes to national symbols change when those in power, be it conservatives or liberals, resort to populism, the former to religious fundamentalism and nationalism, the latter to extreme individualism and cosmopolitanism. National symbolism is then abused, used like a fetish or, conversely, trivialised and commodified. It does not bode well for its authentic existence, as it is stripped of its meaning.[34]

The authenticity of experiencing national symbols is determined by the depth of the relationship between humans and their communities and land, including own history, language, landscape as well as human, animal, plant, water etc. resources. This positive relationship is expressed in environmental patriotism which rises above particular populism in the name of the common good of the human and non-human inhabitants of the homeland.

According to Philip Cafaro, although patriotism has many varieties (some of them very dangerous), it needs to be revisited in the context of the climate crisis in

---

[30] More in Nora P., Realms…

[31] Traba, R., Historia stosowana. Pamięć i krajobraz jako nośniki badań i edukacji historycznej [Applied History: Memory and Landscape as Vehicles for Historical Research and Education], in Purda 1900–2006. Portret wsi [Purda 1900–2006. Portrait of a village], eds. Kardach M. , Pilecki J., Traba E., Wydawnictwo: Stowarzyszenie Wspólnota Kulturowa "Borussia" 2011, p. 7.

[32] More in Schama S., Landscape…

[33] More in Beck U., Risk Society: Towards a New Modernity, Sage, 1992.

[34] Meyer, J. M., "The People" and Climate Justice: Reconceptualizing Populism and Pluralism within Climate Politics, "Polity", 2024/56, pp. 199–354. Cf. Behrens-Nowakowska, E., Symbol…, pp. 54–55.

order to recover its original meaning when people lived close to nature, and patriotism was more than:

> (...) meaningless abstraction for many of us, in part through our mistaking abstractions for our true country. I think we must retrieve the word by rethinking and reliving it. When we live closer to the land, strive to know it better and work to protect all its inhabitants, human and nonhuman—that is patriotism.[35]

This need arises from the millennia-long movement of humanity from the natural world, which inspired it to give it symbolic meanings, through religious, family and crown symbols (which filled the human religious or family imagination) to flags and national emblems, which made extensive use of plant and animal motifs. All this has gradually stripped of meaning not only the symbols themselves, but also their actual references in nature, including their almost complete extinction.

In the meantime, the climate crisis has caught up with us, and with it, to paraphrase Ewa Domańska, the urgent need to save the future by re-imagining the past,[36] which is the sine qua non of rescue history.

As interpreted by Ewa Domańska, rescue history is primarily concerned with "(...) the periphery which increasingly claims the status of important centres where knowledge is accumulated (...) and local, potential, existential and affirmative history, drawing attention to its ecological aspect and formative function".[37] The ecological aspects and formative function of history:

> (...) raises awareness and changes attitudes towards people, nature, animals, plants, and also objects. Moreover, it encourages us to see world processes in terms of economic, social and political cycles but also biological ones (...) [And further, the ecological movement of rescue history aims to change attitudes, I.S.] towards the ideology of the exceptional human nature, a mechanistic vision of the world, and the instrumental treatment of nature as a resource for the satisfaction of human needs.[38]

Therefore, rescue history drives, or would drive, a shift in historical and social awareness towards post-anthropocentric history with the complexity of human-non-human coexistence with a view to the climate security of the next generations of humans, animals and plants as national security remaining the central problem of the contemporaries. Awareness of the inevitability of human and non-human fate in biological terms admonishes that human and natural resources are limited. Under these circumstances, the task of rescue history is a bottom-up and collective action to protect them, i.e. pro-climate human causality shared with the natural world and not above it, much less against it. The causality attributed to humans is distributed between the human and non-human worlds, and environmental patriotism is or can be the common denominator of this causality[39] interpreted as concern beyond the

[35] Cafaro Ph., Patriotism..., p. 182.

[36] Domańska E., Historia... p. 21.

[37] Domańska, p. 12.

[38] Domańska, p. 21, 13.

[39] Domańska, and more: Todd, A. M., Communicating Environmental Patriotism. A Rhetorical History of the American Environmental Movement, Routledge 2013.

present state interests for the homeland in all its wealth of symbolic and natural resources, with their history and present, strongly appealing for climate security for human and non-human inhabitants of the future, including, self-interestedly and biasedly, due to the cases I used, for lovers of palms, oaks, olive groves, roses, or eagles and others, not to be powerless, not to be alone...

And that is my conclusion… as hope.

References

101 STAT. 2038 Proclamation 5574—Nov. 20, 1986. https://www.govinfo.gov/content/pkg/STATUTE-101/pdf/STATUTE-101-Pg2038.pdf

Beck U (1992) Risk society: towards a new modernity. Sage, London

Behrens-Nowakowska E (2021) Symbol w lustrze. Obieg symboliczno-handlowy róży

Blog of International Fund for Animal Welfare (IFAW) (2024) Animals with symbolic meaning around the world, 6 May 2024. https://www.ifaw.org/international/journal/animals-symbolic-meaning-world

Cafaro P (2009) Patriotism as an environmental virtue. J Agric Environ Ethics 23(1–2):3–7

Domańska E (2014) Historia ratownicza [Rescue history]. Teksty Drugie, 5

European Parliament resolution of 4 April 2017 on palm oil and deforestation of rainforests (2016/2222(INI)). https://www.europarl.europa.eu/doceo/document/TA-8-2017-0098_PL.html

Fabrègue BFG (2020) Greece vis-a-vis climate change: tourism and agriculture affected, 20 Nov 2020. https://www.blue-europe.eu/pl/pl-analysis/pelne-raporty/grecja-wobec-zmian-klimatu-wplyw-na-turystyke-i-rolnictwo/

Halbwachs M (1950) La mémoire collective. Presses Universitaires de France

Hallfahrt P (2018) Lasy dziedzictwem kulturowym Niemiec [Forests as a cultural heritage of Germany], 24.09.2018. https://www.deutschland.de/pl/topic/srodowisko/ile-lasow-jest-w-niemczech-liczby-i-fakty

Hans Carl von Carlowitz and "Sustainability", environment & society portal. https://www.environmentandsociety.org/tools/keywords/hans-carl-von-carlowitz-and-sustainability

Jabłońska L (2013) Zmiany na rynku róż ciętych w Unii Europejskiej [Changes in the cut rose market in the European Union] "Zeszyty Naukowe SGGW—Ekonomika I Organizacja Gospodarki Żywnościowej", no 102

Krajobraz narodowy [National landscape], Vocabulary of the landscape studies. http://studiakrajobrazowe.amu.edu.pl/vocabulary_tag/krajobraz-narodowy/

Landscape of memory, vocabulary of the landscape studies. http://studiakrajobrazowe.amu.edu.pl/vocabulary_tag/krajobraz-pamieci/

Meyer JM (2024) "The people" and climate justice: reconceptualizing populism and pluralism within climate politics. Polity 56(2):252–274

National symbol(s), The world factbook. https://www.cia.gov/the-world-factbook/field/national-symbols/

Naukowcy: plantacje palmy olejowej-katastrofą dla azjatyckich lasów [Scientists: Palm oil plantations—A disaster for Asian forests], 12.10.2026. https://naukawpolsce.pl/aktualnosci/news%2C411573%2Cnaukowcy-plantacje-palmy-olejowej-katastrofa-dla-azjatyckich-lasow.html

Nora P (ed) (1996) Realms of memory: rethinking the French past, vol. 1—Conflicts and divisions. Columbia University Press, New York

Nora P (2001) Czas pamięci [Time of the memory]. ResPublica Nowa, no 7 (trans: Dłuski W)

Osojca-Krasiński G (2009) Rybołów—ginący orzeł znad wód i lasów [Osprey—the vanishing eagle of waters and forests]. "Dzikie Życie", no 178/4. On-line article, https://dzikiezycie.pl/archiwum/2009/kwiecien-2009/rybolow-ginacy-orzel-znad-wod-i-lasow
Radlak A. Pożary po grecku [Fires in Greek]. Przegląd Pożarniczy. Online. https://www.ppoz.pl/czytelnia/rozpoznawanie-zagrozen/Pozary-po-grecku/idn:3104
Schama S (1996) Landscape and memory. Vintage, New York
The Eagle Protection Committee. http://www.koo.org.pl/o-koo
Thoreau H (1980) The natural history essays. Peregrine Smith Press, Salt Lake City
Titchmarsh A (2011) As interviewed by Bukalska P, Brytyjczycy kochają ogrody [Britons love gardens]. Polityka, 37.2011 (2824) 18.09. https://www.polityka.pl/tygodnikpolityka/swiat/1519266,1,brytyjczycy-kochaja-ogrody.read
Todd AM (2013) Communicating environmental patriotism. A rhetorical history of the American environmental movement. Routledge, New York
Tomasz [no name], Narodowe kwiaty—roślinne symbole państw [National flowers—plant symbols of countries]. ABC Florysty. https://abcflorysty.pl/kwiatowe-symbole-krajow/
Traba R (2011) Historia stosowana. Pamięć i krajobraz jako nośniki badań i edukacji historycznej [Applied history: memory and landscape as vehicles for historical research and education]. In: Kardach M, Pilecki J, Traba E (eds) Purda 1900–2006. Portret wsi [Purda 1900–2006. Portrait of a village]. Stowarzyszenie Wspólnota Kulturowa "Borussia", Wydawnictwo
Węgiel A. Olej palmowy a lasy tropikalne [Palm oil and tropical forests]. Uniwersytet Przyrodniczy Poznań. http://www1.up.poznan.pl/orangutany/?q=lasy_tropikalne
we współczesnej kulturze [Symbol in the mirror. The symbolic and commercial circulation of the rose in contemporary culture] (2021) Master's thesis written under the supervision of Kasia K, Ph.D., Warszawa (typescript pdf https://upcoming.asp.waw.pl/wp-content/uploads/2021/12/asp-rose-behrens-nowakowska.pdf)
Zwierzęce symbole państw świata [Animal symbols of countries in the world]. https://wikizoo.pl/ciekawostki/zwierzece-symbole-panstw-swiata/

Izabela Skórzyńska, Associate Professor at the Faculty of History at Adam Mickiewicz University, postdoctoral fellowship at the University of Laval (Quebec, Canada), since 2019 executive member of the world Center for Women's Studies, Vice-President of the World-CWS 2024–2025. Co-organizer annual conferences in 2021 (Warsaw, Poland), 2022 (Bucharest, Romania) and organizer 10th Annual Conference in 2025 (Poznań, Poland). Author and co-author with C-F. Dobre, A. Wachowiak, B. Jonda, I. Chmura-Rutkowska, E. Głowacka-Sobiech, A. Chwieduk of research projects on the performative memory and didactics of history and the history and memory of women during communism in the light of autobiographical narratives "(Re) gaining the future by (re)building the past: women's narratives of life under communism in Poland, Romania and the former East Germany"; also the place and role of women in the school textbook narratives to history in Poland (women history/gender history). Co-editor with Maria Anitą Stefanelli "*Scars of War: Migration, Security and Sustainable Future*" *Research in Political Sociology*, 2025/30, Emerald Publishing.

Healing Ravaged Ecologies Through the Concept of Ecosocialism and Eco-Justice: An Ecocritical Reading of Vandana Shiva's Earth Democracy: Justice, Sustainability and Peace

Jagriti Upadhyaya

Abstract **Purpose of the Research Paper:** My paper focuses largely on Vandana Shiva's fight for the rights of Mother Earth through her iconic book 'Earth Democracy: Justice, Sustainability and Peace'. Quoting an 1848 speech attributed to Chief Seattle of the Suquamish tribe, Shiva tries to capture its essence that no one can buy or sell the sky, or air or water because earth does not belong to man, man belongs to the earth. Earth democracy as Shiva states is the awareness of these connections and of the rights and responsibilities that flow from them.

Methodology/Approach: Mine would be a qualitative approach wherein I will refer to the original text mentioned for primary material and other sources for secondary references.

Findings of the Research: We need to establish the health of the Mother Earth through a green political ideology that moves from a global capital economy to a sustenance economy where the rights of ethnic minorities, the poor and women are not marginalized by giant corporations that manipulate knowledge and data and by developing a language' and an environmental praxis that doesn't separate the subject and the object and only then we can usher in biological egalitarianism, and restore justice, peace and sustainability.

Originality: This paper is my own research work and contains all the in-text citations wherever required.

Keywords Ecosocialism · Eco-justice · Sustainability · Peace · Earth democracy · Vandana Shiva · Politics of predatory globalization

---

J. Upadhyaya (✉)
Department of English, Sardar Patel University of Police, Security and Criminal Justice, Jodhpur, Rajasthan, India
e-mail: jagriti@policeuniversity.ac.in

B. Wejnert (ed.), *Democracy in Retreat and the Climate Crisis*, Springer Studies on Populism, Identity Politics and Social Justice,
https://doi.org/10.1007/978-3-032-02881-5_15

Introduction

> In nature's economy, the currency is not money, it is life.
>
> Whenever we engage in consumption or production patterns which take more than we need, we are engaging in violence.
>
> The fight for truth…is not just our right as free citizens of free societies. It is our duty as citizens of the earth (Vandana Shiva)

Mahatama Gandhi had aptly stated: "The earth has enough for everyone's need bur not enough for eveyone's greed." The same sentiment had been echoed by the sage Bhartihari (450–510 CE) when he said "*Trishna na jirna, vayamev jirna*" in his *Niti Shatak* (ethics for behaviour) meaning thereby that we get old but are desires never get old. The same sentiment had been echoed by the sage Bhartihari (450–510 CE). Cheryl Glotfelty *in The Ecocritical Reader: Landmarks in Literary Ecology* quotes the historian Donald Worster (who) explains, "We are facing a global crisis today, not because of how ecosystems function but rather because of how our ethical systems function. Getting through the crisis requires understanding our impact on nature as precisely as possible, but even more it requires understanding those ethical systems and using that understanding to reform them. Historians, along with literary scholars, anthropologists, and philosophers, cannot do the reforming, of course, but they can help with the understanding" (Glotfelty and Fromm 1996, xxi).

Modern day materialism with its focus on unlimited consumption and an unprecedented pursuit of hedonism, and the stress on material production and intricately complex market forces of supply and demand and a raging consumerist culture has led to thoughtless and mindless exploitation of natural resources leading to economic inequality and a skewed development. We live with a false notion that amassing material wealth makes one rich whereas a truly rich man is the one who is devoid of any material desires. As Andrew Heywood in *Political Ideologies: An Introduction* states: "The development of the 'productive forces' roughly, natural resources and the technology used to turn them into humanely useful objects under capitalism has transformed nature beyond all recognition. Humankind currently faces the prospect of environmental disaster precisely because in its passionate and blinkered pursuit of material wealth, it has quite simply, upset the 'balance' of nature and endangered the very ecosystems that make human life possible" (Heywood 2015, 268).

Consumerism is the outcome of unending desires and it is abetted by modern means of production facilitated through technology and the superficially glamorous world of countless advertisements constantly bombarding the subliminal mind creates a perennial want of more and more products and an unprecedented depletion of natural resources. It has resulted in the contamination of water resources and air, the destruction of tropical and temperate rainforests, the extinction of some species and putting some on the list of managed species, the greenhouse effect and the depletion of the ozone layer, global warming and the resultant vagaries of weather, deforestation and disappearance of valuable farmland and wilderness for flyovers and highways and housing colonies creating wide chasms between the haves and the have not's. These interconnected crises are the result of human actions and activities, procreation, recreation and pursuit of wealth, comfort and convenience.

John Stanley and David Loy in their essay *At the Edge of the Roof: The Evolutionary Crisis of the Human Spirit* in *Spiritual Ecology: The Cry of the Earth* by Vaughan Lee contend that indigenous human communities have always had a scared and spontaneous affinity with nature but in the advanced globalized society religious symbolism that earlier put reins on human greed and aggression, now cannot compete with the technology-driven advertising industry manned by giant corporations. They underscore:

> By glorifying self-concern as never before, consumerism generates a mental environment of endless competition. It undermines empathy, altruism and cooperation. (Vaughan-Lee 2013, 30).

But consumerism generates even more desires, leaving consumers as Heywood explicates: "in a constant state of dissatisfaction because, however much they acquire and consume, they always want more. Consumerism thus works not through the satisfaction of desires, but through the generation of new desires, keeping people in an unending state of neediness, want and aspiration" (Heywood 2015, 265).

This avaricious attitude of mankind has resulted in the contamination of water resources and air, the destruction of tropical and temperate rainforests, the extinction of some species and putting some on the list of managed species, the greenhouse effect and the depletion of the ozone layer, global warming and the resultant vagaries of weather, deforestation and disappearance of valuable farmland and wilderness for flyovers, highways and housing colonies creating wide chasms between the haves and the have not's. These interconnected crises are the result of human actions and activities, procreation, recreation and pursuit of wealth, comfort and convenience. Terence Ball in *Green Political Theory* states, "These humans are headless inheritors and practitioners of a perspective variously called 'humanism' or 'anthropocentrism', a hierarchical orientation that stresses the well-being of humans with little or no regard for, and at the expense of, other creatures and the natural environment" (Ball, 536).

Kenneth Boulding is of the opinion that human beings act as if they live in a 'cowboy economy'—an economy with unlimited opportunities as was during the frontier days in America. But Boulding suggests that this encourages "reckless, exploitative, and violent behaviour" as happened in the American West (Heywood 2015, 260). It also leads to the widening of the chasm between the rich and the poor and it is the third-world countries that have to face the brunt of the greed of the developed nations. If the few richest nations, corporations or individuals exploit the limited resources of the world, then a large mass of the marginalized are deprived of even the basic necessities of life.

Ravaged ecologies and carbon footprints are created by:

- Farm and other agricultural land management causes diffuse pollution which can affect water bodies such as rivers, streams and lakes, and the major pollutants are: phosphates, nitrates, pesticides, sediment and faecal bacteria.

- Automobile pollution release carbon monoxide, nitrogen oxides and hydrocarbons. These pollutants damage lung tissue and aggravate respiratory diseases like asthma. Automobile pollution also contributes to acid rain and adds to the greenhouse gases that cause climate change. Acid rain refers to the deposition of wet (rain, snow, sleet, fog, cloudwater and dew) and dry (acidifying particles and gases) acidic components. Acid rain has adverse impacts on forests, freshwaters and soils, killing insects and aquatic lifeforms as well as causing damage to buildings and having impacts on human health.
- More than 75% of the world's usable freshwater comes from forested areas. Rather shockingly, 36 million acres of natural forest are lost each year. More than 300 million people around the world live in forests and more than 1.5 billion people depend on forests for their livelihood (UNEP).
- The depletion of the ozone layer is caused by ODS or ozone-depleting substances like CFCs, HCFCs and halons, emitted from commercial, home and vehicle air conditioners, and refrigerators, foam-blowing agents. Solvents, aerosol spray propellants, fire extinguishing agents, and chemical reactants.

Vandana Shiva is a world-renowned environmental activist and physicist. And a leading voice in struggles for global justice. "In Earth Democracy, Shiva updates the struggles she helped bring to international attention-against genetic food engineering, cultural theft, and Natural resource privatization-uncovering their link to the rising tide of fundamentalisms, violence against women, and planetary death. Starting in the sixteenth century with the initial enclosure of the British commons, Shiva reveals how the commons continue to shrink as more natural resources are patented and privatized. As our ecological sustainability and cultural diversity are at a teetering edge, so too is human life rendered disposable. Through the forces of neoliberal globalization, economic and social exclusion ignite violence across lines of difference, threatening the lives of millions. Yet these brutal extinctions are not the only trend shaping human history. Struggles on the streets of Seattle and Cancun and in homes and farms across the world have yielded a set of principles based on inclusion, non-violence, reclaiming the commons, and freely sharing the earth's resources. These ideals, which Shiva calls Earth Democracy, serve as an urgent call to Peace and as the basis for a society and sustainable future" (Book overview of Earth Democracy: Justice, Sustainability, and Peace. https://www.amazon.in/Earth-Democracy-Vandana-Shiva/dp/8181581490#:~:text=Book%20overview&text=In%20Earth%20Democracy%2C%20Shiva%20updates,against%20women%2C%20and%20Planetary%20death).

Globalization and Its Impact on the Environment and Social Structures

Globalization and corporate capitalism see the world where the planet is a deposit of limitless resources to be exploited for the creation of wealth. The expansion of a global market-based economy has been done at the expense of nature's economy and a sustenance economy where corporations have systematically enclosed the commons. Those resources that are managed and used by all members of communities, that have sustained people and cultures for centuries, are now becoming privatized and commoditized by corporations seeking only to maximize potential profits. Humans have become utility maximizers. Unfettered competition and profits are the engine of corporate globalization, while the people bear the brunt of the externalities created by short-sighted, greed-based projects. As Shiva vocalizes: "one part of the international community cannot destabilize the climate, enclose the atmospheric commons, or ignore the rights of other species and other countries by creating 36 percent of the world's CO2 pollution" (Shiva 2015, 8).

The Politics of Predatory Globalization

Shiva is critical of the myth of prosperity and abundance that Globalization has perpetrated. She underscores how cash-crop production and food processing divert land and water resources away from sustenance needs that leaves more and more people from their entitlement to food. She argues that profit inevitably flows to the giant corporations 'that have no interest in feeding hungry people without money' (Shiva 2015, 33). Market economy becomes ineffective and destructive in the context of the other two economies of nature and sustenance precisely because corporate globalization destroys local and national economies, breeding insecurity, fear and exclusion and knowledge, which ought to allow informed citizens making free choices, becomes a tool to be manipulated by big corporations.

Globalization: A Political Project

Shiva contends that globalization is a political project and it needs a political response which needs to put human beings at the center of economic thought and disallow big corporations to annihilate human rights and practice exclusivist policies, that not only exclude diverse cultures but also women's politics of making change through everyday lives.

Navdanya (A movement) as Shiva states is "a network that attempts to shift from the suicidal/ genocidal economy of agribusiness imposed by the WTO and the World Bank to living economies of food" (Shiva 2015, 67). In one publication it highlights thus: "We are living in a linear, economy-driven world that is inevitably unsustainable. Like colonialization before it, corporate globalization counts on claiming resources outside of the scope of ownership and inside the trust of the commons of people. Through the efforts of the WTO and agribusiness giants, subsistence and small farmers have lost the right to food through the increasing expansion of business-based agricultural methods that benefit the corporations, at the expense of labour and lead to rural despair and hunger in the world. Free trade, industrialization of agriculture, patents and intellectual property rights (IPRs) and the commodification of seed and food have produced an impoverished and toxic planet, ill health and disease, rampant inequality and dying democracies. Unfair trade rules and pseudo-safety laws that dump junk food, fake food, GMFs, synthetic and artificial food on the countries' markets are based on unjust and unscientific laws of extraction, exploitation, domination and corporate colonization, mostly serving the interest of the developed global North." https://navdanyainternational.org/key-issues/earth-democracy/.

The Anthropocene and Slow Violence

Mankind has ushered in the age of Anthropocene by its relentless consumption and exploitation of earth's limited resources. Drawing attention to how in the Anthropocene that human agents have set in motion irreducible climate changes Amitav Ghosh discusses capitalism as the new colonialism in his iconic book *The Nutmeg;s Curse: Parables for A Planet in Crisis.* In agreement with Naomi Klein Ghosh argues that the wheels of modern day capitalist economy run on fossil fuels, those commodities of compacted remains of primordial forests, treated as expendable resources and provides answers to why the switch to 'greener' economy and sustainable energy sources is slow- because greedy corporations and power-wielding individuals in the global order want to continue reaping huge monetary profits and the resultant power dynamics, "even if this should entail the destruction of the world."((Ghosh, 100).

Raising issues about the indelible entanglement of capitalism, geopolitics, empire and power dynamics with today's irreducible climate change Ghosh states that the modern day citizen is now *Homo economicus* where capitalism and resultant rampant consumerism have become the prime movers of modern history but Ghosh finds that 'the era of Western military conquests and imperialism predates the emergence of capitalism by centuries' when the genocide of the Amerindian peoples led to the formation of modern Europe. He expostulates: "No pillage of the Americas, no capitalism, no Industrial Revolution, thus perhaps no Anthropocene either" (Ghosh 2016, 2021, 116).

Serpil Oppermann and Serenella Lovino in their classic studies *in 'Environmental Humanities: Voices from the Anthropocene'* contend: "The idea that natural cycles are not impermeable to capital has inspired others to use the term 'Capitalocene', a definition—proposed somehow independently by Andreas Malm, Jason Moore, and

Donna Haraway—to indicate an age in which capitalism is understood as a world ecology, joining the accumulation of capital, the pursuit of power, and the co-production of nature in dialectical unity." (Oppermann and Lovino 2017, 11).

In his iconic book '*The Anthropocene: The Promise and Pitfalls of an Epochal Era*' Rob Nixon considers a concept in environmental studies that highlights the massive impact that humans have had on the present and future of the planet. The term 'Anthropocene' popularized in 2000 by the atmosphere chemist Paul Crutzen, and the ecologist Eugene Stoermer, refers to a proposed new geological epoch that began in the late eighteenth century and is defined by massive impacts on the earth's geophysical and biophysical systems. Nixon in *The Norton Anthology of Theory and Criticism* highlights the massive inequalities that it has created between the world's rich and poor and "expresses his concerns about the 'anthropocene optimists' who tell a hubristic story about humans emerging as demigods who will increasingly excel at bending the planet to their collective will" (Nixon 2018, 2354). Referring to 'slow violence' Nixon argues: "By slow violence I mean a violence that occurs gradually and out of sight, a violence of delayed destruction that is dispersed across time and space, an attritional violence that is typically not viewed as violence at all." (Nixon 2018, 2356).

Focusing on slow violence Nixon states:

> In doing so we also need to engage the representational narrative, and strategic challenges posed by the relative invisibility of slow violence. Climate change, the thawing cryosphere, toxic drift, biomagnification, deforestation, the radioactive aftermaths of wars, acidifying oceans, and a host of other slowly unfolding environmental catastrophes present formidable representational obstacles that can hinder our efforts to mobilize and act decisively. The long dyings—the staggered and staggeringly discounted casualties, both human and ecological that result from war's toxic aftermaths or climate change—are underrepresented in strategic planning as well as in human memory. (Nixon 2018, 2356)

What Nixon underscores is also raised by Pramod K. Nayar. Commenting upon the discourse of ecoprecarity Nayar finds numerous enmeshed constituents that have rendered the ecosystems of the earth fragile and precarious in the Anthropocene- "the degradation and destruction of habitats, animal populations, climate and Nature…. and we are threatening life itself with toxic landscapes all around, with other life forms 'edging towards the precarious' often resulting in the extinction of species." (Nayar, 8)

Ramchandra Guha in *Environmental: A Global History* is skeptical about the remarks of the US Secretary of the treasury Henry Morgenthau, who was confident that post World War II a dynamic world economy would flourish and the world would enjoy the fruits of material progress 'on an earth *infinitely* blessed with natural riches.' Harvard geologist Kirtley F. Mather deified the age of ecological innocence and claimed that like North America the under-developed world, with the 'endless frontier' of Science harnessed with technology and the 'inexhaustible resources' of the earth, since Mother Earth had enough to spare would embark on the path of material progress. Little could they envisage what Carl Sauer, the great Berkeley geographer found, that this buoyant optimism bound by spatiotemporal limitations who categorically stated that this sort of reckless frontier attitude was nothing more than the residue of days when "north-European freebooters overran the world and put it under tribute" (Guha 2000, 66). Guha also refers to E.F Schumacher who believed that economic expansion has legitimized the 'rapacious exploitation' of non-renewable resources like fossil fuels as also to Lewis Mumford who believed that the 'awful omniscience and the omnipotence of our science and technology', the instruments which we use to ravage nature will lead us to nothing but self-destruction (Guha 2000, 67). Similarly Catriona Sandilands in her essay *Green Things in the Garbage: Ecocritical Gleaning in Walter Benjamin's Arcades* commenting upon exploitative technological development as part of the naturalization of capitalist progress writes that "In capitalism, then, history is garbage. And nature is garbage" (Goodbody and Rigby 2011, 31).

Capitalism, consumerism and rapacious materialism provide the cultural basis for environmental degradation in which nature remains nothing but a commodity or a resource. Thich Nhat Hanh, the famous Buddhist, in his essay *'The Bells of Mindfulness'* in Lee's *'Spiritual Ecology' highlights* human impact on the planet and warns that if humans continue to live the way they have been and keep consuming mindlessly, then 'devastating climate change is inevitable.' (Lee et al, 17)

Ecosocialism and Earth Democracy

Social ecology is a term coined by Murray Bookchin to refer to a range of ideas, chief being the premise that environmental degradation is, in some way, related to existing social structures and that the market economy 'commodifies' nature as also the breakdown of the organic fabric of both society and nature .Andrew Heywood underscores that the core theme of ecosocialism finds capitalism as the enemy of the environment and one solution to ecological problems is to redefine 'the ownership of wealth' (Heywood 2015, 270).

Vandana Shiva promotes the concept of 'Earth Democracy' which redefines 'upholding family values' as respecting the limits on greed and violence. She quotes Chief Seattle who says that 'all things share the same breath, the beast, the tree, the man.... The air shares its spirit with all the life it supports.' Shiva further reiterates that, "one part of the international community cannot destabilize the climate, enclose the atmospheric commons, or ignore the rights of other species and other countries by creating 36 percent of the world's CO2 pollution" (Shivạ 2015, 8).

The Judeo-Christian ideology of God having created man in his own image and man naming all the other creatures establishing his dominance over them essentially negates the ancient pagan and animistic cultures where every tree, every spring, every stream, every hill had its own genius loci, its guardian spirit. Christopher Manes in the chapter 'Nature and Silence' in 'The Ecocritical Reader' quotes a Tuscarora Indian remark that unlike his people's experience of the world, for Westerners, the uncounted voices of nature are dumb. Manes states, "The language that we speak today, the idiom of Renaissance and Enlightened Humanism, veils the processes of nature with its own cultural obsessions, directionalities, and motifs that have no analogous in the natural world" (Glotfelty and Fromm 1996, 15). We are so anthropocentric in our attitude that despite Copernicus, all the cosmos rotates around our little globe; despite Darwin, we in our hubristic attitude consider ourselves superior to nature, are contemptuous of lesser species and exploit nature according to our whims and desires. This unchecked, rampant exploitation driven by a merciless Capitalistic economy has led to a skewed idea of development and the resultant environment crises that we face today.

Quoting the magnitude of the alarming way this power has disrupted the natural resources Rachel Carson in her path breaking book 'Silent Spring' (1962) which spearheaded the environmental movement in America Rachel Carson states: "The most alarming of all man's assaults upon the environment is the contamination of air, earth, rivers, and sea with dangerous and even lethal materials. This pollution is for the most part irrecoverable, the chain of evil it initiates not only in the world that must support life but in living tissues is for the most part irreversible" (Carson 2002, 6).

Shiva is highly critical of Corporate Globalization which she finds, creates new enclosures of the common, which are based on exclusion and violence. She states: "Instead of a culture of abundance, profit-driven globalization creates cultures of exclusion, dispossession, and scarcity. In fact, globalization's transformation of all beings and resources into commodities robs diverse species and people of their rightful share of ecological, cultural, economic, and political space. The 'ownership' of the rich is based on the 'dispossession' of the poor—it is the common, public resources of the poor which are privatized, and the poor who are disowned economically, politically, and culturally" (Shiva 2015, 2–3).

Shiva is critical of patents on life and the ownership of water, biodiversity, cells, genes, animals, plants because she says that such a worldview expresses the fact that life forms have no intrinsic worth of their own and are deprived of their subjecthood because of the concept of an 'ownership society.'

The globalized free market economy denies excess to life and livelihoods by generating scarcity which is brought about by destruction of nature's economy and sustenance economy through over-exploitation of resources, and harboring anti-life practices by disenfranchisement of people by treating them as disposable entities.

Andrew Heywood in *Political Ideologies* in the chapter on '*Ecologism*' of the opinion that the use of technology to use natural resources into utilitarian objects in a world driven by capitalistic forces has development of the 'productive forces' roughly, natural resources has transformed nature beyond all recognition. He states:

> Ecologists argue that humankind currently faces the prospect of environmental disaster precisely because in its passionate and blinkered pursuit of material wealth, it has quite simply, upset the 'balance' of 'nature' and endangered the very ecosystems that make human life possible. (Heywood 2015, 255).

Dipesh Chakravarty in '*The Climate of History in a Planetary Age*' discussing capital, technology, and the Planetary argues that the globe and the planet—as categories standing for the two narratives of globalization and global warming—are connected. He argues: "What connects them are the phenomenon of modern capitalism (using the term loosely) and technology, both global in their reach."(4). He further underscores how this has led to greenhouse emissions and contends that the advocates of climate justice should see global warming as a consequence of uneven capitalist development inflicted by class, gender, and race.

Shiva Highlights the Following Reasons That Have Led to Global Capitalism and Its Resultant Inequalities

The domination of the market—A key to the domination of the market economy is its ability to claim resources from outside its scope and to transform land from public to private ownership.

The economics of enclosure—it worked through the exclusion of people from access to resources that were considered common and created surplus or 'disposable' people by denying them such rights, destroying interdependence, cooperation and collective decision-making.

Terra Nullus—Most sustainable culture view earth as 'terra mater' or mother earth, and have reverence for earth-centered cosmologies, thus receiving nature's bounty with gratefulness and returning the debt with sustainable lifestyles. But colonial constructs brought in the passivity of the earth and thus the concept of *terra nullus* or empty land developed.

From Commons to Commodities: Colonialism as Enclosure

Colonialism not only usurped power but also natural resources and created private property by enclosing the commons, displacing and uprooting indigenous people in the Americas, Asia, and Africa, and destroyed living cultures and ethnic identities through systemic exploitation of natural and human resources.

Both Edward Said and Frantz Fanon have written extensively on how colonialism marginalized the indigenous identity and made it a trope for domination and derision. As Denise Judd argues in his book *Empire: The British Imperial Experience from 1765 to the Present*, "[n]o one can doubt that the desire for profitable trade, plunder and enrichment was the primary force that led to the imperial structure" (Judd 2012, 3). Judd further argues that colonialism was first and foremost part of the commercial venture of the Western Nations that developed from the late seventeenth and early eighteenth centuries, although some date it to the advent of European voyages of discovery. The seizing of 'foreign' lands was in part motivated by the desire to create and control markets abroad for Western goods, and to capture the natural resources and procure the native labour-power through slavery or indentured work. John McLeod states, "Colonialism was a lucrative commercial operation, bringing wealth and riches to Western nations through the economic exploitation of others. It was pursued for economic profit, reward and riches. Hence colonialism and capitalism share a mutually supportive relationship with each other" (McLeod 2010, 7).

Economic Globalization/Corporate Globalization

Unbridled capitalist greed and trade rules have destroyed cultures of compassion and social justice where resources, instead of being shared, are owned and used by a handful of rich corporations. Corporate globalization recognizes water and biodiversity as commodities to be owned and privatized. But certain natural resources are 'not tradable and will be governed by values other than commodity values' (Shiva 2015, 30).

Globalized Agriculture Cash crop production diverts land and water resources from sustenance needs and excludes increasing number of people from their entitlement to food. The displacement of small farmers by giant corporations, the industrialization of traditional methods of farming which leads to soil depletion, the patenting of seeds, the creation of terminator seeds, the destruction of soil, water, and biodiversity, coupled with the vagaries of weather due to massive climate change all contribute to scarcity and human induced poverty.

Privatization of Water It creates the possibility of excluding others from access to that which is necessary for living. Shiva says; "A system which can claim that water can be allocated on this planet according to how capital and control and access it is saying that most species can go extinct. No species seeks its entitlement to its share of water through the market place; they get their access to water through being members of communities and ecosystems" (Shiva 2015, 43).

Intellectual Property Laws Within indigenous communities, innovation is seen as a social and collective phenomenon fostering the use and sharing of new knowledges thus resulting in collective usage of biodiversity. Innovations over time become a common pool of knowledge and this heritage is not viewed as private property. Shiva opines: "The concept of individual 'property' rights to either bioresources or knowledge remains alien to the local community. This undoubtedly exacerbates the usurpation of the knowledge of indigenous people. With serious consequences for them and for biodiversity conservation" (Shiva 2015, 41).

Loss of Workers' Security The rise of capitalism has led to profit-driven capital-run economies replacing producer-run economies, where humans have become cogs in the wheels of the mega-corporations. Shiva contends: "Today, as a result of globalization, workers' rights are being dismantled and instead of growth generating employment, we are witnessing jobless growth" (Shiva 2015, 48). Shiva also underscores the how outsourcing creates wage gaps and helps the corporate giants to avoid the obligation that they have towards providing workers with economic security in their old age.

Ramchandra Guha also refers to E.F Schumacher who believed that economic expansion has legitimized the 'rapacious exploitation' of non-renewable resources like fossil fuels as also to Lewis Mumford who believed that the 'awful omniscience and the omnipotence of our science and technology', the instruments which we use to ravage nature will lead us to nothing but self-destruction (Guha 2000 67). Similarly Catriona Sandilands in her essay *Green Things in the Garbage: Ecocritical Gleaning in Walter Benjamin's Arcades* commenting upon exploitative technological development as part of the naturalization of capitalist progress writes that "In capitalism, then, history is garbage. And nature is garbage" (Goodbody and Rigby 2011, 31).

Shiva highlights: "Nature's economy is the production and reproduction of resources through a complex network of ecological processes. But market-oriented economy jeopardizes nature's economy through the overexploitation of resources and the incomprehended destruction of natural processes." (Shiva 2015, 16). As more and more people get displaced, and insecurities grow negative cultures of terrorism; extremism and xenophobia take virulent forms and humanity begins to define itself through its inhumanity.

Ecologism and ecosocialism seek not only to revise conventional moral thinking but also to reshape our understanding of happiness and well-being. Capitalism, consumerism and rapacious materialism provide the cultural basis for environmental degradation in which nature remains nothing but a commodity or a resource.

States Andrew Heywood in his iconic book 'Political Ideologies.' Perhaps the most influential concept for modern 'Greens' has been developed by looking back to pre-Christian ideas. Primitive religions often drew no distinction between humans and other forms of life, and for that matter, little distinction between living and non-living objects. All things are alive, stones, rivers, mountains, and even the earth itself, often conceived of as 'Mother Earth.' In '*Gaia, A new look at Life on Earth*' (1979) James Lovelock developed the idea that the planet itself was alive and gave it the name 'Gaia' after the Greek goddess of the Earth.

Eco-feminist historians have shown that the mother earth image has been used by western male writers since the Renaissance to underwrite the domination both of nature and of women. The ancient mystification of nature as a "benevolent female," Carolyn Merchant argues in 'Key Concepts in Critical Theory—Ecology' "Contained the implication that nature when plowed and cultivated could be used as a commodity and manipulated as a resource." With the scientific revolution, nature changed "From an active teacher and parent" into "a mindless submissive body', in the hands of the Baconian engineers of the modern world system that is."

Vandana Shiva states: "Earth Democracy movements are the struggles of the disadvantaged and excluded, aimed at conserving nature's balance to preserve their survival. Ecology and justice movements offer the world a future by working to ensure the survival of, and protect the fundamental rights to, the earth's resources."

(Shiva 2015, 61). These movements vocalize the marginalized and help to reverse the trend of deracinated people as disposable people.

Earth Democracy is rooted in the ancient Indian concept of ***Vasudhaiva Kutumbkam***, the earth family. As one family, all beings have equal rights to sustenance through the gifts of the earth. Its three components: economy, environment, and people, must be in balance to maintain sustainability. As Shiva explains: "Earth Democracy is both an ancient worldview and an emergent political movement for peace, justice, and sustainability. Earth democracy connects the particular to the universal, the diverse to the common, and the local to the global" (Shiva 2015, 1).

Shiva referring to Chief Seattle of the Suquamish tribe protests as to how we can sell or buy the warmth of the sky, the freshness of air, and the sparkle of water. It's a transgression of the sanctity of Mother Earth since Chief Seattle says that the earth does not belong to man; man belongs to the earth. He stresses upon the interconnectedness of all things 'like the blood which unites our family' (Shiva 2015, 1). Shiva echoes these sentiments in: "Our world is not for sale. Our water, seeds and biodiversity are not for sale" (Shiva 2015, 2).

Earth Democracy works to promote a shift towards a circular model where the economy is held in no greater esteem than the environment and the resources it provides and the people, for whom justice, fairness, and the right to resources is of equal importance. It's a new vision for a Planetary Citizenship, an alternative worldview, rooted in caring and compassion for the Earth and Society, in which humans are embedded in the Earth Family, where ecological responsibility and economic justice, based on the Law of Return, are central to creating a liveable future for humanity. Shiva reiterates: "In Earth Democracy the concern for human and nonhuman species come together in a coherent, nonconflicting whole that provides an alternative to the worldview of corporate globalization, which gives rights only to corporations and which sees human and other beings as exploitable raw material or disposable waste." (Shiva 2015, 8).

Earth democracy is based on earth centered and community centered knowledge systems. Living knowledge is knowledge that maintains and renews living processes and contributes to health of the planet and people. It is also living knowledge in that it is embedded in nature and society, is not abstract, reductionist and anti-life. Living knowledge is a commons, it belongs collectively to communities that create it and keep it alive. All humans have a duty to share knowledge. No person or corporation has a right to enclose monopolize patent or exclusively own as intellectual property living knowledge. Earth Democracy also stresses that rights are derived from and balanced with responsibility. Those who bear the consequences of decisions and actions are the decision makers.

Principles of Earth Democracy as Given by Shiva in Her Classic Text Earth Democracy: Justice, Sustainability, and Peace

The Principles of Earth Democracy given by Shiva are as follows:

1. "*All species, peoples, and cultures have intrinsic worth*
   All species, humans and cultures have intrinsic worth. They are subjects, not objects of manipulation or ownership. No humans have the right to own other species, other people or the knowledge of other cultures through patents and other intellectual property rights" (Shiva 2015, 9). "Deep ecologists have viewed nature as the source of moral goodness. Nature thus has 'intrinsic' or 'inherent' value, not just instrumental value deriving from the benefits it brings to human beings" states Andrew Heywood (2015, 275).
2. "*The earth community is a democracy of all life*
   In other words we are all members of the Earth community. We all have the duty to protect the rights and welfare of all species and all people. No humans have the right to encroach on the ecological space of other species and other people, or treat them with cruelty and violence" Shiva 2015, 9).
3. "*Diversity in Nature and Culture must be defended*
   Defending biological and cultural diversity is a duty of all people. Diversity is an end in itself, a value, a source of richness both material and cultural" (Shiva 2015, 9).
4. "*All beings have a natural right to sustenance*
   All members of the Earth Community including all humans have the right to sustenance — to food and water, to safe and clean habitat, to security of ecological space. These rights are natural rights, they are birthrights given by the fact of existence on earth and are best protected through community rights and commons. They are not given by states or corporations, nor can they be extinguished by state or corporate action. No state or corporation has the right to erode or undermine these natural rights or enclose the commons that sustain all through privatization or monopoly control" (Shiva 2015, 9–10).
5. "*Earth Economy is based on living economies and economic democracy*
   Earth democracy is based on economic democracy. Economic systems in Earth Democracy protect ecosystems and their integrity, they protect people's livelihoods and provide basic needs to all. In the earth economy there are no disposable or dispensable species or people. The earth economy is a living economy. It is based on sustainable, diverse, pluralistic systems that protect nature and people, are chosen by people, for the benefit of the common good" (Shiva 2015, 10).
6. "*Living Economies are built on local economies*
   Conservation of the earth's resources and creation of sustainable and satisfying livelihoods is most caringly, creatively and efficiently and equitably achieved at the local level. Localization of economics is social and ecological

imperative. Only goods and services that cannot be produced locally, using local resources, local knowledge should be produced non-locally and traded long distance. Earth democracy is based on vibrant, resilient local economies, which support national and global economies. The global economy does not crush and destroy local economies" (Shiva 2015, 10).

7. "*Earth Democracy is a living democracy*

   Earth democracy is based on local living democracy with local communities, organised on principles of inclusion and diversity and ecological and social responsibility having the highest authority on decisions related to the environment and natural resources and to the sustenance and livelihoods of people. Authority is delegated to more distant levels of governance on the principle of subsidiarity. Earth democracy is living democracy. Earth democracy is based on local living democracy with local communities, organized on principles of inclusion and diversity and ecological and social responsibility having the highest authority on decisions related to the environment and natural resources and to the sustenance and livelihoods of people. Authority is delegated to more distant levels of governance on the principle of subsidiarity. Earth democracy is living democracy" (Shiva 2015, 10–11).

8. "*Earth democracy is based on living cultures*

   Vibrant, living cultures nurture peace and harmony and create free spaces for different religions, faiths and identities and embrace cultural diversities based on our common humanity as members of the earth community. Such living cultures are ecological cultures based on the reverence for all life forms as part of a planetary consciousness" (Shiva 2015, 11).

9. "*Living cultures are life-nourishing*

   They promote the dignity of all life, human and nonhuman, people of all genders and cultures as well as present and future generations. They revere life and recognize multiplicity of identities. These cultures represent planetary consciousness—the web that connects individual to the earth and all life" (Shiva 2015, 11).

10. "*Earth democracy globalizes peace, care and compassion*

    Earth democracy connects people in circles of care, cooperation and compassion instead of dividing them through competition and conflict. Earth democracy globalizes compassion, not greed, and peace, not war" (Shiva 2015, 11).

Ancient Concepts of Earth Democracy

Ancient concepts of living peacefully and the ancient concept of the non-separability and interconnectedness is evident in quantum theory, the space-time continuum of general relativity, and the self-organized complexity of living organisms. Andrew Heywood quotes Fritjof Kapra who in '*The Turning Point*' holds "the Cartesian-Newtonian paradigm responsible for contemporary environmental crisis because science treats nature nothing more than a machine which means that it can be toyed with, tinkered with or even be replaced. In his classic work '*The Tao of Physics*' Capra draws parallels between modern physics and ancient Eastern religions, particularly quantum theory and Zen Buddhism that 'have long preached the unity or oneness of all things" (Heywood 2015, 259).

In her path breaking book 'Silent Spring' (1962) which spearheaded the environmental movement in America Rachel Carson states:

> The history of life on earth has been a history of interaction between living things and their surroundings. To a large extent, the physical form and the habits of the earth's vegetation and its animal life have been molded by the environment. Considering the whole span of earthly time, the opposite effect, in which life actually modifies its surroundings, has been relatively slight. Only within the moment of time represented by the present century has one species—man—acquired significant power to alter the nature of his world (Carson, 2002: 5).

George Sessions in the chapter *Ecocentrism and the Anthropocentric Detour* in '*Key concepts in Critical Theory—Ecology*' edited by Carolyn Merchant quotes Stan Steiner who in describing the traditional American Indian philosophy of the sacred 'circle of life' finds a spirit of fraternity and sorority in all creation. Steiner says: "Life is shared with the bird, bear, insects, plants, mountains, clouds, stars, sun" (Merchant 1996, 140).

If there is anything that can "stop the human species from poisoning the Earth or blowing it up". Writes feminist ecotheologian Carol Christ, it is 'a deeply felt connection to all beings in the web of life.'

What modern Western ecologists have now begun to voice their concerns over, was an intrinsic part of the holistic living of ancient India. Ancient texts are replete with examples about the entire biotic as part of the same cosmos, reflected in

> Yat Pinde, tat Brahmande (Upadhyaya 2023, 119).

The concept of Earth as Mother or as the Greek Goddess Gaia, as James Lovelock gave to the earth is a sacred concept in ancient cultures. This concept of the Gaia was used by James Lovelock to emphasize the biosphere as a self-regulating system but for him the earth is not merely a cybernetic system but also a unitary organism. Lawrence Buell in *The Environmental Imagination—Thoreau, Nature writing and the formation of American Culture*, states:

Lawrence Buell in the chapter 'Nature's Personhood' underscores: "Undoubtedly the commonest image, in both Euro-American and Native American traditions, for expressing the idea of an earth –humanity continuum has historically even the metaphor of "mother earth" of "Mother Nature." "Our Mother the Earth", one witness testifies, "is a reality in the cosmologies of virtually every native people in the world." (Buell 1995, 214).

As the Upanishad rightly states that this moving world, and the other, is all permeated by the all-pervasive Ishwara. Man should by renunciation enjoy only what is righteously his and not covet what belongs to others.

> Ishavasyam idam sarvam, yat kinchat jagtyamajat
> Tein tyakten bhunjhithaha, ma gridhah kasyasvidhanam. (Ishvasyopanishad, 1.1)

> Whatever there is in this world, is covered by Ishwar (God) and belongs to him. Man should consume what is righteously his and not covet what by right belongs to others. (My translation).

The Idea of Eco-Justice

Eco-justice seeks not only to revise conventional moral thinking but also to reshape our understanding of happiness and well-being. Capitalism, consumerism and rapacious materialism provide the cultural basis for environmental degradation in which nature remains nothing but a commodity or a resource Vandana Shiva stresses that the concept of 'Earth Democracy' redefines 'upholding family values' as respecting the limits on greed and violence. She quotes Chief Seattle who says that 'all things share the same breath, the beast, the tree, the man.... The air shares its spirit with all the life it supports.' Shiva further reiterates that, "one part of the international community cannot destabilize the climate, enclose the atmospheric commons, or ignore the rights of other species and other countries by creating 36 percent of the world's CO2 pollution" (Shiva 2015, 8).

Environmental ethics and justice mandates that man is not the conqueror or master of 'Mother Nature' but an integral part of the biotic community and is therefore responsible for the welfare and protection of not only the natural environment but the entire biotic community. John Muir questions the very notion nature created plants and animals for the happiness on man alone. Ramchandra Guha quotes Muir, "Why should man value himself as more than a small part of one great unit of creation? And what creature of all that the Lord has taken the pains to make is not essential to the completeness of that unit—the cosmos? The universe would be incomplete without man; but it would also be incomplete without the smallest transmicroscopic creature that dwells beyond our conceitful eyes and knowledge" (Guha 2000, 52). As Shiva highlights: "Earth Democracy connects us through the perennial renewal and regeneration of life—from our daily life to the life of the universe" (Shiva 2015, 8). As Shiva highlights: "Earth Democracy connects us through the perennial renewal and regeneration of life—from our daily life to the life of the universe" (Shiva 2015, 8). Vandana Shiva also stresses upon the fact that we humans share the planet with millions of other species and states that both justice and sustainability can work only if we restrict the use of resources. She says: "Restraint in resource use and living within nature's limits are preconditions for social justice" (Shiva 2015, 50).

And the concept of 'Deep Ecology' of Arne Naess becomes significant in this context. As Naess quotes, "The ecological field worker acquires a deep- seated respect, or even veneration, for ways and forms of life. He reaches an understanding from within, a kind of understanding that others reserve for fellow men and for a narrow section of ways and forms of life. To the ecological field worker, the equal right to live and blossom is an intuitively clear and obvious value axiom. Its restriction to humans is an anthropocentrism with detrimental effects upon the quality of human themselves." (Merchant, 120).

Aldo Leopold's 'Land Ethic' in '*A Sand County Almanac*' envisions "changes in the role of homo sapiens from conqueror of the land community to plain member and citizen of it. It implies respect for his fellow members and also respect for the community as such."

Bill Devall in the chapter 'The Deep Ecology Movement' states, "In Deep Ecology, the wholeness and integrity of person/ planet together with the principle of what Arne Naess calls 'biological equalitarianism' are the most important ideas. Man is an integral part of nature, not over or apart from nature. Man is a 'plain citizen' of the biosphere, not its conqueror or manager. There should be a 'democracy of all God's creatures' according to St. Francis; or as Spinoza said, man is a 'Temporary and dependent mode of the whole of God/Nature" (Devall 1996, 133).

As Shiva insists: "Earth Democracy allows us to reclaim our common humanity and our unity with all life. Erath Democracy relocates the sanctity of life in all beings and all people irrespective of class, gender, religion, or caste" (Shiva 2015, 8). By instilling family values of the earth such a democracy does not allow privatization of water, corporatization of agriculture or patenting of life, 'since all beings have a right to life and well-being' (Shiva 2015, 7).

Shiva insists that: "Our survival demands that we make a transition from vicious cycles of violence to the virtuous cycles of non-violence; from negative economies of death and destruction to living economies that sustain life on earth and our lives; from negative politics of corruption and fascism to living democracies which include concern for and participation of all life; and from negative cultures that are leading to mutual annihilation to positive and living cultures based on caring, compassion, and conservation" (Shiva 2015, 106).

Shiva further states: "Earth Democracy protects the ecological processes that maintain life and the fundamental human rights that are the basis of the right to life, including the right to water, the right to food, the right to health, the right to education, and the right to jobs and livelihood. Earth democracy is based on the recognition of and respect for the life of all species and all people" (Shiva 2015, 8).

Green Political Theory: Terence Ball

Andrew Heywood in Political Ideologies: An Introduction underscores the idea of green political thinkers who are highly critical of any philosophy that views nature only or mainly as a means to an end or a commodity and encourage the idea of an ecocentric theory of value that promotes the 'well-being of nature's myriad species' (Heywood 2015, 541). Andrew Heywood focuses on E. Schumacher's Buddhist economics that is based on the ethics of right livelihood in which individuals, "unlike in conventional economic theories are not 'utility maximizers' but facilitators of personal growth through social bonds and removal of ego centeredness." (Heywood 2015, 262).

Earth Democracy thus posits a paradigm shift of the worldview from one dominated by markets and monocultures, and mechanistic reductionism, to the peaceful and symbiotic co-creation of diverse species enmeshed inextricably in the common web of life. As Shiva truly expostulates "Earth Democracy allows us to overcome artificial scarcity and manipulated and manufactured insecurities by seeing and experiencing connections." (Shiva 2015, 185). In Earth Democracy the concern for human and nonhuman species come together in a coherent, nonconflicting whole that provides an alternative to the worldview of corporate globalization, which gives rights only to corporations and which sees human and other beings as exploitable raw material or disposable waste" (Shiva 2015, 8).

References

Buell L (1995) The environmental imagination—Thoreau, nature writing and the formation of American culture. Harvard University Press, Cambridge

Carson R (2002) Silent spring. Mariner, Boston

Devall B (1996) The deep ecology movement. In: Merchant C (ed) Key concepts in critical theory—ecology. Rawat, New Delhi

Ghosh A (2016) The great derangement: climate change and the unthinkable. Penguin Random House, Gurgaon

Ghosh A (2021) The Nutmeg's curse: parables for a planet in crisis. Penguin Random House, Gurgaon

Glotfelty C, Fromm H (1996) The ecocriticism reader—Landmarks in literary ecology. University of Georgia Press, Athens

Goodbody A, Rigby K (2011) Ecocritical theory: new European approaches. University of Virginia Press, Charlottesville

Guha R (2000) Environmentalism—a global history. Oxford University Press, New Delhi

Heywood A (2015) Political ideologies. Palgrave Macmillan, Basingstoke

Judd D (2012) Empire: The British imperial experience from 1765 to the present. I.B. Tauris, New York

Leopold A (1970) A Sand County Almanac with essays on conservation from round river. Ballantine, New York

McLeod J (2010) Beginning post-colonialism. Viva, New Delhi

Merchant C (1996) Key concepts in critical theory—ecology. Rawat, New Delhi

Nixon R (2018) Slow violence and the environmentalism of the poor. In: Leitch VB, Cain WE, Finke LA, McGowan J, Sharpley-Whiting TD, Williams JJ (eds) The Norton anthology : theory and criticism. W. W Norton, New York

Oppermann S, Lovino S (2017) Environmental humanities: voices from the anthropocene. Rowman Littlefield, London

Shiva V (2015) Earth democracy: justice, sustainability and peace. Natraj, Dehradun in association with Research Foundation for Science, Technology and Ecology, New Delhi. https://www.amazon.in/Earth-Democracy-Vandana-Shiva/dp/8181581490#:~:text=Book%20overview&text=In%20Earth%20Democracy%2C%20Shiva%20updates,against%20women%2C%20and%20Planetary%20death

Upadhyaya M (2023) Interconnectedness of man and ecology in the puranas in literary oracle, vol 7. Issue II. December 2023. Das S et al (ed) Authors Press, Odisha, India

Vaughan-Lee L (2013) Spiritual ecology: the cry of the earth. The Golden Sufi Center, Inverness

Dr. Jagriti Upadhyaya is the Head of the Department of English, Sardar Patel University of Police, Security and Criminal Justice in Jodhpur, India.

Unravelling Sufferings: Ecofeminism and Empowerment in Selina Hossain's *Charcoal Portrait*

Neelam Kedia

Abstract As a social and political movement, ecofeminism seeks to draw attention to how women and nature suffer and are dominated by patriarchy, development, capitalism, and scientific advancement. It examines the resilience and resistance of both women and nature in the face of exploitation. Women's suffering is multi-layered; it includes suffering due to caste, social status, poverty, colonization, and gender among others. Similarly, nature too suffers in the name of progress and development. Women are subjected to many forms of discrimination in developing countries; neither in the domestic nor the economic spheres do they have equal standing. The present paper deals with Bangladeshi writer Selina Hossain's *Charcoal Portrait* which delves into the lives of tea plantation workers, whose seemingly idyllic existence amid lush green surroundings masks the unbearable reality of their conditions. It will involve qualitative and detailed textual analysis with the theoretical framework of ecofeminism. It will analyse the tremendous longing for freedom and identity of the female characters who in one way or the other fall prey to the patriarchal setup, as presented in the novel. Besides that, it will highlight the untold sufferings of third-world women and the effects of deforestation, pollution and violence on nature. The central focus of this paper is to present how the beautiful surroundings have turned into a green prison for its inhabitants and how sustainability and empowerment are needed for smooth survival.

Keywords Ecofeminism · Suffering · Resistance · Identity · Empowerment · Sustainability

---

N. Kedia (✉)
Department of English and Foreign Languages, Central University of South Bihar, Gaya, Bihar, India

B. Wejnert (ed.), *Democracy in Retreat and the Climate Crisis*, Springer Studies on Populism, Identity Politics and Social Justice,
https://doi.org/10.1007/978-3-032-02881-5_16

Introduction

Ecofeminism is an umbrella term that unites and combines women and environmental issues, ranging from oppression and exploitation to claiming empowerment and justice. If we break the term ecofeminism, we find two different concepts, 'eco' and 'feminism'. Eco stands for all the biotic and abiotic components of the earth, and the environment, and feminism is the study of equality and empowerment of women, who have long been dominated and marginalised. Ecofeminism as a movement came out in the 1960s and 70s with women participating in the anti-nuclear movement and the feminist lesbian movement. The term, "ecofeminism" was first coined in 1974 by Françoise d'Eaubonne in her book *Le féminisme ou la Mort* or *Feminism or Death*. In this book, she blames and holds patriarchy responsible for the overuse and exploitation of natural resources as well as the exploitation of women leading to overpopulation. Ecofeminism comes out as a branch of both ecocriticism and feminism and deals with the issues of women and nature together. Apart from the issues of women and nature, the theory brings together the suffering of all the marginalised, as Greta Gaard (1993) says,

> Ecofeminism's basic premise is that the ideology, which authorizes oppression such as those based on race, class, gender, sexuality and physical abilities is the same ideology which sanctions the oppression of nature. Ecofeminist theorists consider the interconnections between sexism, the domination of nature, racism, speciesism and other social inequalities (p. 1).

Any movement begins with the realisation of the lack of rights, equality, and oppression. The feminist movement started when women realised their lack of rights and equality in both the domestic and political spheres. The first thing they fought for was the right to vote. Next, they gradually protested for their rights and established their own identity individually. Similarly, ecocriticism as a movement emerged when people became aware of environmental issues and started taking a stand for them. Given the fact that both nature and women are subjected to oppression from the same oppressive and dominant force, the liberation movements of women and nature combined and thus emerged ecofeminism, also known as ecological feminism. In the first world, the suffering of women is single-layered to some extent as they are dominated by patriarchy only, whereas in developing countries, the scenario is different in the sense that women suffer not only for being women in the hands of their male folk, rather they suffer for their gender, race, caste, and class. They become doubly and triply marginalised in the hands of patriarchy, colonialism, and poverty. Quoting Mies, Paul Guerra (2023) says that a main postulate of ecofeminism is that male domination and modern civilisations are built on the colonisation of women, nature, and nations. Thus, ecofeminism is about promoting the inclusion of nature, or the environment, in understanding the actions of the exploited (4). Ecofeminism stands for highlighting these gender inequalities as well as society's worst attitude towards women and nature, which they look upon as commodities and profit. Coming away from mainstream Ecofeminism centred on Eurocentric views, it expanded its theoretical arena and branched out into several sub-branches, as it failed to fully comprehend the entire struggle and suffering of the women of developing countries. In order to consider the sufferings and experiences of third-world women, there emerged post-colonial ecofeminism. The experiences of all the women of the world are different, and they have several layers of exploitation according to their own situation, background, caste, and class. Ecofeminists have variable opinions regarding the oppression, exploitation, and interconnection of women and nature. According to Elizabeth Carlassare (2000), divergent opinions exist among ecofeminists regarding the means of bringing about social change as well as the ways in which patriarchy, capitalism, Western dualism, and imperialism contribute to ecological degradation and oppression on several fronts, such as class, race, and gender. Despite this difference, the dedication of ecofeminism's proponents to achieving global survival and putting an end to oppression unites them (p. 89). Ecofeminism takes into consideration every form of oppression since all the domination stems from the same place and by unveiling it, it calls for the liberation and freedom of all the marginalised, including women, children, the poor, the weak, the disabled, non-human animals, and nature. According to Vandana Shiva (1988), patriarchy and capitalism are the reasons behind the exploitation of both nature and women. She has considered Western capitalism as "maldevelopment" (p. 4), which destroys everything, including the environment and women. Greta Gaard (Gaard and Murphy 1998) also considers "maldevelopment", patriarchy, global capitalism, and corporation as the reason of the struggle of women and environmental degradation and claims that ecofeminist movement fought "for environmental balance, hierarchal and matrifocal societies, the continuance of indigenous culture and economic value and culture based on subsistence and sustainability" (p. 2). Ecofeminist Karen Warren (2000) talks about the logic of domination as one of the causes of oppression. In this logic, she challenges the Cartesian dualism and says that in this dualistic structure, one group is considered as up and the other down, and the up group has the full authority and liberty to dominate the other one, and it is considered very natural. Whoever is up in the hierarchy always subordinates the poor (p. 47).

Ecofeminism is a philosophical theory yet a political movement. It examines how factors like social, and political have an impact on women's lives. Women and nature are subordinated by various means. The paper deals with Selina Hossain's *Charcoal Portrait*, originally written in Bengali as *Kathkoylar Chhobi* in 2001, and later translated into English by Trina Chatterjee and Jackie Kabir in 2019. The novel centres primarily on the life and suffering of the tea-plantation workers with a little focus on life and suffering of women and nature outside of the plantation too. It unveils various forms of oppression that women and nature face due to patriarchy, colonialism, development, capitalism, and how specially women fall prey both inside and outside their homes. From an ecofeminist lens, the novel uncovers the joint oppression and exploitation of women, nature, and marginalised others and present their interconnection, sustainability and empowerment.

History of the Migration of Tea-Plantation Workers

Britishers first introduced tea to break the monopoly of China over tea production. Sylhet was the first tea plantation of Bangladesh since 1854 and the first tea garden in Bangladesh was Malnichhara of Sylhet. From the book *Story of the Tea Workers: Subsistence against Hegemony*, edited by Dr. Md. Rahmat Ullah (2014), we come to know how in the Sylhet tea plantations, labour migration occurred due to natural disasters, poverty, a need for cheap labour, and the expectation of a better future. There were other factors, though, such as the fact that they were deceived into this servitude (34). The colonials didn't get the natives to work in the plantations, as Ghosh Hazra (2000) mentions in her article,

> the local people were reluctant to join the industry due to the hazardous condition of work as well as its low wage. The labourers had no fixed hour of work neither any fixed type of work. Their wages was as low as three rupees per month as the planters viewed their work as 'easy and light' (p. 60).

Therefore, they brought the poor and distressed people from various regions of India, "mainly from Chotonagpur area of Bihar and its adjoining area of Orissa and Madhya Pradesh. These people were the aboriginal people of Oraon, Murida, Santhal, Barik, Tulsi, Asur, Lahar, Kheria and others, popularly known in India as 'adibasi'" (Ghosh Hazra 61), lying to them for a better future. They were brought in the plantation with the dream of a golden future, but they didn't know they would be imprisoned there lifelong, even their future generation would also not get rid of it. The workers who tried to flee from the drudgery of the plantation, either get shot or imprisoned. They remained stuck there generation after generations with a dream in their eyes to go back to their motherland. In the plantation they live in pathetic conditions with no facilities. They even do not get the basic minimum requirements for survival. Fakhrul Alam (Alam and Sarker 2009) in his article says, "The colonial rulers framed certain rules and regulations to restrict their life in the gardens, keeping them out of the light of all development prospects found outside the domain of tea gardens" (1). The condition of the tea plantation workers remains more or less the same till now. They are still not provided with basic facilities like medical and sanitation, they still live in shanties with the meagre amount they get after a whole day's work. Nothing has changed much since then and the protests are going on till now.

Discussion of Selina Hossain's Charcoal Portrait

Charcoal Portrait deals with the miserable life story of tea-plantation workers of Sylhet in Bangladesh, their fight for identity and freedom, daily hardships and discriminations they experience, and the double exploitation of the female plantation workers. Selina Hossain has given the detailed view of the entire history of the tea plantation, its workers and specially the misery of the women workers in the novel, which is so appropriate in the current situation too. The narrative is set against the backdrop of 1971 on the tea plantation at Sylhet, Bangladesh, where the Britishers long ago brought a community to work in the tea plantation forcefully. The novel prominently centres around one of the tea plantation workers, Chaitirani, it is her tale of her suffering, survival, and resistance to gain her identity by going back to her roots with the sons begotten to her by rape. The workers were brought from several places by the Britishers with the hope of a better future, but once they reached here, they couldn't return to their homeland since then. Whenever they try to flee from there, either they get imprisoned or shot there.

> Pulled out of our roots, we came to this land hoping for better days—four meals a day, a peaceful life. All we got here were twisted iron pots and pans, frayed blankets, straw mattresses, shanties with leaking roofs, rotis and salt tea. And of course this life in captivity, the garden as a prison. (p. 173)

She remembers her parents crying and beating their heads in agony to go back to their homes, "The day we came here we were like a brood of chickens, a flock of ducks, hutched onto a crowded train, not humans. Even after so many years, I don't feel like being human again" (32). The workers live in the most pathetic and unhygienic conditions in shanties inside the plantation, almost like cattle. For generation after generation, they are imprisoned here on the plantation. The condition of the women on the plantation is even more pathetic; they are oppressed and cheated both by the males in their houses and the plantation authority in the workplace. Apart from that, Hossain has presented several layers of exploitation of women even outside the plantation. She has presented how women suffer from the excessive greed of their in-laws for dowry and for giving birth to a girl child. Hossain, in almost all her novels, deals with the ground-level reality; she delves down to the life of the common people, especially women, and puts them down in her writing.

Suffering Due to Patriarchy

Ecofeminists like Francoise d'Eaubonne (2022)), Vandana Shiva (1988)), and Karen Warren (2000) among others consider patriarchy as the root cause of oppression of women and nature. According to Karen Warren (2000), ecofeminists critique male domination and institution of patriarchy (63). Under patriarchal setup, the male member of the house has been bestowed with all the liberty, rights to decision making and controlling others, especially the women of the family. According to Sylvia Walby (1989),

> patriarchy is composed of six structures: the patriarchal mode of production, patriarchal relations in paid work, patriarchal relations in the state, male violence, patriarchal relations in sexuality, and patriarchal relations in cultural institutions, such as religion, the media and education. (p. 214)

Under the clutches of patriarchy, women suffer the most and so is nature. Women inside the plantation are the worst sufferers of these patriarchal relations. Life inside the plantation is no less than hell and when it comes to women, it gets more terrible. They look after the whole family, spend all day in the garden, are beaten by their husbands, are taken advantage of by the authorities, and are imprisoned like animals on a plantation. Women workers inside the plantation are doubly marginalised, first, they are exploited at their homes, then they got exploited by the authority, "three-quarters of labor forces are women and many of them are victimized both inside and outside of family" (Hassan 2014, p. 123). Inside the tea garden, it is the responsibility of the women to pluck the tea leaves because the process needs to be handled with care. They got cheated by the authority as their own husbands in the household who snatched the entire money from them for their booze:

> The women pluck leaves throughout the day; while weighing the bundles the babus cheat them. And when they return home, their men will snatch away all their earning, leaving them with nothing more than bits of gur and a couple of rotis. What happiness! (p. 168)

The males don't bother about running the household or feeding their children, rather they snatch even the hard-earned money of their wives, "After picking leaves, the boys would rush to the shacks and the women to their firepits" (31). Discussing about the condition of the women, Hassan (2014) states,

> It has been found that major portion of them cannot spend money by their own interest and will; even sometimes they are compelled to give full amount of salary to their husband. Otherwise, maximum of them are being beaten and mentally harassed. (p. 125)

The payday is not a joyful day for the women as one of the plantation worker states, "The bastard waits like a hawk for my payday and snatches away the money I earn slogging the whole week" (p. 162).

Apart from the male members of the family, women in the plantation are subjected to the lust of the authority. Previously the women workers used to be exploited, dominated and used by the British colonisers, now they got raped and exploited by the authority of their own country, they are treated as commodities whom the male folk can enjoy. Chaitirani says, "This body that was violated, brutalized, devoured twice in these gardens—first by the Bengali manager of the garden and then a Pakistani soldier" (p. 47). According to Collard and Contrucci (1988), "In patriarchy, nature, animals and women are objectified, hunted, invaded, colonized, owned, consumed and forced to yield and produce (or not). This violation of the integrity of wild, spontaneous Being is rape" (1). The novel presents how, during the liberation war of Bangladesh, the condition of the women workers became worst. They got pathetically exploited by the West Pakistani soldiers. They were raped, abandoned and left pregnant by the soldiers. The soldiers randomly used to pick the female workers from the plantations and rape them for months. They even rape the female workers in front of their families. When Chaitirani returned to her family after the war, she was completely wounded and couldn't even walk properly. Looking at her deplorable condition, her mother couldn't tolerate it anymore and died. Inside the plantation, a woman is no less than a commodity, who is used for labour and pleasure. The Pakistani soldiers not only raped the women, but they set fire to the homes of the workers, and looted their cattle and harmed the serenity of the nature of the plantation.

Since every woman belongs to a different circumstance, class, caste, and background, it is impossible to generalise about women worldwide. The only thing they have in common, despite the many levels of exploitation and oppression they face, is their constant suffering. Selina Hossain has presented about women outside the plantation, who are suffering from the greed of dowry of their husband and in-laws. They are being tormented and assaulted in their households in order to obtain a dowry. Moufuli, Kazim's wife is tortured and dominated by her husband and mother-in-law as her father couldn't provide them with the asked money of the dowry. Her husband beat her and even doesn't consider her as human, "If the wife's father doesn't pay, then hit her" (108). They constantly taunt her for giving birth to a girl child and neither care for her nor her child. She silently works and manages the entire household, but yet her mother-in-law and husband turn a blind eye to all her contributions behind running the family. Another character Deepti is also a subject of her husband, Sattar's excessive greed. He left both Deepti and their daughter Rinku when Deepti's father didn't fulfil the demand for money.

Suffering Due to Capitalism

Quoting Mies, Karen Warren (2000) says that women's body, labour, and natural environment are colonised by a combination of capitalism and patriarchy (26). In order to make the plantation, a larger area of forest and fertile land was cut and destroyed as tea comes under cash crop. Not only did capitalists imprison labourers, but they also imprisoned the natural world. There is significant biodiversity loss when high biodiversity areas such as forests are converted to tea plantations (van der Wal 2008, p. 8). Because of progress and globalisation, human beings are using natural resources carelessly and without hesitation, which has a negative impact on the environment. They are utilising this very environment that aids in their success while simultaneously exploiting nature for their own benefit. Ballard and Parham (2020) in their article state:

> Much of the habitat used for tea cultivation is located in remote areas, areas which would otherwise be teeming with life because of an ability to support high levels of biodiversity. Tea plantations result in the direct loss of habitat, but this impacts the wider environment too. Tea cultivation alters the geographical landscape and effects the natural flow of water, increasing soil erosion and often leading to the loss of unique wetland habitats (p. 101).

The way the land has been cleared in order to make the plantation, in the same way, women labourers also get destroyed and exploited by the authorities. Hassan (2014) in his article states, Bangladesh can earn a lot of foreign currency reservation by exporting tea and this contribution is being made possible mostly by the efforts of female workers of tea garden. But it is pathetic that there is no especial priority in giving facilities to female labourers of tea garden though three quarters of labour forces are women and many of them are victimized both inside and outside of family (p. 123). Both nature and women suffer inside the plantation. The authority only cared about the tea plantation, they didn't care about the upgradation and quality life of the workers, neither did they care about the plantation. They deduct maximum labour from the women workers in return of minimum wages. The condition of the plantation workers is summarised as such—"Being alive like this is no better than the life of a maggot" (Hossain 32). The plantation workers face water scarcity throughout the year but it reaches its worst in summer. They have to suffer from this water crisis as the authority has no take in this matter. Due to environment pollution and unhygienic condition in the tea plantation, the workers even suffer from various deadly diseases like TB and typhoid, among others, and left untreated to die. The workers are not even given a good quality of tea as it has been exported to foreign countries for the sake of money. The people whose hard labour makes tea cultivation possible get only the left-over dust.

In the novel, the stark contrast of the polluted city and the peaceful greenery of the rural area is presented. Dulal, who came from Europe, in search of his original mother, feels choked and suffocated in the city of Dhaka, where there is traffic, noise, and pollution at its peak. Bahar, a villager, whose home location is away from the polluted city has been described as, "Shrouded in greenery, the place is out of the ordinary and does not seem to be part of the overbuilt, congested capital city" (Hossain 13). Because it is not as heavily targeted by anthropocentric exploitative activities, the area is somewhat safe. The effect of pollution and continual deforestation is very clear in the urban areas, "the city is getting increasingly treeless" (190). The way people are recklessly killing trees, in the same way, they are killing people. Inflicting continual torture on nature and women for the sake of profit and pleasure is making people emotionless completely.

Women–Nature Interconnection

Ecofeminists say that women share a special connection with nature because of their biology, similar oppression and their conditioning as caregivers and nurturers. Susan Griffin strongly believes in the relation and interconnection of women and nature. Vakoch (2012) says,

> According to cultural ecofeminism, there is an innate connection between women and nature. By positing an inherent tendency of women to be attuned to nature—to take care for it, to recognize their interrelationship with it—cultural ecofeminists recognize the value of actions and characteristics typically devalued by the dominant (patriarchal) culture (p. 4).

In the tea garden, the tea-picking responsibility is bestowed on the women. Since the tea leaves are so delicate, it requires utmost concentration to pick the leaves in the pattern of two leaves and a bud. In addition to having the patience and nurturing attitude needed to care for tea plants, females are socially conditioned to be extremely tolerant. Because their livelihoods are entwined with the plantation, women have a strong attachment to and care for it, "If the tea plants die, so do we—we know that very well. We don't know any other work. Nor do we have another place to go" (p. 173).

Chaitirani unlike all the plantation workers is very frustrated with the imprisonment inside the plantation, but she gets solace only in the lap of nature. She soothes herself by taking baths in the water of the stream. In order to make her own clan, Chaitirani wants to convert her body to a fertile land, where the sons in the form of crops will be born, "She was no longer a tract of barren land—the boundaries had been stretched—ready for the new crop, another world to begin" (68). Terms like womb, rape, barren, and fertility are sometimes used interchangeably to refer to both natural elements and women. Chaitirani describes her womb as the fertile land, through which she wants to conceive and produce many children in order to make her own clan, to revolt and go back to her motherland. She shares a kind of empathetic relationship with nature, where she shares and parallels her mental state and thoughts with elements of nature. She experiences a comparable loss of identity in the factory because of the way tea leaves burn to change their hue. They have also experienced burns, gone through a difficult procedure, and lost their identities. Although the tea plantation is a prison for them, they connect to the plants and handle them with love and care. They have issues with the imprisonment but they connect with the greenery there.

Many women over there get stuck in their household duties amidst the four walls of their house and they don't get the time or opportunity to see how beautiful the outside world is. The same happened with Deepti and Moufuli, once they go out of their homes, they feel the beauty of the world, and on the lap of nature, they feel the urge to stand for themselves. Moufuli along with Nishi, her daughter, leaves home to take a stand against all the drudgery and realised that she always thought her home to be the centre of peace, but outside nature is more peaceful and beautiful. On a touch of nature, she realised her freedom,

> 'Ah, how beautiful is this village! I never saw it in this light before. I thought it was peace being home, Nishi, I am a free woman and will make sure you too become one'. It's a new Moufuli talking to herself (p. 143).

Women and nature are correlated in a way that allows one to find freedom and comfort from the touch of another; this is a form of healthy equilibrium because both are oppressed and can relate to one other.

Women Empowerment

The ability and entitlement of women to govern their own lives is the essence of women's empowerment. Women need to speak up for themselves and demand their rights after years of being marginalised, mistreated, and discriminated against. Women on the tea plantation are self-sufficient because they possess the courage and strength to manage their family as well as the strength to fight up for their own rights. Chaitirani, a tea planation worker, has gone through many upheavals in her life. Born in the tea-plantation, she has been exposed to tremendous suffering and poverty in her life. She has also been raped twice, first by the manager of the plantation and then by a Pakistani soldier during the liberation war. But she didn't lose hope, she carried in herself the irresistible desire to go back to her homeland and make her own clan with the sons she begat through rape, "I will one day go back to my motherland with my battalion of children" (28). She even didn't marry the love of her life for her tremendous sense of freedom and the urge to go back to her homeland one day, "Tell them I want the sky. How can I run away if I am tied down to another man? Don't I have to get out of the garden?" (70) Almost all the women in plantation are very gutsy, their situation has made them very bold and courageous. They face every situation and stand for what is right without any fear. One of the plantation workers, Budhu's wife Lachmati, is also a very gutsy woman. She single-handedly faced the cops with her spear, when they came to attack her husband. Her husband ran scared out of the back gate, but his wife faced them all by herself and showed no fear. The plantation workers not only raise their voice against the authorities, but are also frustrated with the torture from the males of their household. They wanted to begin a movement for getting freed from the torture, "They are calling for an end to oppression by their men" (162). There are many women like Chaitirani, who are bringing up their child single handedly and they don't need any help from the male members. They got raped several times, but they carry inside their hearts the benevolence and generosity, not to abandon or kill the child, rather to rear him/her up like their own, without any grudges.

Moufuli after tolerating a lot of torture from her husband, finally left his house. She asked for the wages for her daily work in the household. When her husband tried to hit her, she protests, "Before you raise your hand pay for all the work I do around the house" (89). Women work in the household every day without any break or holiday. Yet, everyone takes their selfless work for granted and tortures them. Home is also no less than a work place for her, where her works are seen as leisure. At one point of time, she wanted to end her life, but then she controlled herself and left the house in order to give her child a good and free life. Moufuli is an example to many housewives there who are silently suffering the torture of their in-laws. Finally, Moufuli takes a stand for her and her daughter to give her a free, safe, and happy life.

The Real-Life Situation of the Tea Plantation Workers

We can still claim that, given the existing situation, women's conditions have not significantly changed. Still the women workers are suffering without any basic facilities, even without proper sanitation. They work hard the entire day but still receive only minimum wages, not even sufficient to make a proper meal. Islam, Md

Nazrul, and Md Al-Amin (2024) in their research paper have quoted an interview with Sonali Munda, a tea planation worker from Finlay Tea State, where she stated:

> "I get BDT170 a day. This is quite insufficient as our sole income. We can never get out of distress this way. We get approximately BDT800 a week, after deducting all compulsory fees and subscriptions", with the high marker price of even the essential things, she shares their struggle, "Every day, we must ask ourselves: will we eat, buy clothes or educate our children with this money? Because there is no way to stretch it to cover all of that!" (p. 6)

Philip Gain (2021) in his newspaper article talks about the deplorable condition of the women workers and how even after working so hard, they are not provided with basic requirements. He reports that there are many women workers inside the garden who are not provided with any facilities during pregnancy and child birth, and talking about the problems, he states:

> They also have a tough time throughout their pregnancies, and it is due to the remoteness of the areas, superstition, cultural practices, malnutrition, poor housing, tough and indecent work conditions. The female tea garden workers are often in ill health, with cases of cervical cancer, tuberculosis and other diseases common in their communities (Paragraph 5).

The women workers inform that:

> [T]hey are subject to a variety of physical and mental abuse at work, home, and society. These include verbal abuse by their employers and supervisors, abuse in the hands of their husbands (particularly when they are drunk), coercion, sexual harassment including rape, involuntary abortion, and sexual assault on children and teenage girls (Paragraph 18).

Talking about the basic facilities, Gain reports that how during the lunch break, the workers have to take their lunch under the open sky or a tree as there is "no shade for them to sit under and eat lunch or take rest if someone falls sick—which is a violation of the labour law".(Paragraph 9). The workers are not provided with even a toilet or washing facilities in the working sections inside the garden, "(although one tea garden recently reported setting up toilets). This means women defecate and urinate in the open. This is again a clear violation of the labour law" (Paragraph 10). When women or other workers have to defecate in the open, it harms the surrounding areas too, it pollutes the place. Upasana Borthakur (2023) in her article "Women's Activism and Resistance to Exploitation: A Study of Women Workers in the Tea Estates of Assam" also gives a detailed description of the condition of the tea planation workers in Assam. Although she talks specifically about the tea-plantations in Assam, the scenario can be applicable to all the tea plantation workers. She states that inside the work plantation, there is unequal opportunity between different sexes. Women workers still play no significant part in decision-making inside their household. Women's leadership in the labour movement has not yet materialised, and even after being members of the trade union, they hardly take part in union activities. The women workers are completely arrogant about the role of education in improving their status, and their level of education is relatively low. Although they work in the plantation even more than their male counterparts, they still couldn't still find themselves equal to men. They are subjected to verbal and physical abuse, and domestic violence due to their husband's alcohol consumption and they still silently tolerate all the pain and torment of their lives (490). Talking about the drinking water availability, Acharjee and Ahmed (2016) write, there are very few tube wells and water falls and the wells in most of the cases remain uncovered. "The female workers also complained that due to shortage of tube wells they are required to walk a long distance to collect drinking water, which is a kind of transgression" (102). They also state that, "the water supply system is unsustainable, pathetic and miserable because of the distance of water sources, water unavailability throughout the year and lack of the idea about safe water" (106). Duara and Mallick (2019) state:

> Most of the women plantation workers are anemic and there is high percentage of malnutrition among all the family members of the workers. The estates' management along with the medical staff takes no initiative to provide for health supplements like vitamins and iron tablets. (p. 24)

They also inform that medical staff are not available during "emergencies like delivery, harassment faced at work or labor lines by management, supervisors or any male worker, cases of domestic violence which hardly gets reported, and work hazards" (25). Women who work on plantations are still subjected to harassment and sexual assault; these crimes continue unpunished, and female employees are taken advantage of everywhere they work.

Conclusion

Ecofeminism proposes equity and equality for all. Humanity will be at risk until every region of the earth and every creature has equal access to opportunities. It will be impossible to develop overall. Globally, there are a lot of women who are experiencing some form of hardship. The silent oppression of women and nature are going on and no one is giving a heed to it. Ecofeminism as a movement brings to the fore the silent oppressions and tries to bring them back in a good condition. Ecofeminism calls for a healthy coexistence of all species including those living and non-living with their healthy environment. The current life of the tea-plantation worker is as pathetic as it was before, as Hassan says:

> Their present scenario of social life is very pathetic and miserable because of having been deprived from education, employment and other basic facilities of human life and as a result they are being regarded as a backward community. They are far from touch of modernization, fully separated and alienated from the mainstream of culture (p. 129).

According to Ariel Salleh (1995), "[E]cofeminists are concerned about global sustainability as much as global justice" (26). Sustainability can only be achieved when there will be a healthy relation between each and every species, "the search for sustainability must be discussed in a way that encompasses quality of life and equal access to health, education, employment and political representation, going beyond aspects of resource management and business maintenance" (Kowalczuk Machado 2021, p. 25). Besides consumption and production, one must take care of the labourers and nature. Both women and nature are not safe in the present world scenario. In every step they are faced with exploitation and degradation because of man's excessive greed and lust. Every other day 1 can hear news about rape, murder, and torture as well as ongoing ecological crisis and environmental hazards. By pointing out the ongoing issues and their discussion through the theories of ecofeminism at least they can be brought to focus and equality should be called forth.

References

Acharjee S, Ahmed SI (2016) Access to water supply among the tea pickers in Sylhet, Bangladesh. Space Cult India 4(1):100–108. https://doi.org/10.20896/saci.v4i1.190

Alam MF, Sarker AH (2009) The tea estate labourers in Bangladesh. ResearchGate

Ballard A, Parham J (2020) Ties that bind: International studies in ecocriticism. Green Lett 24(2):101–107. https://doi.org/10.1080/14688417.2020.1830618

Borthakur U (2023) Women's activism and resistance to exploitation: A study of women workers in the tea estates of Assam. Social Change 53(4):486–499. https://doi.org/10.1177/00490857231203415

Carlassare E (2000) Socialist and cultural ecofeminism allies in resistance. Ethics Environ 5(1):89–106. https://doi.org/10.1016/S1085-6633(99)00025-X

Collard A, Contrucci J (1988) Rape of the wild: Man's violence against animals and the earth. Women's Press, London

d'Eaubonne F, Hottell RA, Merchant C, Bahaffou M, Gorecki J, Ramadan E (2022) Feminism or death. Verso, London

Duara M, Mallick S (2019) Women workers & industrial relations in tea estates of Assam. Indian J Ind Relat Econ Soc Dev 55(1):15–26. https://www.jstor.org/stable/10.2307/2712469

Gaard GC (ed) (1993) Ecofeminism: Women, animals, nature. Temple University Press, Philadelphia

Gaard GC, Murphy PD (eds) (1998) Ecofeminist literary criticism: Theory, interpretation, pedagogy. University of Illinois Press, Urbana

Gain P (2021) The strong women of tea gardens. The Daily Star

Ghosh (Hazra) S (2000) Girl child among adibasi plantation labourers of North Bengal: a study of their social situation (University of North Bengal Dt. Darjeeling). Retrieved from http://hdl.handle.net/10603/137485

Guerra P (2023) DIY, fanzines and ecofeminism in the Global South: 'This city is my sister'. DIY Altern Cult Soc. https://doi.org/10.1177/27538702231211062

Hassan ABME (2014) Deplorable living conditions of female workers: A study in a tea garden of Bangladesh. Am J Humanit Soc Sci 2(2):121–132. https://doi.org/10.11634/232907811402522

Hossain S (2019) Charcoal portrait (trans: Chatterjee T, Kabir J). Palimpsest, New Delhi (Original work published 2001)

Islam MN, Al-Amin M (2024) Ethnicity, culture and human development of Bangladesh's tea workers. Labor Hist 1–17. https://doi.org/10.1080/0023656X.2024.2353090

Kowalczuk Machado M (2021) Ecofeminism and sustainability. In: da Cal Seixas SR, de Moraes Hoefel JL (eds) Environmental sustainability sustainable development goals and human rights. CRC, Boca Raton, p 320

Salleh A (1995) Nature, woman, labor, capital: living the deepest contradiction. Capital Nat Social 6(1):21–39. https://doi.org/10.1080/10455759509358619

Shiva V (1988) Staying alive: Women, ecology, and survival in India. Kali for Women, New Delhi

Ullah MR, Reza SM, Baul TK, Empowerment Through Law of the Common People (Project) (eds) (2014) Story of the tea workers: Subsistence against hegemony. Empowerment Through Law of the Common People, Dhaka

Vakoch DA (2012) Feminist ecocriticism: environment, women, and literature. Lexington, Lanham

van der Wal S (2008) Sustainability issues in the tea sector: A comparative analysis of six leading producing countries. SOMO, Amsterdam
Walby S (1989) Theorising patriarchy. Sociology 23(2):213–234. https://doi.org/10.1177/0038038589023002004
Warren K (2000) Ecofeminist philosophy: a western perspective on what it is and why it matters. Rowman & Littlefield, Lanham

Neelam Kedia is a Research Scholar in the Department of English and Foreign Languages at the Central University of South Bihar, Gaya, Bihar, India.

Women and Nature: An Ecofeminist Approach Towards Environmental Conservation and Sustainability

Nibedita Priyadarsini

Abstract According to feminist literature, women have a distinct voice. This means, women have a particular perspective on the relationship between humanity and nature and also have an ethical and moral standpoint to reweave the world. In one aspect, environmental ethics involves identifying and criticizing the flaws in moral perception and thought, that underlie the current harmful practices, which can help to reconstruct our social and ethical relationship with the nonhuman world. Whereas, ecofeminism principles are based upon the views of nature being the central consideration for preservation and protection, requiring efficient and ethical use of natural resources, keeping in mind the consideration of nurturing and community growth and development as important priorities. When we talk about the association of women and nature, it appears with a tragic consequence for humans and nature. Some feminists have given their views on the form of association which can be converted with a positive outlook by affirming certain feminine values like; caring, responsibility, openness, love, and nurturing. Feminists and philosophers have emphasized the importance of ethics and practices that connect humans to the Earth. This paper focuses on the abstractness and generality of deontological and consequentialist theories that proposed a need for the conservation and sustainability of the environment through feminist ethics.

Keywords Ecofeminism · Environment · Deontology · Consequentialism · Utilitarianism

N. Priyadarsini, , Ph.D. (✉)
Department of Philosophy, Ravenshaw University, Cuttack, Odisha, India
e-mail: npriyadarsini@ravenshawuniversity.ac.in

B. Wejnert (ed.), *Democracy in Retreat and the Climate Crisis*, Springer Studies on Populism, Identity Politics and Social Justice,
https://doi.org/10.1007/978-3-032-02881-5_17

Introduction

Women possess a vital role in the ecological well-being of society. Challenges arise when society fails to provide enough opportunities for women and other marginalized groups to contribute to the betterment of our planet. Ecofeminism combines environmentalism and feminism to provide a comprehensive approach to sustainability and justice. Ecofeminists hold the view that the suppression of women and the destruction of the environment are deeply connected. Thus, ecofeminism emphasizes the need for both environmental and gender justice for any potential advancement and progress.

In today's activist and intellectual circles, ecofeminism appears to be less relevant. This framework can be established by understanding how empowering women can align with achieving sustainable and equitable goals and that too, by making a significant contribution to alleviate environmental impacts. The framework delves into the intersection of gender and nature, promote hierarchical thinking.

Ecofeminism has been identified as part of a 'deeper' or more radical approach to the ecological crisis (Merchant 1992; Eckersley 1992; Dobson 1995). Francoise D'Eaubonne first coined the term "ecofeminism" on the occasion of the publication of *Le Feminisme ou la Mort* in 1974 (Estévez-Saá and Lorenzo-Modia 2018). Over the years, ecofeminist values, principles, practices, and orientations have undergone scrutiny, refinement, and criticism. Ecofeminism encompasses women's and feminist viewpoints on the environment by emphasizing the connection between the abuse of women, marginalized communities, and nature. It is claimed that women, who are believed to be closer to nature, play a significant role in nurturing and caring for the environment. Today, understanding nature, managing resources, and ensuring individual well-being have significant gender-environmental implications. Women are the primary users of natural resources like land, forests, and water in many countries because they spend a lot of time working on farms. The responsibility leads to a profound connection between women and their surroundings, which in turn fosters a strong commitment to the protection and preservation of nature and the environment. A number of studies have shown that women have a strong interest in environmental conservation, as proven by their care of natural resources. Women can be specified as instrumental in the management of natural resources such as soil, water, forests, and energy, and often possess both traditional and contemporary knowledge about the natural world around them.

The Development of the Relationship Between Women and the Environment

From ancient times, women have shared an intrinsic relation with the environment as caregivers, cultivators, and preservers of natural resources or biodiversity. Many environmental movements throughout history have been led by women. Various movements and grassroots organizations worldwide are also demonstrating the implementation of ecofeminism and environmental justice on the ground.

1. *Traditional knowledge for the developmental relation*: In traditional societies women have a very profound knowledge and understanding of our ecosystem. Women used their knowledge of plants for medicinal and nutritional purposes. Different forms of worships has been taken cared in order to keep the growth and maintenance of a healthy ecological balance.
2. *The perspective of ecofeminism*: The Ecofeminist concept, which was emerged in the 1970s and 1980s, established a philosophical and political connection between the exploitation that happened between women and environment. The central argument was raised against the capitalist structure that prioritized profit over ecological health and social equity.
3. *Women as the pioneers of change and development*: Women have emerged as leaders in environmental movement, advocating for sustainability, environmental justice and growth. The Chipko Movement in 1970s India, can be taken as a notable example of non-violent protests by women in rural villages to safeguard forests from government-led logging activities, aimed at maintaining ecological balance and tradition. This movement was primarily led by women and involved hugging trees to prevent corporate interests from deforestation. These women, who relied on forests for firewood, water, and food, understood that such kind of destruction of ecosystems would have a devastating effect on their communities. By preserving those trees, they protected their livelihoods and fought against the economic and patriarchal forces that were driving environmental degradation.

   Another similar initiative is the Green Belt Movement in Kenya, established by Wangari Maathai. It is a remarkable initiative that combines both environmental conservation and empowerment of women. A huge number of trees have been planted in Kenya as a result of this movement. This has effectively combated deforestation, soil erosion, and water scarcity. While doing so, it has empowered women through different grounds like, environmental education and sustainable livelihoods. The Green Belt Movement is notably a prime illustration of how ecofeminism and environmental justice can be combined harmoniously to come up with sustainable and socially equitable solutions to environmental challenges.

   The Narmada Bachao Andolan is an Indian social movement which was led by native tribes, farmers, environmentalists and human rights activists against the construction of a number of large dams under the Narmada Dam Project across river Narmada.
4. *Eco measures endorsed by UN Guidelines*: The UN government advocates for women leadership in environmental governance. Programmes like Global Environment Facility (GEF) and the United Nations Framework Convention on Climate Change (UNFCCC) stress importance on educating women on sustainable environment and agricultural practices. The UN Declaration on the Rights of Indigenous People (UNDRIP) emphasizes the contribution of the demilitarization of the lands and territories of indigenous peoples to peace, and economic and social progress. It also advocates for women's equitable access to land and

resources. In order to promote gender equality and women empowerment UNEP has developed a comprehensive programme which provides a complete gender mainstreaming policy statement and operational framework to guide the organization's work. Gender and Climate action ensures that women's as well as men's concerns and experiences are an integral dimension of the design, implementation, monitoring and evaluation of climate action policies and programmes so that women and men benefit equally.

On the whole, women have played a crucial role in preserving the environment and upholding traditions through peaceful protests.

An Exploration of Feminism and Feminist Issues

The main purpose of ecofeminism here, is to underscore the connection between women's treatment and that of nonhuman nature, which is also known as women-nature connections. As it emphasizes the significance and necessity of understanding these connections for a comprehensive, feminist or environmental ethics.

Ecofeminism is commonly seen as a critique of traditional environmentalism, which has traditionally been dominated by males. In connection with this, ecofeminists argue that traditional environmentalists have often prioritized the preservation of nature without taking into account the effects of environmental degradation on the diverse population of our planet. They also argue on the ground, to create a sustainable society, we need to confront all forms of oppression, including those based on gender, race, and class. In order to achieve our sustainable development goals, it is essential to incorporate women into decision-making processes. Women's involvement in decision-making results in their involvement in this sustainable development initiative, which ensures that all community members needs are taken into account, thereby making development initiatives more effective and successful.

It has been observed that ecological feminism's origin can be traced back to various feminist movements, including liberal feminism, marxist feminism, radical and socialist feminisms, and black and third-world feminism. Ecofeminist perspectives on the nature and resolution of current environmental issues are diverse. There are a lot of issues that arise due to their diverse roots. More often, what defines these perspectives as ecofeminist is, their explicit focus on the "women-nature connections." Ecofeminist philosophy expands traditional feminist critiques of systems of domination (e.g., sexism, racism, classism, heterosexism, ageism, antisemitism) to include nature (i.e., naturism). Ecofeminists believe that women's relationship with nature is a feminist issue that deserves a lot of attention and action. Women have a distinct connection to nature that men do not have, particularly the connection they have with their bodies and reproduction. Women, as Salleh puts it, "flow with the system of nature" (Salleh 1984, 340). The importance of a close connection with nature's vital forces can be seen through the front-line participation of women in the peace movement and community actions to protest pollution in the workplace or home community.

Karen Warren argues that ecofeminism is an ethics based on context. It involves a shift from a conception of ethics as primarily a matter of rights, rules, or principles pre-determined and applied in specific cases to entities viewed as competitors in the context of moral standing, to a conception of ethics as growing out of what Jim Cheney calls "defining relationships," i.e., relationships conceived in some sense as defining who one is... Ecofeminism makes a central place for values of care, love, friendship, trust, and appropriate reciprocity-values that presuppose that our relationships to others are central to our understanding of who we are (Warren 1990, 141, 143).

Ecofeminism aims to redefine the concepts of human and nature. Ecofeminism reframes our view of humans as abstract individuals to emphasize the importance of our relationships those with nonhuman nature. According to this perspective, having interconnectedness and relationships with the world around us is more important than possessing reason, will, or effectivity as the sole basis for leading a moral life. Instead, what matters is, "how a moral agent is in relationship to another ... not simply that a moral agent is a moral agent or is bound by rights, duties, virtue, or utility to act in a certain way" (Warren 1990, 142). This leads to emphasizing the values of care, love, friendship, and trust.

Environmental Sustainability Through Ethics and Ecofeminism

In today's global context, environmental sustainability is a crucial concern. It emphasises the responsible use and management of resources to ensure the health and longevity of ecosystems for future generations. Environmental challenges can be addressed differently by different ethical frameworks. Utilitarianism, consequentialist theory, and ecofeminism presents different but complementary perspectives. Our main concern is to explore how each of these ethical theories can be utilized to promote environmental sustainability, focusing on the principles, strengths, and challenges of each approach. Feminists and ecofeminists are among those who have noted that the abstractness and generality of deontological and consequentialist moral principles and their vocabulary of rights, duties, and obligations obscures "the importance of attending to the needs of those with whom we are in relation" (Lauritzen 1989, 32).

For centuries, philosophers have been grappling with one fundamental question: What determines whether an action is morally right or wrong? Utilitarianism is an ethical theory that has been influential and provides an answer to this question. Philosophers like Jeremy Bentham and John Stuart Mill formulated utilitarianism, which is a form of consequentialist ethics. It is rooted in the principle that an action is deemed right if it leads to happiness or pleasure, and wrong if it results in unhappiness or pain. It is not only for the person acting but also for everyone else affected by it. This theory differs from ethical theories that determine the rightness or wrongness of actions based on the agent's motivation.

Utilitarianism has been used to justify a wide range of policies. During the assessment of a policy, it's crucial to consider the impact on the largest number of individuals and weigh the benefits against any potential drawbacks with care. Additionally, utilitarianism can act as a guide for making decisions in daily life. From an Ecofeminist perspective, when making choices about environmental conservation and sustainability, it is crucial to consider not only the personal implications but also how these decisions may impact those around us. Here, the focus of utilitarianism in environmental sustainability is on creating the greatest good for the greatest number of people. It not only encompasses the current but also the future generations. It extends its scope to future generations, by asserting that sustainable practices are crucial for ensuring that future individuals can live in an environment which will enhance their living and well-being. Women have a significant role to play in agriculture, making up a significant portion of the agricultural workforce worldwide and often taking care of crops and livestock. Women farmers have the potential to improve soil health and reduce environmental degradation by adopting sustainable farming practices like organic agriculture and agroforestry. Utilitarianism advocates for the sustainable utilization of resources such as water, forests, and fisheries. So, conserving resources can help societies maximize collective well-being and ensure that future generations also have access to them.

Consequentialist ethics which is also known as consequentialism, is an ethical theory that states that an action's morality is determined by its consequences. The moral value of actions is determined by their environmental impact. It takes into account values such as justice, rights, and the well-being of non-human entities. To assess the morality of these actions, it is important to take into account the long-term environmental consequences of pollution, deforestation, and carbon emissions.

Consequentialism can be extended to the interests of animals, ecosystems, and even inanimate nature, making it suitable for sustainability discissions that are not solely human-centred. Following this ethical approach, environmental policies, and actions it should be evaluated by their capacity to generate sustainable results, not by their intentions or means. Consequentialism also supports the protection of biodiversity by focusing and emphasizing the long-term benefits of healthy ecosystems. The preservation of species and ecosystems is a must and is justified because the consequences of extinction and ecosystem collapse are considered highly negative for both humans and non-human entities.

Deontology is an ethical theory that considers an action to be morally good based on its inherent characteristics. According to this theory, certain actions are morally obligatory regardless of their impact on human welfare. Deontological theories concentrate on adhering to a rule or law. Immanuel Kant is a prominent philosopher from eighteenth-century Germany who was known for his critical philosophy and who has articulated the deontological principles. Kant stressed on the importance of having a strong will and following moral law out of respect for it. He believed that the moral law is a categorical imperative, an unconditional command that could only be established through human reason. Kant's supreme categorical imperative can be summarized as, act only according to that maxim whereby you can, at the same time, will that it should become a universal law. According to Kant, we should act in such a manner that must ensure that we would never treat humanity, whether in ourselves or others, only as a means to an end, but always as an end in itself.

The feminist moral philosophy is important as it illustates how human relationship is characterized by interdependence. "Our relationships with others are inextricably bound and vary at every stage of our lives. We are always someone's daughter or sister, mother, son, brother or father; we are always members of a community" (Lauritzen 1989, 33–34). For ecofeminists, this community extends beyond the human species to our relations with other animals, trees and plants, places and ecosystems (Zimmerman 1987, 34). The purpose of this statement is to advocate for sustainable farming practices that emphasize local knowledge, biodiversity, and food sovereignty. At the same time, it discusses the duty that one should hold for the common good of the environment and society.

Now, if we discuss the moral ground of ecofeminism, then this will emphasize the importance of empowering women and marginalized communities to achieve environmental sustainability. It also recognizes women's traditional ecological knowledge and their leadership in grassroots environmental movements as valuable assets in promoting sustainable practices. Ecofeminism prioritizes love, care, cooperation, kindness and interconnectedness as crucial elements in addressing environmental challenges. It advocates for a shift from exploitative environmental management models to nurturing, community-centred approaches. In this framework this offers a holistic and justice-oriented view, highlighting the interdependence of social and environmental issues.

The ethical frameworks of utilitarianism, consequentialism, deontology and ecofeminism each make valuable contributions to the discussion on environmental sustainability. Utilitarianism offers a practical, outcome-based approach that prioritizes maximizing overall well-being. Consequentialism expands this view to encompass the well-being of all living beings and the environment itself. Through an ethical work culture, a sustainable environment can be defined through a deontological perspective that is based on duty and principles. Ecofeminism advocates for a more equitable and relational approach to environmental sustainability, challenging patriarchal structures and elevating the voices of women and marginalized communities.

In essence, a comprehensive approach to environmental sustainability could benefit from integrating these ethical theories, merging utilitarian, consequentialist foresight with ecofeminist justice to foster a more sustainable and equitable world for both present and future generations.

Conclusion

Ecofeminism acknowledges the differences between human beings and the natural world and aims to build a sense of community while exploring new ways of coexisting on the planet. This involves rethinking our approach to nourishment, shelter, entertainment, education, medicine and science. Ecofeminism starts from recognizing the centrality of nature to human existence. So it will always be a part of the human condition and must be address directly.

The ecofeminism and environmental justice creates a foundation for addressing both environmental degradation and social inequality. Ecofeminism enhances the environmental justice movements by linking women oppression and environmental exploitation, offering distinct insights as to how dynamic issues affect environmental outcomes. The promotion of ecofeminist ideals can be achieved through environmental justice, as this will provide a practical outlook and a practice for a systematic change that will benefit both the people and planet. Together, this will lay the foundation for a future that can be fair, sustainable and just.

References

Card C (ed) (1991) Feminist ethics. University of Kansas Press, Lawrence, KS

Cuomo C (2005) Ethics and the ecofeminist self. In: Zimmerman ME et al (eds) Environmental philosophy: From animal rights to radical ecology. Pearson Prentice Hall, Upper Saddle River, NJ, pp 194–207

Dobson A (1995) Green political thought. Routledge, London

Eckersley R (1992) Environmentalism and political theory. UCL, London

Estévez-Saá M, Lorenzo-Modia MJ (2018) The ethics and aesthetics of eco-caring: contemporary debates on ecofeminism(s). Women's Studies 47(2):123–146. https://doi.org/10.1080/00497878.2018.1425509

Gruen L (1997) Revaluing nature. In: Warren KJ (ed) Ecofeminism: women, culture, nature. University of Indiana Press, Bloomington, IN, pp 356–374

Howard PL (2003) Women and plants: gender relations in biodiversity management and conservation. Zed, London

Kheel M (2007) Nature ethics: an ecofeminist perspective. Rowman & Littlefield, Lanham, MD

King Y (1989) The ecology of feminism and the feminism of ecology. In: Plant J (ed) Healing the wounds: the promise of ecofeminism. New Society, Philadelphia, PA, pp 18–28

Lauritzen P (1989) A feminist ethics and the new romanticism—mothering as a model of moral relations. Hypatia 4(2):29–44

Merchant C (1990) Ecofeminism and feminist theory. In: Diamond I, Orenstein GF (eds) Reweaving the world: The emergence of ecofeminism. Sierra Club, San Francisco, CA

Merchant C (1992) Radical ecology. Routledge, London

Salleh, A. K. (1984). Deeper than deep ecology: The Eco-feminist connection. Environmental Ethics, 6, 340.

Warren KJ (1987) Feminism and ecology: making connections. Environmental Ethics 9:3–20

Warren KJ (1990) The power and promise of ecological feminism. Environmental Ethics 12:125–146

Warren KJ (ed) (1997) Ecofeminism: women, culture nature. Indiana University Press, Bloomington, IN

Warren KJ (2000) Ecofeminist philosophy: a western perspective on what it is and why it matters. Rowman & Littlefield, Lanham, MD

Zimmerman M (1987) Feminism, deep ecology, and environmental ethics. Environmental Ethics 9(1):21–44

Web Sources

https://www.unwomen.org/en

https://www.unep.org/

https://unfccc.int/process-and-meetings/what-is-the-united-nations-framework-convention-on-climate-change

https://www.un.org/development/desa/indigenouspeoples/wp-content/uploads/sites/19/2018/11/UNDRIP_E_web.pdf

Dr. Nibedita Priyadarsini is an Assistant Professor at the Department of Philosophy, Ravenshaw University, Cuttack, Odisha, India. Prior to the current appointment, she served as a lecturer in the Department of Philosophy at S.V.M. Autonomous College, Jagatsinghpur from 2016 to 2022. She received her Ph.D in Philosophy from Indian Institute of Technology (IIT) Bombay, Mumbai in 2021.

Her research interests are in Applied Philosophy especially in Bioethics. She worked extensively on Euthanasia and End-of Life Care. Apart from that she has also worked in other areas like Environmental Ethics, Neuro Philosophy, Philosophical Counselling and Indian Ethics.

Dr. Priyadarsini is the co-author of the book SAṀVĀD: Debates in Indian Philosophy (Red Shine Publication, 2024). She has also several articles in national and international journals in the areas of Applied Ethics and Indian Ethics.

MIX
Papier aus verantwortungsvollen Quellen
Paper from responsible sources
FSC® C105338

FSC
www.fsc.org

If you have any concerns about our products,
you can contact us on
ProductSafety@springernature.com

In case Publisher is established outside the EU,
the EU authorized representative is:
Springer Nature Customer Service Center GmbH
Europaplatz 3, 69115 Heidelberg, Germany

Printed by Libri Plureos GmbH
in Hamburg, Germany